THE PHONOLOGY OF THE WORL

*Series Editor*: Jacques Durand, Université d

# The Phonology of Hungarian

# THE PHONOLOGY OF THE WORLD'S LANGUAGES

*General Editor*: Jacques Durand

Published

# THE
# PHONOLOGY
## OF
# HUNGARIAN

———

*Péter Siptár and Miklós Törkenczy*

OXFORD
UNIVERSITY PRESS

# OXFORD
**UNIVERSITY PRESS**

Great Clarendon Street, Oxford OX2 6DP

Oxford University Press is a department of the University of Oxford.
It furthers the University's objective of excellence in research, scholarship,
and education by publishing worldwide in
Oxford New York

Athens Auckland Bangkok Bogotá Buenos Aires Calcutta
Cape Town Chennai Dar es Salaam Delhi Florence Hong Kong Istanbul
Karachi Kuala Lumpur Madrid Melbourne Mexico City Mumbai
Nairobi Paris São Paulo Singapore Taipei Tokyo Toronto Warsaw
with associated companies in Berlin Ibadan

Oxford is a registered trade mark of Oxford University Press
in the UK and in certain other countries

Published in the United States
by Oxford University Press Inc., New York

British Library Cataloguing in Publication Data

Data available

Library of Congress Cataloging in Publication Data

Data applied for

ISBN 978-0-19-823841-6 (Hbk.)
978-0-19-922890-4 (Pbk.)

1 3 5 7 9 10 8 6 4 2

Typeset by
SPI Publisher Services, Pondicherry, India
Printed in Great Britain on
acid-free paper by
Biddles Ltd., King's Lynn, Norfolk

# ACKNOWLEDGEMENTS

In writing this book we have had the benefit of advice, criticism, and help from many of our colleagues and students.

First of all, it is our pleasure to acknowledge our indebtedness to our colleagues László Kálmán, Ádám Nádasdy, and László Varga, with whom we co-authored another book (Ferenc Kiefer (ed.), *Strukturális magyar nyelvtan, 2. kötet: Fonológia* [The Structure of Hungarian, Volume 2: Phonology], Akadémiai Kiadó, 1994) without which this volume could never have been written.

Special thanks are due to Ferenc Kiefer for having initiated the project the end product of which is the present book. We would also like to thank Anna Kürti, Catherine Ringen, and Péter Szigetvári for their careful reading of and valuable comments on various versions of the manuscript. The discussions of various aspects of Hungarian phonology with László Kálmán, Ádám Nádasdy, Krisztina Polgárdi, Péter Rebrus, and Péter Szigetvári were invaluable and often drove us to despair.

We owe a lot to our students, especially Zsuzsa Bárkányi, Zoltán Kiss, Anna Kürti, and Nóra Wenszky, who had to suffer several trial runs of various chapters of the manuscript and whose comments improved this book considerably.

We would also like to thank Jacques Durand for valuable comments and for the editing of this series.

We thank the Center for Advanced Study in the Behavioral Sciences for supporting the project between September 1996 and December 1996 (Zsigmond Telegdi Grant) and the Hungarian Soros Foundation for the Short Study Grant awarded to us in 1998.

We are grateful to the Holland Institute of Generative Linguistics, University of Leiden, and Harry van der Hulst in particular for the hospitality, help, and the 'atmosphere of intellectual freedom' that surrounded us during the time we spent at HIL working on the final chapters of the book.

# CONTENTS

# LIST OF TABLES

# ABBREVIATIONS AND SYMBOLS

| | | | |
|---|---|---|---|
| abl. | ablative | MSC | morpheme structure constraint |
| acc. | accusative | | |
| adj. | adjective | N | nucleus *or* placeless nasal |
| ant | anterior (feature) | nas | nasal (feature) |
| ATR | advanced tongue root | NPA | nasal place assimilation |
| C | consonant | O | onset |
| CC | cluster of two consonants | OCP | Obligatory Contour Principle |
| CCC | cluster of three consonants | | |
| Co | coda | OP | [+ open₁] |
| comp. | comparative | OVS | object–verb–subject |
| cond. | conditional | P | any place feature |
| cons | consonantal (feature) | pl | plural |
| cont | continuant (feature) | PL | (C-)place node |
| COR | coronal node | poss. | possessive |
| dat. | dative | pres. | present |
| def. | definite conjugation | R | root node *or* rhyme |
| dim. | diminutive | S | sentence |
| DOR | dorsal node | SAA | stop+affricate affrication |
| ECH | Educated Colloquial Hungarian | SCL | Syllable Contact Law |
| | | SFA | stop+fricative affrication |
| FCS | fast cluster simplification | sg | singular |
| FSVS | final stem vowel shortening | SLH | Standard Literary Hungarian |
| GP | government phonology | | |
| imp. | imperative | so | someone |
| ind. | indicative | son | sonorant (feature) |
| indef. | indefinite conjugation | SOV | subject–object–verb |
| iness. | inessive | SPE | classical generative (with reference to *The Sound Pattern of English* (Chomsky and Halle, 1968)) |
| inf. | infinitive | | |
| instr. | instrumental | | |
| intr. | intransitive | | |
| IPA | International Phonetic Alphabet | spr. | superessive |
| | | SSC | syllable structure constraint |
| ISVS | internal stem vowel shortening | SSP | Sonority Sequencing Principle |
| | | | |
| L | laryngeal node | sth | something |
| LAB | labial node | strid | strident (feature) |
| lat | lateral (feature) | SVO | subject–verb–object |
| LVL | low vowel lengthening | SVS | stem vowel shortening |

The OP entry reads: OP $[+ \text{open}_1]$

| | | | |
|---|---|---|---|
| V | vowel *or* vocalic node *or* verb | % | grammaticality judgements differ across speakers |
| $V_d$ | defective vowel | . | syllable boundary |
| $V_f$ | full vowel | { } | syllable boundaries (where |
| $V_{FOP}$ | lowered full vowel | | left and right syllable edges |
| $V_u$ | unstable vowel | | have to be distinguished) |
| VH | vowel harmony | ⟦ ⟧ | analytic domain boundaries |
| VOS | verb–object–subject | ] | (synthetic) morpheme bound- |
| VP | verb phrase | | ary (only in rules) |
| VSO | verb–subject–object | - | concatenation (X-Y = Y is |
| VV | cluster of two vowels | | attached to X) |
| X | timing slot | / / | underlying segment(s) |
| XP | any phrasal category | [ ] | surface forms |
| 1sg$^s$ 2sg$^o$ | 1st sg. subject 2nd sg. object | σ | syllable node |
| * | incorrect or non-existent form | μ | mora |

## DISTINCTIVE VOWEL SEGMENTS

| Transcription symbol | IPA-symbol (if different) | Orthographic symbol | Example | Gloss |
|---|---|---|---|---|
| a [ɔ] | | *a* | *agy* | brain |
| aː | | *á* | *ágy* | bed |
| e [ɛ] | | *e* | *egy* | one |
| eː | | *é* | *ért* | understand |
| i | | *i* | *irt* | eradicate |
| iː | | *í* | *ír* | write |
| o | | *o* | *orr* | nose |
| oː | | *ó* | *ól* | sty |
| ö | ø | *ö* | *öl* | kill |
| öː | øː | *ő* | *őr* | guard |
| u | | *u* | *ujj* | finger |
| uː | | *ú* | *úgy* | like that |
| ü | y | *ü* | *ügy* | affair |
| üː | yː | *ű* | *űr* | space |

## DISTINCTIVE CONSONANT SEGMENTS

| Transcription symbol | IPA-symbol (if different) | Orthographic symbol | Example | Gloss |
|---|---|---|---|---|
| p | | *p* | *por* | dust |
| b | | *b* | *bor* | wine |
| t | | *t* | *tor* | feast |
| d | | *d* | *dal* | song |
| t$^y$ | c | *ty* | *tyúk* | hen |
| d$^y$ | ɟ | *gy* | *gyár* | factory |
| k | | *k* | *kép* | picture |
| g | ɡ | *g* | *gép* | machine |
| f | | *f* | *fal* | wall |
| v | | *v* | *vár* | castle |
| s | | *sz* | *szár* | stalk |
| z | | *z* | *zár* | lock |
| š | ʃ | *s* | *sír* | grave |
| ž | ʒ | *zs* | *zsír* | fat |
| h | | *h* | *hír* | news |
| t$^s$ | ts | *c* | *cél* | aim |
| č | tʃ | *cs* | *csel* | ruse |
| ǰ | dʒ | *dzs* | *dzsem* | jam |
| m | | *m* | *már* | already |
| n | | *n* | *nem* | gender |
| n$^y$ | ɲ | *ny* | *nyár* | summer |
| l | | *l* | *lom* | lumber |
| r | | *r* | *rom* | ruin |
| j | | { *j* | *jár* | walk |
| | | { *ly* | *lyuk* | hole |

## ADDITIONAL TRANSCRIPTION SYMBOLS

[a]   unrounded short low back vowel (in square brackets only)
ɔː   rounded long low back vowel
ç   voiceless palatal fricative
d$^z$   voiced alveolar affricate
[e]   unrounded short mid front vowel (in square brackets only)
ɛː   unrounded long low front vowel
ɦ   voiced glottal glide
i̯   palatal glide
ʝ   voiced palatal fricative

| ɱ | labiodental nasal |
|---|---|
| N | placeless nasal |
| n̠ | palato-alveolar nasal |
| ŋ | velar nasal |
| u̯ | labiovelar glide |
| ʋ | labial approximant |
| x | voiceless velar fricative |
| ɣ | voiced velar fricative |
| ː | length mark |
| ˜ | nasalization (above vowel symbol) |
| ˈ | stress mark |

(for intonation marks, cf. section 2.3)

PART I

—————

# BACKGROUND

# 1

---

# INTRODUCTION

## 1.1. AIMS, SCOPE, AND COVERAGE

This book presents a rule-based account of the phonology of Hungarian. At a time when current phonological research is undergoing a radical shift of emphasis towards non-derivational models (Declarative Phonology, Government Phonology, Optimality Theory), this choice of framework calls for a brief comment. Our decision to use a derivational framework follows from our primary aims rather than our theoretical preferences. As (with the notable exception of vowel harmony) Hungarian phonology is relatively little discussed in the international literature, we wanted to cover as much ground as possible descriptively, and discuss the data in a manner which is coherent and yet transparent in the sense that it is readily accessible for phonologists of a wide range of theoretical affiliations and backgrounds. As the various non-derivational frameworks have very little in common and (some of) their (basic) assumptions are still in a state of flux, this has practically determined our decision. It is perhaps not surprising that while several (sometimes very complex) phenomena from various languages have been given non-derivational treatments, no comprehensive analysis of the sound pattern of a single language has been published in any of these non-derivational theories.[1]

The dialect described is Educated Colloquial Hungarian (ECH), the spoken language of 'educated' people living in Budapest, the capital of Hungary. That dialect (cf. Nádasdy 1985) contrasts with Standard Literary Hungarian (SLH), the speech of conservative or speech-conscious speakers on the one hand and with various types of non-standard speech, including traditional rural dialects (cf. section 2.2.3), on the other. Both authors are native speakers of ECH.

In addition to native speaker judgements that underlie all data and generalizations presented in this book, the description of some of the phonological phenomena discussed here is based on a computerized database (cf. Kornai 1986*a*) comprising phonological (and other types of) information concerning approximately 80,000 lexical items.

---

[1] Harris (1994) may be considered an exception.

## 1.2. OVERVIEW OF PREVIOUS LITERATURE

There is a rich descriptive tradition in Hungarian phonology (see Deme 1961 and references cited therein). Unfortunately, most of these pre-structuralist works are in Hungarian (Simonyi 1907, Tompa 1968, Sauvageot 1971, Benkő and Imre 1972 are notable exceptions). There are also two recent publications in Hungarian that provide a detailed and comprehensive generative (autosegmental) analysis of Hungarian phonology: Kiefer (1994) and Siptár (1998b). For useful overviews of various periods of the Hungarian literature see Vértes (1982), Beöthy and Szende (1985), and Kontra (1995).

A few data-oriented overviews have also been published in English. These are Nádasdy (1985), Nádasdy and Siptár (1989), Cseresnyési (1992), and the chapter by Vago in Kenesei, Vago, and Fenyvesi (1998). The major structuralist works that discuss Hungarian are Lotz (1939), Hall (1944), Austerlitz (1950), Hetzron (1972, 1992). The most important comprehensive classical generative treatments are Szépe (1969) and Vago (1980a).

Vowel Harmony is the only phenomenon of Hungarian phonology that has attracted extensive interest and provoked a host of analyses in post-SPE phonology: see Becker Makkai (1970a), Clements (1976), Vago (1976, 1978a, 1980a, b), Ringen (1977, 1978, 1980, 1982, 1988a, b), Jensen (1978, 1984), Phelps (1978), Zonneveld (1980), Battistella (1982), Booij (1984), Goldsmith (1985), Hulst (1985), Kontra and Ringen (1986, 1987), Farkas and Beddor (1987), Abondolo (1988), Ringen and Kontra (1989), Kornai (1987, 1990b, 1994), Kontra, Ringen, and Stemberger (1991), Olsson (1992), Ritter (1995), Ringen and Vago (1995, 1998a, b), Polgárdi and Rebrus (1996, 1998), Polgárdi (1998). Other phenomena have received but sporadic attention in the international literature (e.g. the phonology/syntax interface: Kenesei and Vogel 1989, Vogel 1989, Vogel and Kenesei 1987, 1990).

Various aspects of Hungarian phonology have been analysed in frameworks outside mainstream generative phonology: e.g. Abondolo (1988), Olsson (1992), Szende (1992); Natural Phonology: Dressler and Siptár (1989); Government Phonology: Törkenczy (1992), Szigetvári (1994), Ritter (1998), Polgárdi and Rebrus (1998). Optimality Theory has also been applied to Hungarian: e.g. Törkenczy (1995) and Ringen and Vago (1998b).

Hungarian intonation is discussed in detail in Fónagy and Magdics (1967), Varga (1983, 1985, 1989, 1994b, 1995, 1996), Fónagy (1989), and Kornai and Kálmán (1989).

There are a few Ph.D. dissertations specifically on Hungarian (only some of which were later published): Austerlitz (1950), Vago (1974), Kornai (1986b), Olsson (1992), Varga (1993), Siptár (1993c), Törkenczy (1993), Zsigri (1994), Dunn (1995), Ritter (1995), Szentgyörgyi (1998).

## 1.3. FRAMEWORK AND THEORETICAL ASSUMPTIONS

In this section we briefly discuss the main theoretical assumptions underlying the description of Hungarian phonology presented in this book. These concern (i) the derivation and the relationship between morphological and phonological domains, (ii) the representation of segments, and (iii) the representation of syllable structure. Further discussion of some details appears in the analytical chapters where they are relevant to the issues at hand.

In this book we assume that—as in other languages (e.g. English)—there are two kinds of morphological domains in Hungarian. We shall refer to the two kinds of domains as 'synthetic' and 'analytic'.[2] The distinction is crucial in (i) the relationship between morphological domains and syllable structure/phonotactics, and (ii) the derivation.

Analytic morphological domain boundaries are opaque to phonotactic constraints, in other words, phonotactic constraints do not apply across them (cf. Kaye, Lowenstamm, and Vergnaud 1990). For instance, in Hungarian there are no phonotactic restrictions that constrain which consonants can be juxtaposed in a cluster $C_\alpha C_\beta$ when $C_\alpha$ is the last consonant of the first half of a compound word and $C_\beta$ is the first consonant of the second half of the compound. The restrictions one may find are purely accidental or non-phonological.[3] Intervocalic /kp/, for example, is only found under the conditions described above (*kerékpár* 'bicycle'), and is in fact not a well-formed interconstituent cluster (i.e. is excluded by a transsyllabic constraint, cf. section 5.3.2). This type of morphological boundary is analytic and is a barrier to syllabification/phonotactic interaction. In Hungarian, compounds[4] (⟦⟦*kerék*⟧ ⟦*pár*⟧⟧) and preverbs (⟦⟦*meg*⟧ ⟦*dob*⟧⟧ 'throw at') are analytic. Suffixes may be analytic (e.g. *-ban/ben* 'in': ⟦⟦*fény*⟧*ben*⟧ 'in (the) light', *-d* 'imp. def.': ⟦⟦*nyom*⟧*d*⟧ 'push!') or synthetic (e.g. *-t/-ot/-et/-öt* 'acc.': ⟦*nyom-ot*⟧ 'trace' (acc.)). Note that analytic suffixes are in an analytic domain separate from the stem, but—unlike compounds and preverbs—they do not form an independent one.[5] In Hungarian, the phonotactic pattern of monomorphemic stems is similar to though not always identical with that of stem + synthetic suffix

---

[2] The terms are borrowed from Government Phonology (cf. e.g. Kaye, Lowenstamm, and Vergnaud 1990, Kaye 1995), but the distinction is traditional in different varieties of Generative Phonology. It is the same as that between '+' boundary and '#' boundary, or Level 1 and Level 2 affixation (cf. Harris 1994).

[3] The few non-accidental regularities that can be found are due to postlexical assimilations such as Voice Assimilation (cf. sections 4.1.1 and 7.3) and Nasal Place Assimilation (cf. sections 4.2 and 7.4).

[4] Of course, words always form their own analytical domains: ⟦*Légy*⟧⟦*bátor*⟧ 'Be brave!'.

[5] This difference will be important in Vowel Harmony (cf. sections 3.2 and 6.1) because (most) analytic suffixes harmonize, but preverbs and compound members do not.

combinations.[6] The boundary between the stem and a synthetic suffix is thus transparent to syllabification/phonotactic interaction. However, it is not completely invisible to phonology since there are regularities that can only be expressed if it can be referred to (e.g. Hiatus (section 8.1.4.2), Low Vowel Lengthening (section 6.2.1), Lowering (section 8.1.4.3)).

We follow Lexical Phonology in assuming that (i) there is a lexical and a postlexical phase of the derivation, (ii) there is a modular difference between (potentially partially overlapping) sets of lexical rules, and (iii) this difference is related to the morphological domains within which rules apply. (We shall refer to the two lexical modules as Block 1 and Block 2.) The relationship between morphological domains and modules has been interpreted in various ways. Classical Lexical Phonology (e.g. Kiparsky 1982*b*) assumed an interleaving of morphology and phonology and thus both phonological processes and morphological operations were said to take place in the module they 'belong to'. As the modules are ordered with respect to one another, both the phonological processes and the morphological operations in Block 1 have to precede those in Block 2. Because of the problem of violations of the affixal order predicted by level ordering and that of 'bracketing paradoxes' (cf. Aronoff 1976, Fabb 1988, Cole 1995), a different interpretation was proposed in Halle and Vergnaud (1987) and Halle and Kenstowicz (1991). In their view it is only the phonological processes that are assigned to the modules. All morphology happens before phonology and each suffix is simply marked according to which block of rules it triggers. Thus the order of morphological operations does not have to mirror the order of the modules. There is evidence of violations of level ordering in Hungarian. The suffix -*hat/-het* 'may' is a case in point. It can be attached without a linking vowel to any stem that ends in a single consonant: *lop-hat* 'may steal' (3 sg indef.), *döf-het* 'may thrust' (3 sg indef.), *lát-hat* 'may see' (3 sg indef.), *rak-hat* 'may put' (3 sg indef.), etc. The lack of phonotactic interaction between the stem-final and the suffix-initial consonant suggests that it is an analytic (Block 2) suffix. Yet it can be followed by a suffix such as the past tense suffix -*(V)t(t)*, which is synthetic (Block 1) since the occurrence of its initial linking vowel depends on the last consonant of the stem (cf. section 8.1.4.4): *rohan-hat-ott* 'may run' (3sg past indef.)—compare *rohan-t* 'run' (3sg past indef.). We adopt the view that morphology precedes phonology rather than being interleaved with it, and that the phonological rules belong to (ordered) lexical modules, but otherwise we shall interpret derivation in a somewhat different way.[7]

---

[6] For instance, identical coda clusters are permitted monomorphemically and when the coda consists of a stem-final consonant and a consonant that belongs to a synthetic suffix. However, hiatus is possible monomorphemically (and across an analytic boundary) but not when one of the vowels is stem-final and the other is initial in a synthetic suffix. See section 8.1.4.2.

[7] This interpretation owes very much to Government Phonology (cf. Harris 1994, Kaye 1995), but is very different from it in many respects (e.g. Government Phonology does not permit rule ordering, let alone blocks of phonological rules).

We shall assume that the suffixes are marked according to whether they are analytic or synthetic. Analytic suffixes must be in a (dependent) domain which is different from that of the stem they are attached to. This domain may be monomorphemic or may contain synthetic suffixes as well. Block 1 rules will apply only *within* (dependent or independent) analytic domains (thus in a structure [[[X]Y]], they may apply (independently) to X and Y).[8] Block 1 rules show derived environment effects, but the derivation is not (necessarily) cyclic within the domain (cf. section 8.1.4.3). Following Cole (1995) we assume that derived environment effects (i.e. that a given rule does not apply within the morpheme, but does when the triggering environment is the result of affixation (of certain affixes)) are not (exclusively) the property of cyclic rules, so we shall refer to the Derived Environment Constraint instead of the Strict Cycle Condition. When all the Block 1 rules have applied, the whole word is subjected to the rules of Block 2. An extended syllable template (cf. Chapter 5 and section 8.1.4.5) is available when this happens and Block 2 rules are assumed not to be subject to the Derived Environment Constraint. A given rule may occur in both blocks or only one of them.[9]

The feature geometry assumed in this book will be a slightly adapted version of that proposed in Clements and Hume (1995), as shown in (1):

(1)    *a. consonants*                    *b. vowels*

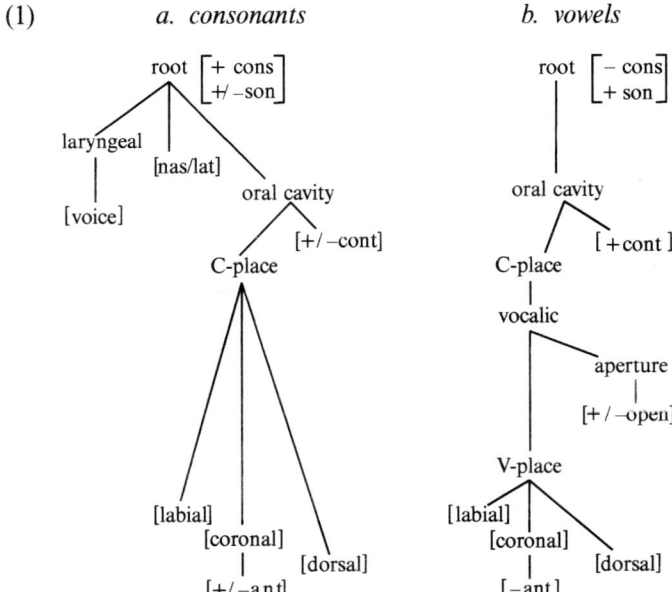

---

[8] It is necessary to allow Block 1 rules to apply within a dependent analytic domain because it may contain a synthetic suffix.

[9] We take no stand as to whether rules can be 'turned on' or only 'turned off' (cf. Mohanan 1986, Halle and Mohanan 1985, Borowsky 1986).

Root nodes are associated to timing slots or skeletal points, represented as X. In short segments, roots are associated to single X's, in long segments to two X's. The features [cons] and [son] are assigned to root nodes and are unable to spread independently of the root.

Note that some features are binary whereas others are unary (single-valued). The latter type includes the place features [labial], [coronal], and [dorsal], as well as [nas], [lat], and [voice]. Of the three unary place features, each segment has (at most) one (note that [+/–anterior] is dependent on [coronal], hence it is not specified for labials and dorsals). Similarly, nasals bear [nas], laterals [lat], and all other segments have neither. The feature [voice] is present in voiced obstruents and is absent elsewhere (together with the laryngeal node that dominates it). In voiceless obstruents, the lack of this feature indicates voicelessness, whereas the spontaneous vocal cord vibration that characterizes sonorants (including vowels), being phonologically irrelevant, is assigned to them in the phonetic implementation module (cf. Lombardi 1995*a*, *b* and section 7.3 below). The class node labelled 'vocalic' dominates the place and aperture features of vowels (the latter correspond to the traditional notion of 'vowel height'). Secondary articulations of consonants can be accounted for by spreading of the vocalic node (this is why it is dominated by C-place, even in the case of vowels).

Crucially, the place features characterizing vowels and consonants are the same, as opposed to more traditional feature systems in which consonant places were characterized by various combinations of values of the binary features [+/–coronal] and [+/–anterior], whereas vowel articulations were defined in terms of [+/–back] and [+/–round]. The insight that front vowels are best characterized in terms of a single-valued (privative) feature COR that also defines coronal (dental/alveolar or palatal) consonants, whereas back vowels share the feature DOR with dorsal (velar) consonants, is due to Hume (1992).[10] For the characterization of rounded vowels, LAB suggests itself: the feature that also identifies labial (bilabial/labiodental) consonants. The features LAB, COR, DOR—just like all class nodes and nearly all features in the diagrams in (1)—are located on separate tiers, but in a way that e.g. the LAB of a rounded vowel is on the same tier as the LAB of a labial consonant; similarly for the other two place features. However, the place features of vowels are not immediately dominated by C-place but only via the V-place and vocalic nodes. Therefore, COR harmony (the spreading of the COR of a vowel onto the vowel in the following syllable) is not blocked by the COR of the consonant between them, if it has one. This is because an association line between a COR node and a V-place node and an association line between another COR node and a C-place node run on two different

---

[10] Since front vowels are not dental/alveolar but palatal, all of them are—redundantly—[–ant].

planes even though the two CORs are on the same tier, hence no crossing of association lines ensues.[11]

Our view of syllable structure and syllabification will be fairly traditional. We assume that syllable structure is not present underlyingly, but is built up by syllabification in the course of the derivation.[12] Syllabification is seen as a template-matching algorithm (Itô 1986, 1989)—cf. section 8.1.4.1.

We assume that the segments belonging to a syllable are organized into the subsyllabic constituents onset, nucleus, rhyme, and coda. We also make the assumption that the constituents are hierarchically organized:[13]

(2)

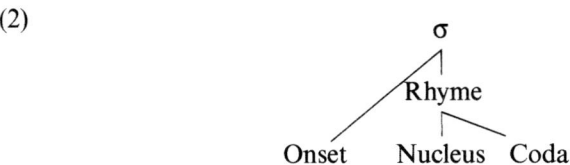

Under this view, syllable well-formedness derives from the well-formedness of the subsyllabic constituents. Given the hierarchical structure in (2), no restrictions (or at least weaker ones) are expected to apply between the constituents' onset and rhyme than between the nucleus and the coda or within each (sub)constituent. This is sometimes referred to as the Principle of Free Co-occurrence (Kaye 1995) and appears to hold true of Hungarian.[14] Furthermore, (in Hungarian and universally) constraints on syllable well-formedness seem to apply to subsyllabic constituents and not to the constituent 'syllable' itself. This has led some researchers (Aoun 1979, Kaye, Lowenstamm, and Vergnaud 1990) to deny the existence of the syllable as a constituent altogether. As nothing seems to hinge on this matter, we take no theoretical stand and retain the syllable as a convenient way of referring to the combination of an onset and a rhyme.

---

[11] Hume (1992: 95–101) illustrates this (i.e. the fact that palatal consonants are transparent with respect to COR spreading between palatal vowels) with a Hungarian example (among others). For instance, in *megy-ek* [mɛdʲɛk] 'I go' the COR of the *e* in the stem spreads onto the vowel of the suffix bypassing the COR of the *gy*, while in *vagy-ok* [vɔdʲok] 'I am' the COR of *gy* does not start a new domain of harmony, i.e. it does not turn the vowel of the suffix into a front vowel (because it is directly dominated by C-place, not by the vocalic node). See sections 3.2 and 6.1 for the details of vowel harmony.

[12] For a contrary view cf. Kaye, Lowenstamm, and Vergnaud (1990).

[13] For other views of syllable structure and subsyllabic organization cf. Clements and Keyser (1983), Kahn (1980), Kaye, Lowenstamm, and Vergnaud (1990), Hyman (1985), Hayes (1989).

[14] Some authors deny the validity of the principle as a universal (Clements and Keyser 1983, Davis 1985) and there are known counterexamples. However, it appears that the unmarked case is when the principle holds (e.g. Fudge 1987). Note that this does not mean that there may be no phonotactic constraints holding between a vowel and a preceding consonant; it only means that if such a constraint obtains, it is not a syllable structure constraint—it can easily be a constraint on morpheme shape, for instance (cf. Davis 1991, Booij 1999).

We assume that all segments that are phonetically interpreted must be prosodically licensed (Itô 1986, 1989). The question is whether this assumption necessarily means that each segment that appears at the surface is affiliated to one of the subsyllabic constituents. The answer is very important in the analysis of the so-called edge effects, i.e. the special character of (certain) clusters at the edges of (certain) morphological domains. There are strict and permissive approaches to this problem. Under the strict view, edge effects must be accounted for by normal syllable *structure* (i.e. the answer to the question above is yes). Thus, no special syllable structures are postulated that are limited to domain edges. Government Phonology exemplifies this approach.[15] In the permissive approach edge effects are accounted for by special syllable structures that can only appear at domain edges. There are several variations: in some analyses the special syllable structures in question may contain an additional subsyllabic constituent such as the appendix (e.g. Fudge 1969, Fujimura 1979, Hulst 1984), other approaches permit direct licensing (i.e. unmediated by a subsyllabic constituent) by the syllable node in the special syllables (e.g. Steriade 1982, Clements and Keyser 1983), still others allow direct licensing of segments by prosodic nodes higher than the syllable at domain edges (Rubach and Booij 1990, Törkenczy 1994a). It is difficult (and not always possible) to find empirical differences between the various approaches.

In this book we adopt the permissive approach and allow an extended syllable, i.e. one containing an appendix, in Block 2 (cf. Chapter 5 and section 8.1.4.5). Only the core syllable template shown in (2) is available for syllabification in Block 1.

Phonotactic constraints are often explicable with reference to sonority and the Sonority Hierarchy (e.g. Clements 1988, Vennemann 1988, Rice 1992). Despite the difficulties with the phonetic definition of the Sonority Hierarchy (Clements 1990, Laver 1994), we take it to be a well-established phonological relationship between classes of segments. We also assume that the Sonority Hierarchy is universal[16] and is the following:

(3) *Sonority Hierarchy*
    stops, affricates < fricatives < nasals < liquids < glides < vowels

Although the Sonority Hierarchy is universal, we allow for some language particular variation: sonority 'reversals' are not permitted (e.g. a language may not classify obstruents as less sonorous than nasals), but different language-particular settings of sonority distance between segment classes are possible (e.g. a language may determine that the sonority distance between stops and fricatives is smaller than that between fricatives and nasals; cf.

---

[15] In GP, instead of special structures, special segmental material (empty vowels) may appear at the edges of domains. See Kaye, Lowenstamm, and Vergnaud (1990), Kaye (1990). See also Burzio (1994) on the relationship between allowing special structures vs. special segments.

[16] For arguments against language particular Sonority Hierarchies, cf. Clements (1990).

Steriade 1982, Hulst 1984). We assume that phonotactic constraints can refer directly to the Sonority Hierarchy.[17] In order to account for sonority-based asymmetries of segment combination we shall borrow the term 'government' from Government Phonology (e.g. Kaye, Lowenstamm, and Vergnaud 1990), and Rice (1992)[18] and state:

(4) *Government*
    A segment X governs an adjacent segment Y if X is less sonorous than Y.[19]

We take government to be asymmetrical, but not intrinsically strictly directional in all governing domains, i.e. it is always directional, but its direction may be fixed in some structural positions but free in others.[20] We assume that government is universally left to right in onsets and right to left in codas. In Hungarian transsyllabic clusters, however, the directionality of government is not fixed (it may be left to right or right to left), cf. section 5.3. We follow Kaye, Lowenstamm, and Vergnaud (1990) and assume that government applies between timing slots.

Following Rice (1992) we assume that there may be another asymmetrical relationship between adjacent segments, i.e. the relationship of 'binding'. We follow (and generalize) Rice's definition (compare Rice 1992):

(5) *Binding*
    A bound segment contains dependent structure.

Thus, a bound segment contains structure that does not differ from that of the segment that binds it (e.g. in a homorganic nasal + stop cluster the nasal is bound by the stop). Binding can apply to various nodes of the feature tree, e.g. to the root node ('root-binding') or the place node ('place binding') for instance. (In the example above the nasal is 'place-bound'; in a (true) geminate the first consonant is 'root-bound', i.e. it has the same structure from the root down as the second consonant.) We assume that binding is strictly directional and is right to left.

The mora is not a primitive in the present treatment, but is considered to be derivative of syllable structure. It is only used as a unit of measuring syllable weight (which, incidentally, plays very little role in Hungarian phonology,[21] cf. section 5.4.1) and does not function as a subsyllabic

---

[17] Note that this does not mean that the Sonority Hierarchy is a primitive (a scalar feature, for instance). We assume that the Sonority Hierarchy is derived. We take no stand whether it is to be defined in terms of features (cf. Clements 1990) or structurally (cf. Kaye, Lowenstamm, and Vergnaud 1990, Harris 1990, Rice 1992).

[18] There are important differences between GP's and Rice's interpretation of government. Our interpretation here is closer to Rice (1992).

[19] We do not take a stand as to the interpretation/derivation of sonority. For the sake of simplicity (4) can be interpreted as directly referring to (3).

[20] Compare Kaye, Lowenstamm, and Vergnaud (1990) who assume that government is strictly directional in all governing domains.

[21] For a different view, cf. Vago (1980a, b, 1992).

constituent/timing unit. We shall use 'bimoraic' as a convenient label to refer
to syllables that have a branching rhyme or/and a branching nucleus. For
arguments against moraic syllable structure (as proposed in Hyman 1985,
Hayes 1989) cf. Brentari and Bosch (1990), Davis (1990), Sloan (1991), Tranel
(1991), Rialland (1993).


## 1.4. CHAPTER LAYOUT

This book consists of three parts, entitled 'Background', 'Systems', and
'Processes', respectively. The first part (Chapters 1–2) presents a general
introduction and summarizes preliminary information on the Hungarian language, its segmental and suprasegmental phonology, morphology, and syntax. The second part (Chapters 3–5) contains detailed discussion of the
vowel system (including length alternations and vowel harmony), the consonant system (with each manner-of-articulation class considered in its turn),
and the phonotactics of the language (in terms of syllable structure constraints, transsyllabic constraints, and morpheme structure constraints).
Finally, the third part (Chapters 6–9) analyses the various phonological
processes that the vowels, consonants, and syllables undergo and/or trigger.
Chapter 6 gives a new analysis of vowel harmony, the most celebrated aspect
of Hungarian phonology (6.1) and includes brief accounts of low vowel
lengthening (6.2.1) and stem vowel shortening (6.2.2). Chapter 7 presents
processes involving consonants, including palatalization (7.1), strident assimilations (7.2), voice assimilation (7.3), and various processes targeting nasals
and liquids (7.4). Chapter 8 analyses processes conditioned by syllable structure, paying special attention to vowel ~ zero alternations (8.1) and briefly
considering two types of consonantal alternations as well (8.2). Finally,
Chapter 9 surveys various surface phenomena, including surface vacillation
of vowel length, compensatory lengthening, hiatus filling, degemination, and
fast-speech cluster simplification.

# 2

## PRELIMINARIES

### 2.1. THE HUNGARIAN LANGUAGE

Hungarian (*magyar*) is a Uralic language spoken in Central Europe. In terms of number of speakers, it is the twelfth largest language of Europe. Approximately ten million people speak it within Hungary, and another three million in the surrounding countries: Austria, Slovakia, the Ukraine, Romania, Serbia, Croatia, and Slovenia. There are, furthermore, over a million Hungarian speakers elsewhere, mostly in the United States and Canada.

Being a Uralic (more specifically, Finno-Ugric) language, Hungarian is typologically unlike the majority of European languages. However, it is also atypical among the members of the Uralic family. More than half of all speakers of Uralic languages are Hungarians. Unlike Finnish, the other major language of the family, Hungarian has no close relatives; the Ob-Ugric languages (Vogul and Ostyak), traditionally bundled together with Hungarian into the Ugric branch of Finno-Ugric languages, are radically different from Hungarian in their phonology, syntax, and vocabulary.[1]

Typologically, Hungarian is a language of agglutinating morphology (cf. Kornai 1994), non-configurational syntax (cf. É. Kiss 1987, Kiefer and É. Kiss 1994), and syllable-timed prosody (cf. Roach 1982, Crystal 1995). The dialect described here is Educated Colloquial Hungarian (ECH, cf. Nádasdy 1985), as distinct from both Standard Literary Hungarian (SLH) and various types of non-standard speech, including traditional rural dialects. (A list of the major dialect areas is given in section 2.2.3 below; cf. Benkő and Imre 1997: 299–326 for details.) Educated people living in one of these dialect areas tend to speak a mixture of ECH and their traditional accent, referred to as Regional Standard (of the area involved).

The word stock of Hungarian includes a high number of loanwords, in addition to the most ancient layer of vocabulary of Finno-Ugric origin (*szem* 'eye', *fej* 'head', *kéz* 'hand', *vér* 'blood', *ló* 'horse', *holló* 'raven', *hattyú* 'swan',

---

[1] For a concise characterization of Uralic languages in general, see Austerlitz (1990) and references cited there, especially Collinder (1960), Comrie (1981), Décsy (1965), and Hajdú (1975). For a very general overview of some salient characteristics of Hungarian, see Abondolo (1990); for a detailed, data-oriented description of this language, see Kenesei, Vago, and Fenyvesi (1998).

*hal* 'fish', *háló* 'net', *nyíl* 'arrow', *víz* 'water', *kő* 'stone', etc.). Some of the oldest loanwords entered the language during the migration of the Proto-Hungarian tribes from the primeval Uralic/Finno-Ugric homeland to the Carpathian Basin, whereas the majority of such vocabulary was borrowed after their settlement in the present territory (also known as the Hungarian Conquest, AD 896).[2]

Hungarian words of Iranian (Ossete) origin include *asszony* 'woman', *híd* 'bridge', *gazdag* 'rich', *kard* 'sword', *méreg* 'poison', *tölgy* 'oak', *üveg* 'glass'.

Due to repeated contacts with Turks, a large number of Turkic loanwords entered the language. These include *búza* 'wheat', *árpa* 'barley', *borsó* 'pea', *szérű* 'barnyard', *sarló* 'sickle', *gyümölcs* 'fruit', *alma* 'apple', *körte* 'pear', *szőlő* 'grape', *komló* 'hop', *bor* 'wine', *sör* 'beer', *barom* 'cattle', *ökör* 'ox', *csikó* 'foal', *disznó* 'pig', *tyúk* 'hen', *gyapjú* 'wool', *túró* 'cottage cheese', *sajt* 'cheese', *tengely* 'axle', *karó* 'stake', *teknő* 'trough', *korsó* 'jar', *gyűrű* 'ring', *gyöngy* 'pearl', *dolmány* 'dolman', *saru* 'sandal'.

After the Conquest, a large number of Slavonic loanwords were borrowed. Examples are *puszta* 'lowland plain', *barázda* 'furrow', *róna* 'flat open country', *patak* 'brook', *széna* 'hay', *szalma* 'straw', *pince* 'cellar', *tornác* 'porch', *udvar* 'yard', *kémény* 'chimney', *kemence* 'oven', *lóca* 'bench', *asztal* 'table', *abrosz* 'tablecloth'.

The layer of early German loanwords includes *föld* 'land', *rét* 'meadow', *határ* 'field', *herceg* 'prince', *polgár* 'citizen', *torony* 'tower', *kehely* 'chalice', *példa* 'example', *törköly* 'marc', *céh* 'guild', *málha* 'luggage', *selejt* 'rejects', *pék* 'baker'. For further details concerning the early layers of loanwords in Hungarian, cf. Benkő and Imre (1972: 171–92).

An important factor contributing to the word stock as it is today was the eighteenth/nineteenth-century movement known as 'language reform' (*nyelvújítás*) that added literally thousands of newly coined (or revitalized) words to replace foreignisms and to extend and update the lexicon so that it could better meet the requirements of the age. Many extreme proposals have since gone to well-deserved oblivion, but a very large number of words are now indispensable items of present-day Hungarian. The major methods of extending the word stock included the revitalization of obsolete words (*fegyelem* 'discipline', *iker* 'twin', *szobor* 'statue', *verseny* 'competition'), often with the 'wrong' meaning (either by mistake or consciously, e.g. *alak* 'figure' (< 'puppet'), *agy* 'brain' (< 'skull'), *börtön* 'prison' (< 'executioner'), *hős* 'hero' (< 'lad'), *kór* 'illness' (< 'ill person'), *váz* 'framework' (< 'scarecrow')); bringing dialect words into standard use (*betyár* 'highwayman', *burgonya* 'potato', *csapat* 'team', *csuk* 'close', *doboz* 'box', *hanyag* 'negligent', *kelme* 'cloth', *pata* 'hoof', *szikár* 'lanky', *zamat* 'flavour'); morphological derivation

---

[2] Note that such old loanwords are by no means identified by speakers as 'non-native' or 'foreign' vocabulary, as opposed to more recent loans of English or German origin, that have clear subregularities that set them apart from 'native' lexical items. (The loanword data listed in this section come from Antal, Csongor, and Fodor 1970: 119.)

using productive, obsolete, and invented suffixes (some of the favourite verb-forming derivational suffixes were causative *-szt* as in *fagyaszt* 'freeze', *törleszt* 'pay off', frequentative *-g*, especially in the combination *-log/leg* as in *társalog* 'converse', *tiszteleg* 'salute', frequentative *-ng* as in *dühöng* 'rage', *forrong* 'revolt', *ömleng* 'gush', and denominal *-ít* as in *alapít* 'found', *egyesít* 'unite'; among nominalizing suffixes, *-alom/elem* as in *fogalom* 'notion', *győzelem* 'victory', *-ász/ész* as in *nyomdász* 'printer', *régész* 'archaeologist', *-mány/mény* as in *állítmány* 'predicate', *festmény* 'painting', *-vány/vény* as in *látvány* 'sight', *növény* 'plant' should be mentioned; examples of invented suffixes are *-c* as in *bohóc* 'clown', *élc* 'joke', *-nc* as in *fegyenc* 'prisoner', *újonc* 'freshman', *-da/de* as in *szálloda* 'hotel', *képezde* 'school'); and truncation (responsible for quite a number of final clusters that would not otherwise be attested in this language: *rajz* 'drawing', *szomj* 'thirst', *taps* 'applause', *törzs* 'trunk', etc.) For more on the 'language reform', see Benkő and Imre (1972: 276–83).

## 2.2. ORTHOGRAPHY AND SOUND SYSTEM

### 2.2.1. The vowel inventory

The (lexical) vowel inventory of Hungarian consists of the following fourteen items:

(1)

| Transcription symbol | IPA-symbol (if different) | Orthographic symbol | Example | Gloss |
|---|---|---|---|---|
| ɔ | | *a* | *agy* | brain |
| aː | | *á* | *ágy* | bed |
| ɛ | | *e* | *egy* | one |
| eː | | *é* | *ért* | understand |
| i | | *i* | *irt* | eradicate |
| iː | | *í* | *ír* | write |
| o | | *o* | *orr* | nose |
| oː | | *ó* | *ól* | sty |
| ö | ø | *ö* | *öl* | kill |
| öː | øː | *ő* | *őr* | guard |
| u | | *u* | *ujj* | finger |
| uː | | *ú* | *úgy* | like that |
| ü | y | *ü* | *ügy* | affair |
| üː | yː | *ű* | *úr* | space |

In addition, four 'marginal vowels' ([a], [e], [ɔː], [ɛː]) are sometimes discussed in the literature; see section 9.1 for some of the processes responsible for these surface segments and cf. Siptár (1991*b*) for a more extensive discussion. It has also been suggested that Hungarian has diphthongs, too (cf. Kylstra and de Graaf 1980, Kylstra 1984). This issue will be briefly considered here.

Compare Hungarian *ajtó* 'door' and *autó* 'car' with English *item* and *outer*, or with German *Eiter* [aı̯tə] 'pus' and *Autor* [au̯to:g] 'author'. Phonetically, i.e. at the level of surface representation, the portions preceding the [t] are identical (except for the slight rounding of Hungarian *a*). Compare further German *Euter* [ɔı̯tə] 'udder' where the first segment is strongly rounded: Hungarian *aj(tó)* is roughly halfway between German *Eu(ter)* and *Ei(ter)*; similarly, Hungarian *baj* 'trouble' is intermediate between English *boy* and *buy*. Thus, surface diphthongs do occur in Hungarian speech. However, as we will see presently, this does not mean that Hungarian has underlying diphthongs as well.[3]

Let us discuss *aj-* and *au-* separately, since their phonological behaviour is not parallel. With respect to *j*-final 'diphthongs', notice that *j* can occur after any vowel in Hungarian (*baj* 'trouble', *táj* 'landscape', *fej* 'head', *kéj* 'pleasure', *mily* 'what kind of', *díj* 'prize', *moly* 'moth', *rój* 'carve' (imp.), *bögöly* 'gadfly', *lőj* 'shoot' (imp.), *paszuly* 'bean', *fúj* 'blow', *süly* 'scurvy', *fűlj* 'get heated' (imp.); note that orthographic *ly* and *j* both represent /j/). If all of these were to be diphthongs, the number of Hungarian vowels would be doubled. (By the same token, *j* + vowel combinations could be analysed as rising diphthongs, cf. *javít* 'repair', *játék* 'toy', *jelen* 'present', *jég* 'ice', *jiddis* 'Yiddish', *jog* 'right', *jó* 'good', *jön* 'come', *jő* 'come', *jut* 'get', *június* 'June', *jüt* 'Jute'; this would almost triple the vowel inventory—Hungarian would have forty distinct vowel types—without any descriptive gain at all.) This is not a knock-down argument in itself but it shows that something is wrong with this idea. What is more important is that in languages where diphthongs are phonological objects, one cannot just combine any vowel with any semi-vowel (e.g. in English, we have [eı̯], [aı̯], [ɔı̯] but no [æı̯], [ʌı̯], [uı̯], [ɔ̯i], [ɒı̯], etc., we have [au̯], [ou̯], but no [iu̯], [eu̯], [æu̯], [ʌu̯], [ɔu̯], etc.).[4]

The consonantal character of Hungarian *j* is further demonstrated by a number of facts. First, *jV*-initial words select the preconsonantal allomorph of the definite article *a*, not its prevocalic allomorph *az*, e.g. *a játék* 'the toy', *\*az játék*. Second, the initial consonant of the suffix *-val* 'with' (whether its underlying representation is just an empty consonant slot, cf. Vago 1989, or

---

[3] Some regional dialects exhibit further types of diphthongs, e.g. *ló* [lo:u̯] (in another dialect: [lu̯o:]) 'horse', *szép* [se:i̯p] (in another dialect: [si̯e:p] 'beautiful'). The following discussion is not concerned with such dialectal diphthongs: rather, it is restricted to Standard Hungarian where this kind of diphthong never occurs: *ló* 'horse' is [lo:] and *szép* 'beautiful' is [se:p]. In other words, the following discussion is restricted to cases like *aj(tó)* and *au(tó)*.

[4] Also, in such languages, diphthongs may alternate with monophthongs (cf. *crime/criminal*), whereas in Hungarian this never happens. This, however, is not a very strong argument **against** the diphthong interpretation whereas the existence of such alternation in English is a fairly compelling argument **for** one.

a full-fledged /v/, cf. section 8.2.1 for discussion) fully assimilates to stem final consonants but appears as [v] after vowel-final stems (cf. *lábbal* 'with foot', *\*lábval*, vs. *szóval* 'with a word', *\*szóal*). Therefore, the fact that e.g. 'with butter' is *vajjal* rather than *\*vajval* suggests that *j* is a consonant. And finally, the example *vajjal* also shows that intervocalic *j* can be geminated (long): this in itself is enough to render any kind of diphthong interpretation impossible.

In sum: Hungarian /j/ is not a semivowel but a consonant (cf. section 4.2 for some more discussion); hence, neither *j*-initial rising diphthongs nor *j*-final falling diphthongs exist in Hungarian phonology either underlyingly or at any shallower level (as opposed to surface pronunciation in which the realization of *jV* and *Vj* sequences—as well as the casual-speech reduction of *Vi* and *iV* vowel sequences, e.g. *kaleidoszkóp* 'kaleidoscope' [-lɛi̯-] (= [-lɛj-]), *Mária* <proper name> [-ri̯ɔ] (= [-rjɔ]), *Garay utca* 'G. Street' [-ɔi̯u-] (= [-ɔju-])—may produce phonetic diphthongs).

Let us turn now to the portion spelt *au* in *autó* 'car', *augusztus* 'August', *bauxit* 'bauxite', *kaucsuk* 'caoutchouc', *mauzóleum* 'mausoleum', *tautológia* 'tautology', *trauma* 'trauma', *kalauz* 'conductor'. This time, we have no reason to analyse the [u] as a coda consonant (as we did for (postvocalic) [j] above). The question here is whether we have one or two syllables (a diphthong or hiatus). Two tests come to mind.

(i) The intonation of yes/no questions involves a rise–fall pattern (LHL) which spreads over the last three syllables provided that the last major stress occurs on the antepenultimate (or earlier) syllable of the utterance (cf. section 2.3 below). Thus, given a question whose focus is well before the third-last syllable, a bisyllabic final word will have a pitch peak on its initial syllable, whereas a trisyllabic word will have one on its medial syllable: *Megjött már a* ↑*ta-*↓*xi?* 'Has the taxi arrived yet?' vs. *Megjött már a vil-*↑*la-*↓*mos?* 'Has the tram arrived yet?' Applied to our problem, this test suggests that *autó* 'car' (at least for a large number of speakers) is bisyllabic (cf. *Megjött már az* ↑*au-*↓*tó?* 'Has the car arrived yet?' and not ... *a-*↑*u-*↓*tó?*). However, the same test shows *kalauz* 'conductor' to consist of three syllables (cf. *Jó ka-*↑*la-*↓*uz?* 'Is he a good conductor?' and not *\*Jó* ↑*ka-*↓*lauz?*), and is inapplicable to *augusztus* etc., as well as to longer words starting *autó-* or *auto-* (*autójavító* 'car repair shop', *automatizálás* 'automation'; although it is not unreasonable to assume that these have the same representation as *autó* has).

(ii) The language game known as 'bird language' (e.g. *Tuvudsz ivígy beveszévélnivi?* < *Tudsz így beszélni?* 'Can you speak like this?') can be described as follows. Expand each syllable into a bisyllabic foot by inserting two skeletal slots between the onset (if there is one) and the nucleus. Fill the second inserted slot by [v]; and fill the first inserted slot by a copy of the first or only mora of the nucleus (i.e. copy the vowel if it is short and supply a short realization—not the short counterpart!—of the vowel if it is long). Thus *be.szél.ni* [bɛse:lni] 'to speak' gives *beve.szévél.nivi* [bɛvɛse:lnivi] (the dots in the output form are there to facilitate reading: they indicate foot boundaries corresponding to the input syllable boundaries). Given that an [ɔu̯] diphthong would be two morae rather than a single mora (cf. the optional monophthongized version [ɔːtoː]), this predicts that the language game should give *avau.tovó, avau.guvusz.tuvus* for the diphthong interpretation but *ava.uvu.tovó, ava.uvu.guvusz.tuvus* for the hiatus interpretation. If the test is done, we find that (with some inter-speaker variability, but mostly) *avau.tovó* etc. but, for *kalauz*, invariably *kava.lava.uvuz* (not *kava.lavauz*) is produced.

The results of both tests show that *kalauz* has three syllables, i.e. an /a/+/u/ hiatus, whereas *autó* has a diphthong (at the surface). This latter can be phonologically interpreted either as an /au̯/ diphthong, i.e. a nucleus-internal sequence of the (head) vowel /a/ and the semivowel (or, non-head vowel) /u̯/; or as an /a/+/u/ vowel sequence (two nuclei) with an /u/ → [u̯] realization rule which some speakers have, others lack, and still others have as an optional rule. The left environment of the rule will be /a/ or, more generally, a short low vowel (cf. *Európa* [ɛ.u.roː.pɔ] ~ [ɛu̯.roː.pɔ] ~ [ɛː.roːpɔ] 'Europe'). Depending on the exact range of lexical items in which the rule is to apply (speakers may differ in this respect), the rule must be restricted either by requiring the low vowel to be stressed (this will exclude *kalauz*) or else by requiring that the low vowel be immediately preceded by a word boundary (to admit *autó* etc., *augusztus*, and *Európa* but to exclude all other examples given above: *bauxit* etc.).

Which analysis is the correct one? In view of the more or less marginal lexical load and the variability of the whole phenomenon, the second solution appears to be better; hence, we can state that the phonology of standard Hungarian has no lexical diphthongs at all.

### 2.2.2. The consonant inventory

The (lexical) consonant inventory of Hungarian consists of the following twenty-four items:

| (2) | Transcription symbol | IPA-symbol (if different) | Orthographic symbol | Example | Gloss |
|---|---|---|---|---|---|
| | p | | p | *por* | dust |
| | b | | b | *bor* | wine |
| | t | | t | *tor* | feast |
| | d | | d | *dal* | song |
| | tʸ | c | ty | *tyúk* | hen |
| | dʸ | ɟ | gy | *gyár* | factory |
| | k | | k | *kép* | picture |
| | g | ɡ | g | *gép* | machine |
| | f | | f | *fal* | wall |
| | v | | v | *vár* | castle |

| s | | sz | szár | stalk |
|---|---|---|---|---|
| z | | z | zár | lock |
| š | ʃ | s | sír | grave |
| ž | ʒ | zs | zsír | fat |
| x | | h | hír | news |
| tˢ | ts | c | cél | aim |
| č | tʃ | cs | csel | ruse |
| ǰ | dʒ | dzs | dzsem | jam |
| m | | m | már | already |
| n | | n | nem | gender |
| nʸ | ɲ | ny | nyár | summer |
| l | | l | lom | lumber |
| r | | r | rom | ruin |
| j | | ⎰ j | jár | walk |
| | | ⎱ ly | lyuk | hole |

According to some analyses (cf. e.g. Szende 1992: 113–16), the Hungarian consonant inventory consists of fifty, rather than twenty-four, segment types; in addition to those in (2), they include */dᶻ/, as well as long /pː/, /bː/, /tː/, /dː/, . . ., /lː/, /rː/, /jː/. The problem of */dᶻ/ will be looked at in section 4.1.4; the issue of long consonants will be briefly considered here.

Superficially, all Hungarian consonants can occur short and long; however, most geminates are derived (either by concatenation or by various assimilation rules; [vː] and [žː] exclusively occur as derived segments). Genuine (underlying) geminates are relatively infrequent and tend to be restricted to various marginal lexical classes like onomatopoeic vocabulary and interjections, proper names, and recent loanwords (cf. Obendorfer 1975 for a detailed overview). All geminates, underlying and derived, are subject to an intricate set of degemination processes (see Nádasdy 1989a and section 9.4 below).[5]

With respect to the classification of the consonantal segments appearing in (2), a number of controversial issues have to be clarified. The oral palatal non-continuants /tʸ dʸ/ are taken to be affricates in part of the literature (cf. Szende 1992: 119 ff. and references cited there) but have been argued to be stops in Siptár (1989a) and in Nádasdy and Siptár (1989: 19–20). We accept the latter analysis: this problem will be briefly reconsidered in section 4.1.2. The prevocalic realization of /x/ is [h] as in hír 'news'; the reasons why the underlying segment is nevertheless taken to be a velar fricative are discussed in section 8.2.2. The class of affricates is usually said to include

---

[5] The long-standing debate concerning whether Hungarian long consonants should be analysed as single [+long] segments or as pairs of identical short consonants has now come to a satisfactory conclusion: given the current non-linear representation of length (two skeletal points/timing units associated to a single melody), the problem simply disappears: the proliferation of segment types inherent in the former approach and the superfluous repetition of tokens (feature matrices) entailed by the latter can both be avoided.

*/d$^z$/; however, no phonological argument seems to support the existence of that segment at any level other than the surface (cf. É. Kiss and Papp 1984 and Siptár 1989*a* for details; the main line of argument will be reviewed in section 4.1.4 below). The palatal liquid /j/ is classified traditionally as a fricative and in most current analyses as a glide; both positions have been argued against by Nádasdy and Siptár (1989: 15–16) and Siptár (1993*a*). See section 2.2.1 above for some of the details. The asymmetrical behaviour of /x/ and /v/ with respect to voice assimilation and to certain other aspects of Hungarian phonology will be considered in section 4.1.1.

### 2.2.3. Dialect variation

In what follows, the major dialect areas of Hungarian will be listed and briefly characterized in terms of selected phonological discrepancies between them and ECH, the dialect described in this book (as well as SLH, cf. the opening paragraphs of section 2.1 above). The source of data listed here is Antal, Csongor, and Fodor (1970: 117–19).

(i) *Western dialect area.* Comprises primarily inhabitants of Vas and Zala counties. One characteristic feature of this dialect is the use of short mid front unrounded [e] (conveniently indicated in orthography-based transcriptions as *ë*) in addition to the seven short vowels of SLH/ECH, shown in section 2.2.1 above. For instance, forms like [sem] 'eye', [ember] 'man', [tesem] 'I do it' are heard here (cp. SLH/ECH *szem* [sɛm], *ember* [ɛmbɛr], *teszem* [tɛsɛm]). Another feature of this dialect is that some [j]'s (those indicated in the spelling by *ly* or derived from underlying /lj/ and spelt *lj*) are pronounced [l] or [lː], e.g. [milːɛn] (cp. SLH/ECH *milyen* [mijɛn] 'like what'), [ülːön] (cp. SLH/ECH *üljön* [üjːön] 'let him sit'). The incidence of short high vowels is higher than in ECH (where, in turn, it is higher than in SLH): [fü] (for ECH/SLH *fű* [füː] 'grass'), [fiu] (for *fiú* 'boy', ECH [fiu], SLH [fiuː]).

(ii) *Transdanubian dialect area.* Except for the parts discussed above, most of Transdanubia (the territory of Hungary west of the river Danube) belongs here. Mid [e] and the relative scarcity of long [uː üː] are characteristic here, too. *Ly* is pronounced as [l], but *lj* as [j]: [hɛl] (for SLH/ECH *hely* [hɛj] 'place') but [üjön le] (for *üljön le* [üjːönlɛ] 'let him sit down'). In the northern part of this area, [iː] replaces standard [eː], as in [viːn] (for SLH/ECH *vén* [veːn] 'old'), [niːp] (for *nép* [neːp] 'people'), [tɛrmiːš] (for *termés* [tɛrmeːš] 'crop').

(iii) *Alföld dialect area.* Covers the middle part of the Great Hungarian Plain (*Alföld*). The most conspicuous feature of dialects belonging here is that [ö] replaces the mid [e] found in some others: [söm] 'eye', [tösöm] 'I do it', [embör] 'man'. In some cases, word-final *l* is deleted: [ɔbːu] (for *abból* 'from that', ECH [ɔbːol], SLH [ɔbːoːl]).

(iv) *Duna–Tisza dialect area.* Comprises most of the territories between the rivers Danube and Tisza. Mid [e] is found here, too, but long [u: ü:] are retained. Orthographic *ly* is pronounced [j] as in SLH/ECH.

(v) *North-western dialect area*. The *palóc* dialect and related varieties belong here. The most conspicuous features are fronted/unrounded [a] and fully back, slightly rounded [ɔ:] for SLH/ECH [ɔ] and [a:], respectively, as in *apám* [apɔ:m] 'my father' (SLH/ECH [ɔpa:m]), and a palatal lateral [ʎ] for *ly* (e.g. *milyen* [miʎɛn] 'like what' vs. SLH/ECH [mijɛn] vs. Western [mil:ɛn]). The plural and possessive morphemes are non-lowering here ([hɔ:zakot] 'houses' (acc.), [hɔ:zamot] 'my house' (acc.), cp. SLH/ECH *házak-at* [ha:zɔkɔt], *házam-at* [ha:zɔmɔt], see section 8.1.3).

(vi) *North-eastern dialect area*. The Upper Tisza region and adjacent counties. There is no mid [e] here; all orthographic *e*'s are pronounced [ɛ] as in SLH/ECH. On the other hand, [i:] often replaces standard [e:], as in Northern Transdanubia (examples in (ii) above).

(vii) *Trans-Királyhágó dialect area*. Covers Transylvania (*Erdély*) in present-day Romania (except for *Székelyföld*). No mid [e]; [o] is lowered to [ɔ] in some contexts: [ɔzɔk] (for SLH/ECH *azok* [ɔzok]) 'those', [vɔdʸɔk] (for SLH/ECH *vagyok* [vɔdʸok]) 'I am'.

(viii) *Székely dialect area*. Like the *palóc*, the *székely* also fail to lower their [o]'s and [ö]'s in plural/possessive accusatives: [ha:zɔkot] 'houses' (acc.), [dʸü:rüŋköt] 'our ring' (acc.), cp. SLH/ECH *házak-at* [ha:zɔkɔt], *gyűrűnk-et* [dʸü:rü(:)ŋkɛt]). Vowel-length alternation in disyllabic stems is eliminated in favour of the short alternant, e.g. [tɛfiɛn] 'cow', [sɛkɛr:ɛl] 'with a cart', [lɛvɛl-bɛn] 'in a letter' (cp. SLH/ECH *tehén* [tɛfie:n], *szekér-rel* [sɛke:r:ɛl], *levél-ben* [lɛve:lbɛn] but plural *tehen-ek* [tɛfiɛnɛk], *szeker-ek* [sɛkɛrɛk], *level-ek* [lɛvɛlɛk]). In some stems, [ü] replaces standard [i]: [küš] 'small', [hüt] 'faith', [mü] 'we' (SLH/ECH *kis, hit, mi*).

## 2.3. STRESS AND INTONATION

In its citation form, a Hungarian word typically has a single primary stress, which falls on its initial syllable, no matter whether the word is simple (e.g. 'iskola 'school') or derived (e.g. 'forrósodik 'grows hot') or a compound (e.g. 'szénanátha 'hay fever').[6] In the metrical literature, it has been repeatedly

---

[6] There are several types of exception to this general stress rule. First, a marginal class of interjections has a single primary stress but not on its first syllable, e.g. a'há 'I see'. Second, a class of compound words (mostly but not exclusively numerals) have primary stresses on both compound members, e.g. 'kilencszáz'kilencven 'nine hundred and ninety'; depending on the phrasal context, these forms regularly undergo rhythmic stress alternation (roughly as in English, cf. Varga 1994b). Third, in exclamations, a primary stress is mechanically placed on every odd-numbered syllable, e.g. 'Ponto'san! 'Exactly!', 'Dehogy'is! 'Not at all!', with the possible exception of the last syllable of the sequence as in 'Termé'szetesen! 'Of course!'. Fourth, primary (contrastive) stress will fall on the last syllable of a word in corrective answers like the following:

claimed that Hungarian words exhibit a regular trochaic pattern of secondary stresses (cf. Hayes 1994: 330, Kager 1995: 374), and even that pairs of such binary feet are organized into 'cola' such that the overall pattern is 103020302030... where the numbers stand for primary, secondary, and tertiary stresses, respectively (cf. Hammond 1987). Native intuition does not support the assumption of such regular physical patterning superimposed on Hungarian strings. At any rate, this putative rhythmic intensity alternation is phonologically irrelevant as it does not interact in any way with the rest of the phonology.[7] Opinions differ, furthermore, concerning whether a (contrastive) degree of stress intervenes between primary and zero; Varga (1994*b*, 1996), Hetzron (1992), and Vogel and Kenesei (1987), for instance, posit a three-level stress system, whereas Kálmán and Nádasdy (1994) argue for just two levels: stressed and stressless.

Turning to sentence stress, the lexical primary stress referred to above is but a potential locus for stress: whether or not the syllable concerned is actually assigned primary stress in the sentence depends on syntactic structure. There are two major types of stresslessness: spontaneous enclisis as in (3) and stress eradication as in (4). Consider the following examples (from Kálmán and Nádasdy 1994: 398):

(3)  *a.*  'Géza      'táncolni      *akar*
                       to-dance      want
            'Géza wants to dance'

     *b.*  'Géza 'táncolni *akar*      a       'magas      'fekete      'lánnyal
                                       the      tall         black       girl-with
            'Géza wants to dance with the tall black(-haired) girl'

     *c.*  'Géza *bácsi*
            'Uncle Géza'

     *d.*  'Géza *bácsi* 'táncolni *akar* a 'magas 'fekete 'lánnyal
            'Uncle Géza wants to dance with the tall black girl'

(4)  *a.*  'Jenő      'táncolni      *imád*
                       to-dance      love
            'It is to dance that Jenő loves'

A: *Az öccse katona volt?* 'Was his brother a soldier?'
B: *Nem, kato'na.* 'No, he IS a soldier.'
This is due to the fact that in present indicative sentences there is no overt 3sg. copula to carry the contrastive stress (cf. Varga 1979 for extended discussion; see also Varga 1985: 213–14 for a short summary in English). Finally, a small set of function words is always unstressed: this set includes the definite articles *a, az* 'the', the particle *is* 'also', the conjunction *meg* 'and', the preposition *mint* 'in the quality of', and a few others.

[7] This is of course the reason why native intuition is unaware of the pattern, even if it indeed exists (which we claim is not the case).

b. 'Jenő 'táncolni *imád* a magas fekete lánnyal
   'It is to DANCE with the tall black girl that Jenő loves'

c. 'Jenő 'táncolni *akar*
   'It is to dance that Jenő wants'

d. 'Jenő 'táncolni *akar* a magas fekete lánnyal
   'It is to DANCE with the tall black girl that Jenő wants'

In (3), the italicized words are (lexically specified as) enclitic: they join the stress domain of the previous word. By contrast, in (4), the stress on *táncolni* eradicates the rest of the lexical stresses in its whole domain.[8] Unlike *akar* 'want', the example *imád* 'love' is a kind of verb that cannot be unstressed unless its stress is eradicated (by the focus of the sentence, *táncolni* in the present case, that precedes it). A non-focused counterpart of (4a) would be *'Jenő 'imád 'táncolni* 'Jenő loves to dance'. Notice that the surface stress patterns of (3a) and (4c) are identical: the only way to tell which is which is to append further words as in (3b) and (4d). If those further words are also unstressed, the sentence exhibits an eradicating stress pattern. Two important facts about eradicating stress are that it need not be stronger than a non-eradicating stress; and that it cannot be followed by another stress within the same sentence unless that other stress is also of the eradicating type. A sentence with no eradicating stress is said to have flat prosody, corresponding to neutral interpretation; a sentence with eradicating prosody has a contrastive or emphatic interpretation (for further details, see Kálmán and Nádasdy 1994 and references cited there; cf. also Hetzron 1992).

Primary stress is defined with reference to character contours, the linguistically significant pitch contours in terms of which Hungarian intonation patterns are described (see Varga 1994a, 1996). That is, an intensity peak that does not initiate a character contour is either regarded as secondary stress (Varga 1994b) or else ignored as phonologically irrelevant (Kálmán and Nádasdy 1994). Actual intensity differences are therefore not important: two primary-stressed syllables need not have identical intensity values, nor is a primary-stressed syllable necessarily stronger in intensity than a secondary-stressed one.

Varga (1996) distinguishes eleven character contours (or 'characters' for short), an appended contour and a preparatory contour.[9] Each contour is realized in several phonetic variants conditioned by the number of syllables on which they are spread out. Three major variants are distinguished: the one-syllable, the two-syllable, and the three-or-more-syllable variant. In addition to these, there is a certain amount of free variation in some cases.

---

[8] The domain of stress eradication extends to the following eradicating stress if there is one in the sentence; otherwise, to the end of the sentence.
[9] A simpler system is offered in Kornai and Kálmán (1989); both analyses are couched in autosegmental terms.

The front-falling characters comprise the full fall, the half fall, and the fall–rise. In their multisyllabic variants, the pitch radically drops down between the first and the second syllables. The full fall ends on the baseline (the lower limit of the speaker's normal voice range) and signals end of utterance in statements and in question-word questions. The half fall does not reach the baseline and is the most common character in non-utterance-final position. The fall–rise steps up at its end or rises steadily after the initial fall. It carries a 'conflicting' meaning component as in ˇNem alszom! 'I'm not sleeping', implying a situation like 'You must believe I am asleep, as you are walking on tiptoe; but I am not asleep'.

The second group of characters, called sustained characters, comprises the rise, the high monotone, and the descent. All three of them signal an explicitly incomplete preparation for something complete and significant to follow. They are typically used on certain non-final sentence constituents and on 'broken questions'. To use Varga's example, if one goes to see Aunt Angela in her home but, somewhat unexpectedly, the doctor opens the door, one might ask Angéla néni? 'Aunt Angela?' with a rise, a high monotone, or a descent, depending on the degree of unexpectedness of the doctor's presence in her flat.

The third group consists of three end-falling characters, the rise–fall, the monotone–fall, and the descent–fall. The falling part invariably starts on the penultimate syllable when the carrier phrase is more than three syllables long. All three are used for yes–no interrogatives, with an additional overtone of genuine questioning (rise–fall) or else the expression of the speaker's surprise or disbelief (descent–fall). The actual form that the rise–fall takes requires some comment (cf. Gósy and Terken 1994 for more phonetic detail). In a monosyllabic utterance (equivalently, in a longer utterance whose last primary stress falls on the last syllable), the contour goes up and down on the single syllable available but the falling part may be physically missing, especially if the syllable is light (short-vowelled and open) or ends in a voiceless consonant. When the contour has two syllables to spread out on, the pitch steps up between the two and then slides back on the second (this falling movement can, again, be physically missing). When there are more than two syllables at its disposal, the melody rises (gradually, early, or late) until it reaches the penultimate syllable and then it drops abruptly between the penultimate and the last syllable.

In addition to yes–no questions, these contours are also used for echoed question-word questions as in (5) and in repetitive question-word questions as in (6). The two mini-dialogues also exemplify the fact mentioned above that normal question-word questions exhibit a full fall.[10]

---

[10] In both B-sentences the single primary stress is on the syllable *Ki* but the pitch peak is on the syllable *koz*. Echoed questions take a rise–fall because they suggest the matrix yes–no question 'Did you ask ... ?'; repetitive questions are implicitly embedded in the yes–no question 'Would you repeat ... ?'.

(5) A: ˋKivel         találkoztak?    (= ˈWho did they ˋmeet?)
      who-with    met-they

    B: ˆKivel találkoztak?       (= ˈWho did they ´meet?)

(6) A: ˋAngélával    találkoztak.    (= They ˈmet ˋAngela.)
      Angela-with  met-they

    B: ˆKivel találkoztak?       (= ´Who did they meet?)

The rest of the contours in Varga's analysis are what he calls second-type descent (used for evaluative exclamations), the stylized fall (used for calling someone's attention from a distance, see Varga 1989, 1995 for details), as well as the appended contour (a low monotone—practically on the baseline—which is melodically separate from any meaningful intonation contour that happens to precede it and which is not associated with any degree of stress on its first syllable) and the preparatory contour (a lowish monotone or one of the sustained contours without stress, characterizing the material that precedes the first character contour of the utterance or serves as a buffer between two character contours). The examples in (7) are here to illustrate the last two types.

(7) *a.* ˆMaga     az,   ₀Pál?       'Is that you, Paul?'
       you      that

    *b.* ˆMaga az?—   ₀kérdezte Pál.  '"Is that you?", Paul asked'
                    asked

    *c.* ₀És   ha      ´eljön?     'And if he comes?'
       and  if     away-come

    *d.* ˋGyere!   ₀Ide ˋne  firkálj!   'Come. Don't scribble here'
       come-imp. here not  scribble-imp.

    *e.* ˋGyere ide!     ˋNe firkálj!  'Come here. Don't scribble'

(7*a*) shows a vocative, and (7*b*) a quoting clause, both appended to a rise–fall. (7*c*) illustrates an utterance-initial preparatory contour. *Ide* in (7*d*) carries a preparatory contour sandwiched between two full falls, whereas the same word in (7*e*) is part of the first contour and continues the falling pitch movement that began on *Gyere*.

This concludes our cursory overview of the stress and intonation patterns of Hungarian. For further details, as well as a thorough autosegmental analysis, the reader is invited to consult Varga (1996) directly (see also Kornai and Kálmán 1989 and Hetzron 1992).

## 2.4. MORPHOLOGY AND SYNTAX

In this section we give a short summary of the major morphological patterns of Hungarian, and make a few points about the word order of simple sentences. For a more comprehensive treatment, cf. Kornai (1994) with respect to morphology and É. Kiss (1987), Kiefer and É. Kiss (1994) with respect to syntax.

As an agglutinative language, Hungarian builds its word forms by juxtaposing a number of suffixes, each of which represents a single morphological function. The root morpheme (or 'absolute stem') may be followed by a number of derivational suffixes (where the addition of each suffix produces a new stem, 'relative stem' as it is conventionally referred to), then by a number of inflectional suffixes (again, each time—except the last—a new 'relative stem' is produced). Consider the following examples:

(8) *a.* barát -ság -os -abb -an
     friend -ship adj. comp. adv.
     'in a more friendly manner'
   *b.* ház -as -ul -andó -k -nak
     house adj. verb participle pl. dat.
     'for those intending to get married'
   *c.* te -het -ség -es -ebb -ek -et
     do -able -ness adj. comp. pl. acc.
     'the more talented ones' (acc.)

In (8*a*), the noun stem *barát* is followed by the derivational suffix *-ság* to form an abstract noun ('friendship') which is then followed by the derivational suffix *-os* to form an adjective ('friendly'), by the comparative marker *-abb* ('more friendly'), and finally by the adverb-forming suffix *-an*. In (8*b*), the noun stem *ház* is first converted into an adjective meaning 'married', then into a verb meaning 'get married'. The participial suffix *-andó* adds futurity to the meaning; the resulting participle is used as a noun and receives a plural marker followed by a dative case ending. In (8*c*), the verb root *te-* (citation form: *tesz* 'do') is followed by the derivational suffix *-het* ('can do'), the nominalizing suffix *-ség* (a vowel-harmony alternant of *-ság* in (8*a*), see section 3.2) to produce the abstract noun meaning 'talent', the adjectivizing suffix *-es* ('talented'), the comparative *-ebb* ('more talented'), the plural marker, and the accusative case ending.

Inflectional suffixes are traditionally subdivided into two major categories on the basis of how they can combine with stems and with other suffixes. The first set (traditionally called *jel* 'sign') includes possessive suffixes, the plural marker of nouns, the comparative marker of adjectives, and mood and tense markers of verbs; the defining feature of this set is that more than one of its members can appear in the same word form and that it can be followed by a member of the other set (called *rag*, a back-formation from *ragaszt* 'stick on'). This other set includes verbal person/number endings and case endings; only one *rag* can occur in a word form and it is always the terminal morpheme of the string.

In this book, we will not follow this traditional classification. Rather, we introduce a new distinction between 'analytic' and 'synthetic' suffixes, based on their respective phonological properties and cross-cutting both the traditional *jel/rag* distinction and (even) the morphologically valid—but phonologically not necessarily relevant—distinction between derivational and inflectional suffixes. See Chapters 5 and 8 for discussion and extensive exemplification; see Harris (1994), Kaye (1995) for the notion of 'analytical' morphological boundaries in general.

Prefixes are much less common in Hungarian morphology: in addition to superlative *leg-* (*legbarátságosabban* 'in the most friendly manner', *legtehetségesebbeket* 'the most talented ones' (acc.)) and a number of loan prefixes like *anti-*, *pre-*, *extra-*, etc., the only eligible category is that of 'verbal prefixes' or 'preverbs' but these are better analysed as separate words (or compound members); see below.

The fact that Hungarian is an agglutinating language implies that the overwhelming majority of stems do not exhibit any morphophonological alternation. Their shapes are either literally constant or else they do alternate but only in terms of very general phonological processes like voicing assimilation (cf. sections 4.1.1 and 7.3), nasal place assimilation (cf. sections 4.2 and 7.4.1) or low vowel lengthening (cf. sections 3.1.1 and 6.2.1) if they happen to end in an obstruent, a nasal, or a low vowel, respectively. We will call these stems 'major' stems, as opposed to various 'minor' stem classes that do exhibit stem alternants particular to these (usually rather small) classes. All minor stems constitute closed, unproductive classes: new members are not added and existing ones often have regularized 'major stem' counterparts or variants. It is to be noted that the incidence of members may vary from dialect to dialect or even from one speaker to the next; it is often the case that some suffixes select the minor stem but others the corresponding major stem, again with considerable interspeaker variation. The types of verbal minor stems will be briefly introduced in section 2.4.2, those of nominal minor stems in section 2.4.3.

How much of the behaviour of minor stems falls within the realm of phonology depends on the framework one adopts. For instance, classical generative phonology offered a battery of 'minor rules', ordering devices, exception mechanisms and other means to come to grips with possibly all minor patterns in a language. An example of this with respect to Hungarian is Vago (1980*a*). Various recent phonological theories, on the other hand, claim that all of this is outside phonology and should not be covered in a systematic manner at all. A third possibility is that such minor patterns are not ignored altogether but are treated as part of morphology (or accounted for by morphological, rather than phonological, means). A recent example of this latter solution (cast in an Optimality Theory framework) is Stiebels and Wunderlich (1999).

In this book, certain minor patterns will be discussed quite extensively, whereas others will be ignored as they reveal nothing of interest about the phonological system of Hungarian as a whole.

### 2.4.1. Derivation and compounding

The most important derivational suffixes will be presented here in a tabular form. For a more extensive treatment of the derivational morphology of Hungarian, see Kenesei, Vago, and Fenyvesi (1998: 351–81). Table 1 shows deverbal verb-forming suffixes; vowel-harmony alternants are listed separately for convenience.[11]

TABLE 1. *Deverbal verb-forming derivational suffixes*

| Base | Gloss | Causative | Gloss | Suffix |
|------|-------|-----------|-------|--------|
| mos | wash | mosat | make wash | -at |
| küld | send | küldet | make send | -et |
| olvas | read | olvastat | make read | -tat |
| nevet | laugh | nevettet | make laugh | -tet |

| Base | Gloss | Reflexive | Gloss | Suffix |
|------|-------|-----------|-------|--------|
| mos | wash | mosakodik | wash (oneself) | -kod(ik) |
| emel | lift | emelkedik | rise | -ked(ik) |
| fésül | comb | fésülködik | comb (one's hair) | -köd(ik) |
| táplál | feed | táplálkozik | take food | -koz(ik) |
| jelent | mean | jelentkezik | present oneself | -kez(ik) |
| töröl | wipe | törölközik | dry oneself | -köz(ik) |
| húz | pull | húzódik | drag on | -ód(ik) |
| vet | throw | vetődik | throw oneself | -őd(ik) |
| takar | cover | takarózik | cover oneself | -óz(ik) |
| kerget | chase | kergetőzik | chase about | -őz(ik) |

[11] The forms of suffixes as cited in Tables 1–4 are not meant to stand for underlying representations or to make any claim about the phonological structure of the suffixes concerned. They are merely listed in their surface-observable shape (or set of shapes) for ease of reference.

Parenthesized (ik) is a third person singular ending, restricted to a small set of root verbs but regularly appearing in the derived verb types listed in the table; see section 2.4.2.

Some reflexive suffixes not listed in the table are: -kóz (zárkózik 'lock oneself'), -ód (csavarodik 'wind itself'). Reflexive suffixes may also express mutual action as in kergetőzik 'chase (one another)', verekedik 'beat (one another)', ölelkezik 'embrace (one another)' or have passive meaning (where the instigator is unknown) as in bepiszkolódik 'get dirty', becsukódik 'get closed', leleplеződik 'get revealed'. 'May'-verbs can be based on derived stems as in mosathat 'may have sth washed', emelkedhet 'may rise'. Additional frequentative suffixes include -g (füstölög 'emit smoke'), -kál (járkál 'keep walking up and down'), -csál (rágcsál 'keep chewing'), -del (tördel 'keep breaking'), -décsel (nyögdécsel 'keep moaning'), -károz (futkározik 'keep running about'), etc.

TABLE 1. *Cont'd*

| Base | Gloss | Passive | Gloss | Suffix |
|------|-------|---------|-------|--------|
| ad | give | adatik | be given | -atik |
| kér | beg | kéretik | be begged | -etik |
| táplál | feed | tápláltatik | be fed | -tatik |
| nevel | rear | neveltetik | be brought up | -tetik |

| Base | Gloss | 'May do' | Gloss | Suffix |
|------|-------|----------|-------|--------|
| mos | wash | moshat | may wash | -hat |
| kér | beg | kérhet | may beg | -het |

| Base | Gloss | Frequentative | Gloss | Suffix |
|------|-------|---------------|-------|--------|
| olvas | read | olvasgat | read now and then | -gat |
| beszél | speak | beszélget | converse | -get |
| szalad | run | szaladgál | run up and down | -gál |
| keres | search | keresgél | search here and there | -gél |
| vág | cut | vagdos | cut into pieces | -dos |
| tép | tear | tépdes | tear into pieces | -des |
| lök | push | lökdös | keep pushing | -dös |
| kap | catch | kapkod | keep catching (at) | -kod |
| lép | step | lépked | amble along | -ked |
| köp | spit | köpköd | spit about | -köd |
| áll | stand | álldogál | stand about | -dogál |
| néz | look | nézdegél | look around | -degél |
| ül | sit | üldögél | sit about | -dögél |

| Base | Gloss | Inchoative | Gloss | Suffix |
|------|-------|------------|-------|--------|
| él | live | éled | come to life (again) | -d |
| szól | speak | szólal | start speaking | -l |
| kon(g) | toll | kondul | begin to toll | -dul |
| csen(g) | ring | csendül | begin to ring | -dül |

Table 2 shows denominal (including deadjectival) verbalizing suffixes.[12]
Table 3 exhibits major types of deverbal noun and adjective forming suffixes.[13]

---

[12] The meanings of the sets of suffixes listed in Table 2 are 'use some instrument', 'provide with something', 'collect something', 'behave in some manner', 'think of something as', 'provide something with some quality', 'acquire some quality', respectively.

[13] Further suffixes of this type include *-ék* (*festék* 'paint', instrument of action), *-et* (*kelet* 'east', place where something happens, where the sun rises in this case, from *kel* 'rise'), *-at* (*nyugat* 'west', place where something happens, from *nyug(szik)* 'set'), *-i* (*maradi* 'old-fashioned', characteristic property, from *marad* 'stay'). The groups in the table share the meaning components 'process', 'result of action', 'ability/property', 'lack of ability/property', respectively; the last three groups comprise infinitives, participles, and deverbal adverbials. Present participles (*író, kérő*) are often converted/zero-derived into nouns: the examples, as nouns, mean 'writer' and 'suitor', respectively. Similarly, past participles are often converted into adjectives.

TABLE 2. *Denominal verb-forming suffixes*

| Base | Gloss | Derived verb | Gloss | Suffix |
|---|---|---|---|---|
| gyalu | plane | gyalul | trim (wood) | *-l* |
| fésű | comb | fésül | comb (hair) | |
| gereblye | rake | gereblyéz | (use a) rake | *-z* |
| pipa | pipe | pipázik | smoke a pipe | |
| talp | sole | talpal | (re-)sole (shoes) | *-l* |
| fej | head | fejel | (provide with) head | |
| folt | patch | foltoz | (put a) patch (on) | *-z* |
| fűszer | spice | fűszerez | season (food) | |
| hal | fish | halászik | catch fish | *-ász(ik)* |
| egér | mouse | egerészik | catch mice | *-ész(ik)* |
| málna | raspberry | málnázik | pick raspberries | *-z(ik)* |
| bohóc | clown | bohóckodik | fool around | *-kod(ik)* |
| ügyes | skilful | ügyeskedik | act skilfully | *-ked(ik)* |
| őr | guard | őrködik | keep guard (over) | *-köd(ik)* |
| csoda | wonder | csodál | admire | *-l* |
| helyes | right | helyesel | approve (of) | |
| sok | plenty | sokall | find sth too much | *-ll* |
| rossz | bad | rosszall | disapprove (of) | |
| szabad | free | szabadít | set free | *-ít* |
| szép | pretty | szépít | make pretty | |
| szabad | free | szabadul | get free | *-ul* |
| szép | pretty | szépül | get pretty | *-ül* |
| fiatal | young | fiatalodik | get young(er) | *od(ik)* |
| öreg | old | öregedik | get old(er) | *-ed(ik)* |

TABLE 3. *Deverbal nouns, adjectives, participles, and adverbials*

| Base | Gloss | Derived form | Gloss | Suffix |
|---|---|---|---|---|
| tanul | learn | tanulás | (the process of) learning | *-ás* |
| ég | burn | égés | (the process of) burning | *-és* |
| gondol | think | gondolat | thought | *-at* |
| felel | answer | felelet | (an) answer | *-et* |
| ad | give | adomány | gift | *-(o)mány* |
| vet | sow | vetemény | vegetable sown | *-(e)mény* |
| lát | see | látvány | sight (sth seen) | *-vány* |
| önt | mould | öntvény | mould (sth moulded) | *-vény* |
| tanul | learn | tanulékony | teachable | *-ékony* |
| hisz | believe | hiszékeny | credulous | *-ékeny* |

TABLE 3. *Cont'd*

| fal | devour | falánk | greedy | *-ánk* |
| fél | fear | félénk | timid | *-énk* |
| tanul | learn | tanulatlan | uneducated | *-(a)tlan* |
| keres | search | keresetlen | unsought for | *-(e)tlen* |
| ír | write | írni | to write | *-ni* |
| kér | ask | kérni | to ask (for sth) | |
| ír | write | író | one who writes | *-ó* |
| kér | ask | kérő | one who asks (for sth) | *-ő* |
| ír | write | írott | written | *-tt* |
| kér | ask | kért | asked for | *-t* |
| ír | write | írandó | (sth) to be written | *-andó* |
| kér | ask | kérendő | (sth) to be asked for | *-endő* |
| ír | write | írva | (while) writing | *-va* |
| kér | ask | kérve | (while) asking | *-ve* |

Finally, Table 4 shows a selection of denominal/deadjectival noun/adjective forming suffixes.[14]

TABLE 4. *Denominal/deadjectival noun/adjective-forming suffixes*

| Base | Gloss | Derived form | Gloss | Suffix |
|------|-------|--------------|-------|--------|
| kocsi | car | kocsis | carman | *-s* |
| lakat | padlock | lakatos | locksmith | |
| juh | sheep | juhász | shepherd | *-ász* |
| kert | garden | kertész | gardener | *-ész* |
| katona | soldier | katonaság | army | *-ság* |
| hegy | mountain | hegység | mountain range | *-ség* |
| barát | friend | barátság | friendship | *-ság* |
| szép | beautiful | szépség | beauty | *-ség* |
| leány | girl | leányka | little girl | *-ka* |
| egér | mouse | egérke | little mouse | *-ke* |
| fiú | boy | fiúcska | little boy | *-cska* |
| könyv | book | könyvecske | little book | *-cske* |
| Szabó | \<name\> | Szabóné | Mrs Szabó | *-né* |

[14] The group meanings are as follows: 'occupation', 'collective noun', 'abstract noun', 'diminutive', 'wife of', 'having some property', 'lacking some property', 'belonging somewhere', 'measure', 'fraction', and 'ordinal number', respectively.

TABLE 4. *Cont'd*

| | | | | |
|---|---|---|---|---|
| só | salt | sós | salty | -s |
| erő | strength | erős | strong | |
| barna haj | brown hair | barna hajú | brown-haired | -ú |
| kék szem | blue eyes | kék szemű | blue-eyed | -ű |
| só | salt | sótlan | saltless | -tlan |
| erő | strength | erőtlen | strengthless | -tlen |
| szag | smell | szagtalan | scentless | -talan |
| íz | taste | íztelen | tasteless | -telen |
| iskola | school | iskolai | school (adj.) | -i |
| Pécs | \<town\> | pécsi | inhabitant of P. | |
| hüvelyk | thumb | hüvelyknyi | as small as a thumb | -nyi |
| tenger | sea | tengernyi | very many/much | |
| öt | five | ötöd | one-fifth | -d |
| hét | seven | heted | one-seventh | |
| öt | five | ötödik | (the) fifth | -dik |
| hét | seven | hetedik | (the) seventh | |

Compounds are formed much like in English (see Kiefer 1992). One important aspect of compound formation (from a phonological point of view) is that compounds form a single domain for word stress assignment but each compound member is a separate domain with respect to vowel harmony. Vogel (1989) resolves this apparent contradiction by claiming that each compound member constitutes a phonological word, whereas the compound as a whole is a prosodic constituent called a 'clitic group'. Other instances of clitic groups include preverb + verb combinations as in (9a) and combinations of bare noun object + verb as in (9b). Furthermore, Hungarian has a number of directional clitics, too: proclitics as in (9c) and enclitics as in (9d). Note that proclitics are special in that (unlike in all other cases where the single primary stress of the whole clitic group falls on its initial syllable) in proclitic + host combinations the first syllable of the host is stressed.

(9)   a.   'fel-     darabol        'cut up'
           up       cut
           'oda-     küldenek      'they send (so) there'
           there     send-3pl

       b.   'kenyeret   eszik         'he eats bread'
           bread-acc.   eat-3sg
           'fát        vág           'he chops wood'
           wood-acc.   cut-3sg

| *c.* | az | 'ablak | 'the window' |
| | the | window | |
| | egy | 'ablak | 'a window' (vs. '*egy* '*ablak* 'one window') |
| | a | window | |
| | és | 'János | 'and John' |
| | and | John | |
| | hogy | 'elmész | 'that you will leave' |
| | that | away-go-2sg | |
| | és ha | 'írta | 'and if he wrote it' |
| | and if | wrote-3sg | |

| *d.* | 'János | is | | 'John, too' |
| | John | too | | |
| | 'Mari | meg | | 'and Mary' |
| | Mary | and | | |
| | 'számtanból | sem | | 'neither in maths' |
| | maths-from | neither | | |

Phonologically, then, preverb-verb combinations (cf. (9*a*)) act like compounds. However, syntactically the behaviour of preverbs parallels that of preverbal bare noun objects (cf. (9*b*)) rather than that of regular compounds in that they can be postposed (cf. *nem darabol fel* 'does not cut up', *nem vág fát* 'does not chop wood') or separated from the verb by an auxiliary[15] (cf. *fel kell darabolni* 'it must be cut up', *fát kell vágni* 'wood must be chopped') or by other small words (cf. *fel is darabol* 'cuts up, too', *fát sem vág* 'does not even chop wood'). Semantically, the meaning of a preverb-verb combination is often non-transparent or 'figurative', i.e. it cannot be derived from the meanings of the components.

Preverbs are listed in (10). Wherever possible, a primary (locative/directional) and one or two more figurative examples are both provided.

| (10) | *abba-* | abbahagy 'stop doing sth' |
| | *agyon-* | agyonüt 'strike dead', agyondicsér 'praise to the skies' |
| | *alá-* | alátesz 'put under', |
| | | alábecsül 'underestimate', aláír '(under)sign' |
| | *át-* | áttesz 'put over (to)', |
| | | átgondol 'think over', átolvas 'skim (through)' |
| | *be-* | betesz 'put in', |
| | | beindul 'start up', beijed 'get frightened', belép 'enter' |
| | *bele-* | beletesz 'put into it', |
| | | belegondol 'think over', belemarkol 'grab at' |

---

[15] On the auxiliary system of Hungarian, see Kálmán *et al.* (1984, 1989); on the syntax of preverbs, see Farkas and Sadock (1989); more on prosodic constituents: Kenesei and Vogel (1989). Vogel and Kenesei (1987, 1990), and Vogel (1989).

| | |
|---|---|
| *benn-* | bennmarad 'stay in', bennfoglal 'include' |
| *egybe-* | egybeolvad 'merge', egybevet 'compare' |
| *el-* | eltesz 'put away', |
| | elcsíp 'catch', elintéz 'put straight', eljut 'get somewhere' |
| *ellen-* | ellenáll 'resist', ellenőriz 'control' |
| *elő-* | elővesz 'produce (from pocket)', előír 'prescribe' |
| *előre-* | előretesz 'put to the front', előremutat 'point ahead' |
| *fell föl-* | feltesz 'put up, suppose', |
| | felgyújt 'set fire to', felismer 'recognize' |
| *félbe-* | félbehajt 'fold in two', félbeszakít 'interrupt' |
| *félre-* | félretesz 'put aside', |
| | félreért 'misunderstand', félrebeszél 'rave' |
| *felül fölül-* | felülfizet 'overpay', felülbírál 'supervise' |
| *fenn fönn-* | fenntart 'maintain', fennáll 'be valid' |
| *hátra-* | hátratesz 'put to the back', |
| | hátratántorodik 'stagger back' |
| *haza-* | hazamegy 'go home', hazabeszél 'have an axe to grind' |
| *helyre-* | helyretesz 'put in its place', |
| | helyrehoz 'remedy', helyreállít 'restore' |
| *hozzá-* | hozzátesz 'add', |
| | hozzászokik 'get used to', hozzámegy 'be married to' |
| *ide-* | idetesz 'put here', |
| | ideküld 'send here', idevág 'be appropriate' |
| *keresztül-* | keresztülszúr 'pierce through', keresztülhúz 'cross out' |
| *ketté-* | kettéhasad 'split in two', kettéágazik 'bifurcate' |
| *ki-* | kitesz 'put out', |
| | kiépít 'build up', kifejez 'express', kiöl 'kill off' |
| *körül-* | körülvesz 'surround', körülnéz 'have a look round' |
| *közbe-* | közbeszúr 'interpolate', közbelép 'intervene' |
| *közre-* | közrefog 'surround', közreműködik 'contribute' |
| *külön-* | különtesz 'put aside', |
| | különír 'write as two words', különválaszt 'separate' |
| *le-* | letesz 'put down', |
| | lefoglal 'reserve', leköszön 'resign', lead 'hand in' |
| *meg-* | megtesz 'do' (perfective), |
| | megelőz 'prevent', meglát 'catch sight of' |
| *mellé-* | mellétesz 'put beside sth', |
| | mellérendel 'co-ordinate', melléfog 'blunder' |
| *neki-* | nekitámaszt 'lean sth against sth', nekilát 'set about doing sth' |
| *oda-* | odatesz 'put somewhere', |
| | odaad 'hand over', odaég 'get burnt', odacsap 'smite' |
| *össze-* | összetesz 'put together', |
| | összedől 'collapse', összeforr 'heal' |

| | |
|---|---|
| *rá-* | rátesz 'put on top of', |
| | rájön 'find out', ráolvas 'cast a spell on' |
| *rajta-* | rajtaveszt 'fare ill', rajtaüt 'take so unawares' |
| *széjjel-* | széjjelválaszt 'separate', széjjeltép 'tear to pieces' |
| *szembe-* | szembefordul 'turn against', szembeállít 'contrast with' |
| *szerte-* | szerteszór 'scatter about', szerteágazik 'ramify' |
| *szét-* | széttesz 'sprawl', |
| | szétszéled 'disperse', szétfő 'boil to a pulp' |
| *tele-* | teletesz 'fill', teleszór 'bestrew', telezsúfol 'cram' |
| *tova-* | tovatűnik 'fade away', tovaterjed 'spread' |
| *tovább-* | továbbad 'pass on', továbbmegy 'go further' |
| *tönkre-* | tönkretesz 'spoil', |
| | tönkrever 'beat hollow', tönkremegy 'go bankrupt' |
| *túl-* | túltesz 'surpass', túllép 'exceed', túlteng 'superabound' |
| *újjá-* | újjáépít 'rebuild', újjászületik 'be born again' |
| *újra-* | újrakezd 'recommence', újratermel 'reproduce' |
| *utána-* | utánacsinál 'do sth after so', utánanéz 'go into the matter' |
| *végbe-* | végbemegy 'take place', végbevisz 'carry out' |
| *végig-* | végiggondol 'think over', végigmér 'look so up and down' |
| *vissza-* | visszatesz 'put back', |
| | visszalép 'backtrack', visszatart 'restrain' |

### 2.4.2. Verbal inflection

Verbs are inflected for mood (indicative, conditional, imperative/subjunctive), tense (past, present, future), number (singular, plural), and person (first, second, third). The indicative, the present, and the third person singular (except for *-ik*-verbs, see below) have no overt marker; past conditional and future indicative are expressed periphrastically.[16] Thus, the Hungarian verb has four distinct paradigms: present (indicative), past (indicative), (present) conditional, and imperative/subjunctive. In each case, a definite and an indefinite paradigm are differentiated: the definite conjugation is used in transitive sentences where the direct object is definite and third person (singular or plural), whereas the indefinite forms are used in all other cases: intransitively, with

---

[16] The future is formed by the infinitive (*várni* 'to wait', *kérni* 'to ask') and the auxiliary *fog*, thus *várni fogok* 'I will wait', *kérni fogsz* 'you will ask', etc. The past conditional is made up by the appropriate past-tense form of the verb plus the form *volna* 'would be', thus *vártam volna* 'I would have waited', *kértél volna* 'you would have asked', etc. Notice that in past conditional forms *volna* is not an auxiliary since its form is invariant; person/number suffixes are attached to the verb stem, unlike in the future forms where the stem is invariable (infinitive) and *fog* is conjugated.

indefinite direct objects, and with first or second person direct objects.[17] These eight patterns are shown in Table 5 for *vár* 'wait for', *kér* 'ask (to do sth)', and *tűr* 'endure, suffer'.[18] In other words, a Hungarian verb form has three parts: a stem, a mood/tense marker, and a person/number ending. The marker of the past tense is *-t-* (*-tt-* in other cases, not shown in Table 5), the marker of the conditional is *-na/ne/ná/né-* (the alternation is subject to vowel harmony, see section 3.2 and low vowel lengthening, see section 3.1.1), and the marker of the imperative is *-j-*. The person/number endings are shown in (11).

(11)    *Indefinite*    *Definite*
  1sg   -k              -m
  2sg   -sz/l           -d
  3sg   -Ø              -(j)a/e/i
  1pl   -unk/ünk        -(j)uk/(j)ük
  2pl   -tok/tek/tök    -(j)átok/étek/itek
  3pl   -nak/nek        -(j)ák/ék/ik

The small class of *-ik*-verbs takes special person/number suffixes in the present indicative singular: *-m* (1sg, both definite and indefinite), *-ik* (3sg indefinite); in the conditional and the imperative, the special suffixes are obsolete and are normally replaced by the corresponding general suffix. Thus, 'I would sleep' is either *aludnám* or (usually) *aludnék*, 'he would sleep' is either *aludnék* or (usually) *aludna*; 'let me sleep' is *aludjam* or (usually) *aludjak*, and 'let him sleep' is *aludjék* or (usually) *aludjon*.

*Ik*-verbs are an instance of 'minor stems' in the sense defined in section 2.4 above. Other minor stem classes include 'epenthetic' stems, '*v*-adding' stems, and '*sz/d* stems', as well as a few verb stems whose conjugation is totally irregular (on these, see Törkenczy 1997: 38–42, Vago 1980*a*: 84–5).

'Epenthetic' stems end in -$CV_uC$ where $V_u$ stands for an unstable vowel, i.e. one that alternates with zero (cf Vago 1980*a*: 79–84, Stiebels and Wunderlich 1997: 28–9):

(12)  sodor   'roll'      sodor-ja (3sg pres. def.)    sodr-om (1sg pres. def.)
      seper   'sweep'     seper-ted (2sg past def.)    sepr-ek (1sg pres. indef.)
      gyötör  'pester'    gyötör-nék (3pl cond. def.)  gyötr-i (3sg pres. def.)

---

[17] If the subject is first person singular and the object is second person (singular or plural), a special form is used: *vár-lak* 'I wait for you-sg.', *kér-te-lek* 'I asked you-sg.'; *vár-ná-lak* (*titeket/benneteket*) 'I would wait for you-pl.', *kér-je-lek* (*titeket/benneteket*) 'I ask-imp. you-pl.'. *Titeket* or *benneteket* (both: 'you-pl.-acc.') is added to disambiguate the number of the object which the verb inflection itself leaves undetermined, though the singular reading is the unmarked one.

[18] The paradigm of *tűr* is given for the present forms only; in all other cases *tűr* takes exactly the same endings as *kér* does.

For a more extensive presentation of the verbal morphology of Hungarian, cf. Kenesei, Vago, and Fenyvesi (1998: 282–330).

TABLE 5. *Conjugation paradigms*

|  |  | Indefinite | Present | | Definite | |
|---|---|---|---|---|---|---|
| Sg. | 1 | várok | kérek | tűrök | várom | kérem | tűröm |
|  | 2 | vársz | kérsz | tűrsz | várod | kéred | tűröd |
|  | 3 | vár | kér | tűr | várja | kéri | tűri |
| Pl. | 1 | várunk | kérünk | tűrünk | várjuk | kérjük | tűrjük |
|  | 2 | vártok | kértek | tűrtök | várjátok | kéritek | tűritek |
|  | 3 | várnak | kérnek | tűrnek | várják | kérik | tűrik |

|  |  | Indefinite | | Past | Definite | |
|---|---|---|---|---|---|---|
| Sg. | 1 | vártam | kértem | vártam | kértem |
|  | 2 | vártál | kértél | vártad | kérted |
|  | 3 | várt | kért | várta | kérte |
| Pl. | 1 | vártunk | kértünk | vártuk | kértük |
|  | 2 | vártatok | kértetek | vártátok | kértétek |
|  | 3 | vártak | kértek | várták | kérték |

|  |  | Indefinite | | Conditional | Definite | |
|---|---|---|---|---|---|---|
| Sg. | 1 | várnék | kérnék | várnám | kérném |
|  | 2 | várnál | kérnél | várnád | kérnéd |
|  | 3 | várna | kérne | várná | kérné |
| Pl. | 1 | váránk | kérnénk | várnánk | kérnénk |
|  | 2 | várnátok | kérnétek | várnátok | kérnétek |
|  | 3 | várnánk | kérnénk | várnák | kérnék |

|  |  | Indefinite | | Imperative | Definite | |
|---|---|---|---|---|---|---|
| Sg. | 1 | várjak | kérjek | várjam | kérjem |
|  | 2 | várj(ál) | kérj(él) | vár(ja)d | kér(je)d |
|  | 3 | várjon | kérjen | várja | kérje |
| Pl. | 1 | várjunk | kérjünk | várjuk | kérjük |
|  | 2 | várjatok | kérjetek | várjátok | kérjétek |
|  | 3 | várjanak | kérjenek | várják | kérjék |

Some of these stems are -*ik*-stems as well: *fürdik* 'bathe'–*fürdeni/fürödni* (inf.), *lélegzik* 'breathe'–*lélegezni* (inf.), *ugrik* 'jump'–*ugrani/ugorni* (inf.), etc. Note that the infinitive suffix usually attaches to the longer stem form (*sodor-ni, seper-ni, gyötör-ni*) but may exhibit variation, especially for stems that are both 'epenthetic' and -*ik*- stems. 'Epenthetic' stems will be discussed in section 8.1.1.

'*V*-adding' stems are the only Hungarian verb stems that end in a vowel in isolation and before consonant-initial suffixes; before vowel-initial suffixes, a /v/ is inserted while the stem final vowel gets shortened (cf. Vago 1980*a*: 76–8, Stiebels and Wunderlich 1997: 29–30):

(13)  ró   'scribble'   ró-nak (3pl pres. indef.)   rov-om (1sg pres. def.)
      lő   'shoot'      lő-nek (3pl pres. indef.)   löv-öm (1sg pres. def.)

In the past tense, these stems end in a vowel to which the past marker is attached in its long form as -*tt*-: *ró-tt, lő-tt* (3sg past indef.) *ró-tt-ak, lő-tt-ek* (3pl past indef.) This class will not be dealt with any further in this book.

'*Sz/d* stems' are all -*ik*-verbs. Their stem-final /s/ alternates with *Vd* as follows (cf. Vago 1980a: 85–6):

(14)  alsz-ik    'sleep'     alud-tam (1sg past)   alsz-om (1sg pres.)
      feksz-ik   'lie'       feküd-tem             feksz-em
      öregsz-ik  'get old'   öreged-tem            öregsz-em

Some stems in this class have regular variants in -*Vdik* along with -*szik*:

(15)  mosaksz-ik   'wash' (refl.)   —               mosaksz-om
      mosakod-ik                    mosakod-tam      mosakod-om
      dicseksz-ik  'boast'          —               dicseksz-em
      dicseked-ik                   dicseked-tem     dicseked-em

Others exhibit -*usz/üsz* vs. -*ud/üd* allomorphy:

(16)  alkusz-ik   'bargain'   alkud-tam   alkusz-om
      esküsz-ik   'swear'     esküd-tem   csküsz-öm

This class will be ignored in what follows.

The verbal morphophonology of Hungarian is analysed, in various frameworks, by Hetzron (1972), Vago (1980*a*), Abondolo (1988), Olsson (1992), Kornai (1994). In the present book, no comprehensive account is attempted; however, a number of purely phonological generalizations that emerge from the above data (and additional data not summarized here) will be treated in the appropriate places below.

### 2.4.3. Nominal inflection

The possible endings of nouns can be represented diagrammatically as in (17) (adapted from Kálmán 1985*c*: 255):

(17)

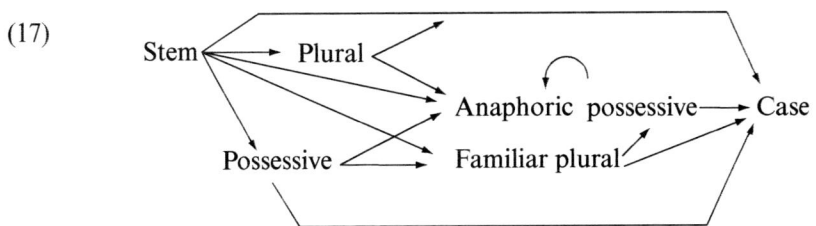

The plural ending is *-k/ok/ek/ök/ak* (see Chapter 8 on the alternation). The possessive endings are tabulated in (18); note the partial overlap with verbal person/number endings as in (11).

(18)  1sg   -m/om/öm/em/am
      2sg   -d/od/öd/ed/ad
      3sg   -a/e/ja/je
      1pl   -nk/unk/ünk
      2pl   -tok/tek/tök/otok/etek/ötök/atok
      3pl   -uk/ük/juk/jük

If the possessed entity is itself plural, this is reflected by a combination of *-jai/jei/ai/ei/i* with *-m*, *-d*, *-Ø*, *-nk*, *-tok/tek*, *-k*, respectively.[19] Table 6 presents examples of the combinations discussed so far.

TABLE 6. *Plural, possessive, and plural possessive forms*

| | | Singular | | | Plural | | |
|---|---|---|---|---|---|---|---|
| | | 'castle' | 'picture' | 'trough' | 'castles' | 'pictures' | 'troughs' |
| | | vár | kép | teknő | várak | képek | teknők |
| 1sg | 'my' | váram | képem | teknőm | váraim | képeim | teknőim |
| 2sg | 'your' | várad | képed | teknőd | váraid | képeid | teknőid |
| 3sg | 'his/her' | vára | képe | teknője | várai | képei | teknői |
| 1pl | 'our' | várunk | képünk | teknőnk | váraink | képeink | teknőink |
| 2pl | 'your' | várotok | képetek | teknőtök | váraitok | képeitek | teknőitek |
| 3pl | 'their' | váruk | képük | teknőjük | váraik | képeik | teknőik |

The 'familiar plural' ending is *-ék*, e.g. *Szabó* <surname>, *Szabóék* 'the Szabó family'; *János* 'John', *Jánosék* 'John and his people (friends, family, team, etc.)'. The familiar plural can be added after a possessive suffix, e.g. *barátomék* 'my friend and his family, a group that includes my friend'; *barátaimék*

---

[19] Detailed descriptions of the possessive in Hungarian are given in Antal (1963) and in Kiefer (1985).

'a group that includes several of my friends'. The 'anaphoric possessive' is
-*é* 'that of' (singular), -*éi* 'those of' (plural). It can follow the stem: *emberé*
'that of (the) person', the plural suffix: *embereké* 'that of (the) people', or any
of the possessive suffixes: *emberemé* 'that of my man', *embereinkéi* 'those of
our men', etc.[20]

The end point of the diagram in (17) is a case marker; if it is assumed that
the nominative has a zero case marker (rather than no case marker) then all
noun forms have to terminate in the category of case.[21] The cases are listed in
Table 7.[22]

TABLE 7. *The Hungarian case system*

| Case | Marker | Gloss |
|------|--------|-------|
| Nominative | -*Ø* | (subject) |
| Accusative | -*t* | (object) |
| Dative | -*nak/nek* | to/for |
| Instrumental | -*val/vel* | with |
| Illative | -*ba/be* | into |
| Sublative | -*ra/re* | onto |
| Allative | -*hoz/hez/höz* | to |
| Inessive | -*ban/ben* | in |
| Superessive | -*on/en/ön* | on |
| Adessive | -*nál/nél* | at |
| Elative | -*ból/ből* | out of |
| Delative | -*ról/ről* | of/about/from top of |
| Ablative | -*tól/től* | from |
| Causal/Final | -*ért* | for |
| Translative | -*vá/vé* | (turn) into |
| Essive/Formal | -*ként, képp, -ul/ül* | like sth |
| Terminative | -*ig* | up to |

[20] As the diagram in (17) suggests, the anaphoric possessive morpheme can even be added
recursively (in practice, at most twice): *gazda* 'master', *gazdáé* 'that of the master', *gazdáéé* 'that
of that of the master', *gazdáéi* 'those of the master', *gazdáééi* 'those of that of the master'.

[21] Thus, the possibilities depicted by the diagram are as follows: [Stem + Case], [Stem + Plural
+ Case], [Stem + Plural + Anaphoric possessive + Case], [Stem + Familiar plural + Case], [Stem
+ Familiar plural + Anaphoric possessive + Case], [Stem + Possessive + Case], [Stem + Possessive + Familiar plural + Case], [Stem + Possessive + Anaphoric possessive + Case], [Stem +
Familiar plural + Anaphoric possessive + Case]. 'Stem' in each case can stand for a simplex or
derived stem (with possibly several derivational suffixes); Anaphoric possessive, when it occurs,
may (marginally) occur twice. The total number of possible forms is thus rather astronomical.

[22] A number of further 'cases' are sometimes posited in the literature, including Locative -*tt*
(e.g. *Vácott* 'in Vác', *Pécsett* 'in Pécs', *Győrött* 'in Győr'), Multiplicative -*szor/szer/ször* (e.g. *hat-*
*szor* 'six times', *hétszer* 'seven times', *ötször* 'five times'), Temporal -*kor* (e.g. *hatkor* 'at six
(o'clock)', *hétkor* 'at seven', *ötkor* 'at five'), Distributive -*nként* (e.g. *egyenként* 'one by one'), Distributive-Temporal -*nta/nte* (e.g. *naponta* 'daily, once a day', *hetente* 'weekly, once a week'), and
Associative -*stul/stül* (e.g. *ajtóstul (ront be)* '(burst into the room) "together with the door" ').
For discussion, see Antal (1961).

On the *v* of the instrumental and the translative, see Vago (1989) and section 8.2.1.

The various case forms of personal pronouns are formed by adding possessive suffixes to roots that are mostly identical with the above case suffixes (but sometimes not quite). Compare Table 7 with Table 8.

TABLE 8. *Case markers used as roots for personal suffixes*

| Case | Marker | Case forms of personal pronouns |
|------|--------|--------------------------------|
| Nom. | -Ø | én, te, ő, mi, ti, ők |
| Acc. | -t | engem, téged, őt, minket, titeket, őket |
| Dat. | -nak/nek | nekem, neked, neki, nekünk, nektek, nekik |
| Instr. | -val/vel | velem, veled, vele, velünk, veletek, velük |
| Ill. | -ba/be | belém, beléd, bele, belénk, belétek, beléjük |
| Subl. | -ra/re | rám, rád, rá, ránk, rátok, rájuk |
| All. | -hoz/hez/höz | hozzám, hozzád, hozzá, hozzánk, hozzátok, hozzájuk |
| Iness. | -ban/ben | bennem, benned, benne, bennünk, bennetek, bennük |
| Sup. | -on/en/ön | rajtam, rajtad, rajta, rajtunk, rajtatok, rajtuk |
| Adess. | -nál/nél | nálam, nálad, nála, nálunk, nálatok, náluk |
| Elat. | -ból/ből | belőlem, belőled, belőle, belőlünk, belőletek, belőlük |
| Delat. | -ról/ről | rólam, rólad, róla, rólunk, rólatok, róluk |
| Abl. | -tól/től | tőlem, tőled, tőle, tőlünk, tőletek, tőlük |
| Caus. | -ért | értem, érted, érte, értünk, értetek, értük |

Nominal minor stems include 'lowering' stems, 'epenthetic' stems, '*v*-adding' stems, 'shortening' stems, and 'unrounding' stems.

After 'lowering' stems, the suffix-initial unstable vowel is low -*a/e* instead of the regular -*o/e/ö* (except in the superessive), and the unstable vowel of the accusative is realized even after stem-final consonants that otherwise do not require a linking vowel before -*t* (cf. Vago 1980*a*: 110–12, Stiebels and Wunderlich 1997: 10 13). The phenomenon of lowering will be given extensive treatment in section 8.1.3; at this point, a list of the most frequent nominal lowering stems is provided for reference:

(19)

| ág | 'branch' | agy | 'brain' | ágy | 'bed' |
|----|----------|-----|---------|-----|-------|
| ár | 'price' | árny | 'shadow' | fal | 'wall' |
| fej | 'head' | férj | 'husband' | fog | 'tooth' |
| föld | 'earth' | fül | 'ear' | gally | 'twig' |
| gyár | 'factory' | hal | 'fish' | has | 'stomach' |
| ház | 'house' | héj | 'peel' | hely | 'place' |
| híd | 'bridge' | hold | 'moon' | könny | 'tear' |
| könyv | 'book' | láb | 'leg' | levél | 'leaf' |
| ló | 'horse' | lyuk | 'hole' | madár | 'bird' |
| máj | 'liver' | mell | 'breast' | méz | 'honey' |
| nyak | 'neck' | nyár | 'summer' | öv | 'belt' |

| száj | 'mouth' | szárny | 'wing' | szög | 'nail' |
|------|---------|--------|--------|------|--------|
| szörny | 'monster' | szűz | 'virgin' | tál | 'dish' |
| talp | 'sole (of a shoe)' | tárgy | 'object' | társ | 'partner' |
| tehén | 'cow' | tej | 'milk' | tél | 'winter' |
| tó | 'lake' | toll | 'feather' | ügy | 'affair' |
| ujj | 'finger' | vágy | 'desire' | vaj | 'butter' |
| váll | 'shoulder' | víz | 'water' | völgy | 'valley' |

'Epenthetic' stems end in -CV$_u$C where V$_u$ stands for an unstable vowel, i.e. one that alternates with zero (cf. Vago 1980a: 116–18, Stiebels and Wunderlich 1997: 16–19):

(20)
| bokor | 'bush' | bokor-ban (iness.) | bokr-ok (pl.) |
|-------|--------|--------------------|---------------|
| eper | 'strawberry' | eper-ben | epr-ek |
| ökör | 'ox' | ökör-ben | ökr-ök |

This stem class will be discussed in section 8.1.1. Note that in three 'epenthetic' stems the consonants flanking the unstable stem vowel are metathesized in the vowelless alternant (cf. Vago 1980a: 118–19; Stiebels and Wunderlich 1997: 19–20; Kenesei, Vago, and Fenyvesi 1998: 449):

(21)
| teher | 'weight' | teher-ben (iness.) | terh-ek (pl.) |
|-------|----------|--------------------|---------------|
| kehely | 'chalice' | kehely-ben | kelyh-ek |
| pehely | 'fluff' | pehely-ben | pelyh-ek |

'V-adding' stems end in a vowel but add a final /v/ when followed by a vowel-initial synthetic suffix (but not before analytic suffixes like terminative -ig, causal -ért, or anaphoric possessive -é). Ló 'horse', fű 'grass', nyű 'maggot', tő 'stem', cső 'pipe', and kő 'stone' shorten their vowels when they take a /v/ (i.e. they are 'shortening' stems as well, see below), whereas mű 'work of art' does not (cf. Vago 1980a: 112–13; Vago 1989: 296–304; Stiebels and Wunderlich 1997: 21–3):

(22)
| ló | 'horse' | ló-ban (iness.) | lov-ak (pl.) |
|----|---------|-----------------|--------------|
| cső | 'pipe' | cső-ben | csöv-ek |
| mű | 'work of art' | mű-ben | műv-ek |

In the three stems in (23), the stem-final /o:/ changes into /av/ before a vowel-initial (synthetic) suffix (cf. Vago 1980a: 113–14; Vago 1989: 303; Stiebels and Wunderlich 1997: 23):

(23)
| hó | 'snow' | hó-ban (iness.) | hav-ak (pl.) |
|----|--------|-----------------|--------------|
| szó | 'word' | szó-ban | szav-ak |
| tó | 'lake' | tó-ban | tav-ak |

Note, however, that the accusative of szó is szót (not *szavat).

In the three stems in (24), the stem-final *vowel* changes into /v/ before a vowel-initial (synthetic) suffix (cf. Vago 1980*a*: 114–15; Vago 1989: 304–5; Stiebels and Wunderlich 1997: 23–4):

(24) falu 'village'   falu-ban (iness.)   falv-ak (pl.)
     daru 'crane'     daru-ban            darv-ak
     tetű 'louse'     tetű-ben            tetv-ek

Note that all nominal '*v*-adding' stems are also 'lowering' stems (e.g. *lovak* 'horses', *csövek* 'pipes'). '*V*-adding' stems will not be discussed in this book as such, but they will crop up in the discussion of 'lowering' stems (section 8.1.3) and in that of 'shortening' stems (section 3.1.2).

'Shortening' stems (or FSVS stems, as they will be called in section 3.1.2.1) shorten their last (or only) stem vowel when followed by vowel-initial (synthetic) suffixes (cf. Vago 1980*a*: 121–3; Ritter 1995: 9–10; Stiebels and Wunderlich 1997: 20–1):

(25) nyár 'summer'   nyár-ban (iness.)   nyar-ak (pl.)
     kéz  'hand'     kéz-ben             kez-ek
     tűz  'fire'     tűz-ben             tüz-ek

All 'shortening' stems are 'lowering' stems (e.g. *nyarak, tüzek*). Stem vowel shortening will be discussed more extensively in section 3.1.2.

Finally, 'unrounding' stems (cf. Vago 1980*a*: 120–1; Stiebels and Wunderlich 1997: 14–15) may change their final /ö:/ or /o:/ into /e/ and /a/, respectively, before some possessive suffixes (in particular, all the plural-possessed suffixes and both 3sg and 3pl singular-possessed suffixes, see Table 6):

(26) erdő 'forest'   erde-je 'his/her/its forest'   erdeitek 'your (pl.) forests'
     ajtó 'door'     ajta-ja 'his/her/its door'     ajtaitok 'your (pl.) doors'

These will be ignored in this book. (The plural-possessed items are rather obsolete today, and are replaced by forms like *erdőitek, ajtóitok*.)

This concludes our short summary of nominal morphology. For more details, as well as for diverse analyses of the above and other facts of Hungarian morphology, see Hetzron (1972), Vago (1980*a*), Abondolo (1988), Olsson (1992), Kornai (1994), as well as Kenesei, Vago, and Fenyvesi (1998: 191–282, 330–41) and references cited there.

## 2.4.4. Word order

Hungarian is often naïvely characterized as a 'free word order' language. This is of course not true—word order in this language is not less strictly

grammatically determined than in what are called 'configurational languages' (like English). However, there are at least two senses in which the
above simplistic statement contains an element of truth.

First, in terms of the usual SVO/SOV/etc. typology, Hungarian cannot be
classified in an unambiguous manner. Observe the following examples (partly
based on É. Kiss 1987: 24–5):

(i) SVO is the unmarked order if both subject and object are definite and not
interrelated (see (vi) below):

(27) *a.* Imre   ismeri        Erzsit
            know-3sg.def.  Erzsi-acc.
           'Imre knows Erzsi'

     *b.* A   fiú   ismeri          a    lányt
          the  boy  know-3sg.def.  the  girl-acc.
          'The boy knows the girl'

(ii) SVO and SOV are both equally unmarked if the object is indefinite:[23]

(28) *a.* Imre   ismer         egy   lányt
            know-3sg.indef.  a     girl-acc.
           'Imre knows a girl'

     *b.* Imre   egy   könyvet   olvas
             a     book-acc.  read-3sg.indef.
          'Imre reads/is reading a book'

(iii) Only SOV is possible if the object is a bare (determinerless, incorporated)
noun (cf. (9a–b) in section 2.3 above):

(29) Imre   könyvet    olvas
         book-acc.   read-3sg.indef.
     'Imre is (engaged in) book-reading'

(iv) OVS is the normal order for definite object and indefinite subject:

(30) Az   igazgatót     felhívta      egy   újságíró
     the   director-acc.  call-3sg.past  a    journalist
     'The director was called up by a journalist'

---

[23] Even a definite object can precede the verb in a neutral (non-focused) sentence if the verb
is more or less 'old information' (predictable from the object), as in *Imre a könyvét olvassa* 'Imre
is reading his book'.

(v) If the object is a proper name and the subject is a class noun, the neutral order is, again, OVS:

(31) Jánost     megbüntette     a     rendőr
     John-acc.  perf-fine-3sg.past  the  policeman
     'John was fined by the policeman'

(vi) OVS is furthermore used if the reference of the subject is defined with respect to the object (e.g. by means of a possessive suffix):

(32) Az    igazgatót      figyelmeztette   a     titkárnője
     the   director-acc.  warn-3sg.past    the   secretary-his
     'The director was warned by his secretary'

(vii) There are OVS sentences with a human object and a non-human subject:

(33) Jánost     elütötte           a     vonat
     John-acc.  run over-3sg.past  the   train
     'John was run over by the train'

(viii) An OV sentence with a 3rd person plural (subject) marker on the verb is the Hungarian impersonal construction:

(34) Jánost     keresték
     John-acc.  seek-3pl.past
     'John was looked for'

(ix) VSO and VOS are also possible neutral orders under various circumstances (including, but not restricted to, idiomatic expressions like (35a); in such sentences, the subject is not topicalized):

(35) *a.* Veri            az    ördög   a     feleségét
          beat-3sg.def.   the  devil    the   wife-his-acc.
          'It is raining and the sun is shining at the same time'

     *b.* Hozott          egy   könyvet    a     barátom
          bring-3sg.past  a     book-acc.  the   friend-my
          'My friend has brought (me) a book'

On the basis of certain typological characteristics that are usually shared by SOV languages, like 'auxiliaries follow verbs' (*várni¹ fog²* 'he will² wait¹', *tudnod¹ kell²* 'you must² know¹', etc.), 'adverbs precede verbs' (*gyorsan¹ fut²* 'he runs² quickly¹', *otthon¹ tanul²* 'he learns² at home¹', etc.), 'postpositions rather than prepositions' (*a ház¹ mögött²* 'behind² the house¹', *egy perc¹ alatt²* 'in² a minute¹', etc.), Hungarian has been claimed to be basically an SOV language, or one developing from SOV to SVO (Dezső 1980). There have also been attempts to describe Hungarian along the lines of generative grammars elaborated for English, attributing an [s NP [vp V NP]] structure to it, e.g. Dezső (1965), Kiefer (1967), Horvath (1986). However, É. Kiss (1981, 1987) has convincingly argued that the basic structure of Hungarian simple sentences is as shown in (36):

(36)

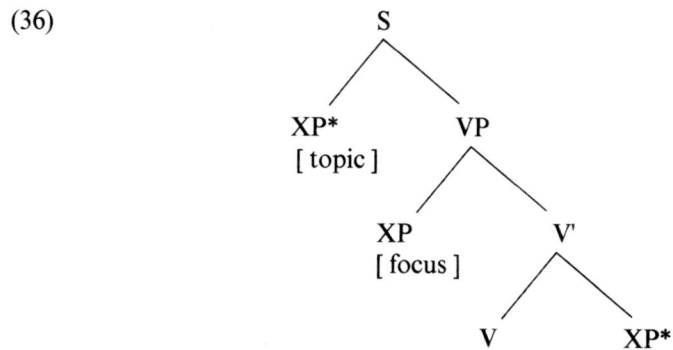

(where the asterisk means 'one or several instances of'; XP means 'some phrasal category').

The VP following the topic is conventionally referred to as 'comment'. The examples listed under (i)–(ix) above turn out to be remarkably uniform if we analyse them into topic and comment (as opposed to their structural diversity in terms of subject, verb, and object):

(37)

| topic | comment | |
|---|---|---|
| Imre | ismeri Erzsit | (27*a*) |
| A fiú | ismeri a lányt | (27*b*) |
| Imre | ismer egy lányt | (28*a*) |
| Imre | egy könyvet olvas | (28*b*) |
| Imre | könyvet olvas | (29) |
| Az igazgatót | felhívta egy újságíró | (30) |
| Jánost | megbüntette a rendőr | (31) |
| Az igazgatót | figyelmeztette a titkárnője | (32) |
| Jánost | elütötte a vonat | (33) |
| Jánost | keresték | (34) |
| | Veri az ördög a feleségét | (35*a*) |
| | Hozott egy könyvet a barátom | (35*b*) |

The second sense in which Hungarian word order is 'free' is that e.g. (27*a*) remains a grammatical Hungarian sentence in all possible permutations: *Imre ismeri Erzsit, Erzsit ismeri Imre, Imre Erzsit ismeri, Erzsit Imre ismeri, Ismeri Imre Erzsit, Ismeri Erzsit Imre*. Of course, not all of these mean exactly the same thing (that would be a truly free or non-configurational word order); rather, all of these are different structures with different stress patterns and interpretations. Actually, most orders correspond to several structures as can be seen in (38) where obligatory stress is shown by ', obligatory stress-lessness (enclisis) is shown by – (no mark suggests that the constituent may or may not be stressed, depending on whether it does or does not constitute 'new information', with no structural consequence):

(38) *a.* Imre 'ismeri Erzsit    'Imre knows Erzsi'
   *b.* 'Imre –ismeri Erzsit   'It is Imre who knows Erzsi'
   *c.* Erzsit 'ismeri Imre   'Erzsi is known to Imre'
   *d.* 'Erzsit –ismeri Imre   'It is Erzsi whom Imre knows'
   *e.* Imre Erzsit 'ismeri   'Talking of Imre and Erzsi: he knows her'
   *f.* Imre 'Erzsit –ismeri   'Talking of Imre: it is Erzsi whom he knows'
   *g.* Erzsit Imre 'ismeri   'Talking of Erzsi and Imre: she is known to him'
   *h.* Erzsit 'Imre –ismeri   'Talking of Erzsi: it is Imre who knows her'
   *i.* 'Ismeri Imre Erzsit   'It is a fact that Imre knows Erzsi'
   *j.* 'Ismeri Erzsit Imre   'It is a fact that Erzsi is known to Imre'

In terms of the structure given in (36), the above strings can be assigned to the four slots as follows:

| (39) | *topic* | *focus* | *verb* | *rest* |
|---|---|---|---|---|
| *a.* | Imre | | 'ismeri | Erzsit |
| *b.* | | 'Imre | –ismeri | Erzsit |
| *c.* | Erzsit | | 'ismeri | Imre |
| *d.* | | 'Erzsit | –ismeri | Imre |
| *e.* | Imre Erzsit | | 'ismeri | |
| *f.* | Imre | 'Erzsit | –ismeri | |
| *g.* | Erzsit Imre | | 'ismeri | |
| *h.* | Erzsit | 'Imre | –ismeri | |
| *i.* | | | 'Ismeri | Imre Erzsit |
| *j.* | | | 'Ismeri | Erzsit Imre |

Whether the above constituents are directly generated in the given structural positions or get there by movement transformations is irrelevant for our purposes.[24]

---

[24] É. Kiss (1987) actually suggests that all examples start out with the three constituents under V'; (*i*) and (*j*) directly mirror the underlying orders. In (*a*) and (*f*) *Imre* is topicalized

However, calling the preverbal slot 'focus' is something of an oversimplification; this slot can be filled by a focused element as in (39*b, d, f, h*) but by a number of other things, too. In particular, it can be filled by a question word (*ki* 'who', *hol* 'where', *mikor* 'when', etc.), by a negative particle (*nem* 'not', *ne* 'don't'), or by various types of 'verb modifiers', see (40). Whatever occupies this position, the verb itself will be stressless; however, it is only if a focused element takes this position that the whole comment (the part of the sentence that follows the topic) will necessarily constitute a domain for eradicating stress.[25]

(40) *a.* preverb (cf. (9*a*), (10)):
    *meg*-számol 'count' (*perf.*), *föl*-megy 'go *up*', *át*-néz 'look *through*'
  *b.* determinerless object (cf. (9*b*), (16)):
    *fát* vág 'chop *wood*', *matekot* tanul 'learn *maths*', *részt* vesz 'take *part*'
  *c.* goal adverbial:
    *Pécsre* utazik 'travel *to Pécs*', *kenyérért* indul 'set out *to get some bread*'
  *d.* result adverbial:
    *keményre* vasal 'iron *stiff*', *laposra* ver 'beat *hollow*',
    *fényesre* csiszol 'polish (until it is) *shiny*', *ripityára* tör 'break *to shivers*'
  *e.* manner adverbial:
    *keményen* bánik 'be *hard* on', *hűtlenül* kezel '*mis*appropriate'
  *f.* infinitive:
    *menekülni* igyekszik 'try *to escape*', *enni* szeretne 'would like *to eat*'
  *g.* noun/adjective before copula:
    *katona* lett 'became a *soldier*', *fáradt* vagyok 'I am *tired*'
  *h.* theme/patient subject:
    *víz* ment (a szemébe) '*water* got (into his eye)'.
    *gyereke* született 'a *child* was born by her'
  *i.* miscellaneous:
    *okosnak* tartják '(he is) thought to be *clever*', *rosszul* volt 'felt *ill*'

In sum: it is not strictly true that Hungarian word order is free; but the organizing principle is not the usual NP–VP type. Rather, the immediate constituents of a Hungarian simple sentence are the topic and the comment; the latter consists of a preverbal slot often (but not always) occupied by the focus of the sentence (if it has one), the verb slot, and a postverbal slot.[26]

(moved into topic position); in (*c*) and (*h*) *Erzsit* is topicalized; and in (*e*) and (*g*) both *Imre* and *Erzsit* are (in the orders corresponding to (*i*) and (*j*), respectively). On the other hand, in (*b*) and (*h*) *Imre* is focused (moved into focus position), and in (*d*) and (*f*) *Erzsit* is (both constituents cannot be moved as the focus position can only accommodate a single constituent).

[25] The examples in (40) are mostly from Kálmán and Nádasdy (1994: 437).

[26] On Hungarian syntax, see further Kiefer and É. Kiss (1994), Alberti (1997), Bartos (1997), Laczkó (1997), É. Kiss (1998), Horvath (1998), Kenesei (1998), Molnár (1998), Puskás (1998), as well as Kenesei, Vago, and Fenyvesi (1998), and references cited there.

# PART II

## SYSTEMS

---

# 3

# THE VOWEL SYSTEM

A surface phonetic classification of the Hungarian vowel system is shown in (1):[1]

(1)

| | FRONT | | CENTRAL | BACK |
|---|---|---|---|---|
| | UNROUNDED | ROUNDED | UNROUNDED | ROUNDED |
| HIGH | i  i: | ü  ü: | | u  u: |
| UPPER MID | e: | ö: | | o: |
| LOWER MID | | ö | | o |
| UPPER LOW | ɛ | | | ɔ |
| LOWER LOW | | | a: | |

This classification involves five heights, three points of articulation along the sagittal axis, plus the rounded/unrounded distinction. Obviously, a number of phonetic details can be filtered out of this representation on grounds of predictability. The difference between upper mid and lower mid might be taken to be a matter of tense/lax (or else, [+/–ATR]); but even that is predictable (redundant) on the basis of long vs. short (i.e. XX vs. X in terms of skeletal positions/timing slots). On the other hand, the two lows may be simply taken to be the same height phonologically: the exact height of [a:], as well as its centrality, is a matter of phonetic implementation since in the phonological pattern of Hungarian (e.g. with respect to vowel harmony, long/short alternations, etc., see below) [a:] behaves as a low back vowel. Hence, the simplified pattern in (2) emerges.

(2)

| | [–back] | | [+back] | |
|---|---|---|---|---|
| | [–round] | [+round] | [–round] | [+round] |
| [+high, –low] | i  i: | ü  ü: | | u  u: |
| [–high, –low] | e: | ö  ö: | | o  o: |
| [–high, +low] | ɛ | | a: | ɔ |

The system of Hungarian orthography, as well as traditional descriptions, suggest that these vowels constitute seven short/long pairs. For reference, the

---

[1] Transcription: ü = IPA [y], ö = IPA [ø].

orthographic symbols for the above vowels are given in (3), arranged in the
same way as in (2).

(3)   i      í      ü      ű                u      ú
            é      ö      ő                o      ó
      e                          á      a

Two questions arise with respect to this classification. First, is vowel length
contrastive in this language? Second, are all pairs symmetrical (differing in
length only) as the spelling suggests or is the phonetic asymmetry shown in
(2) phonologically valid?

High vowels, as we saw in (1), may differ in length without any quality difference; although fully satisfactory minimal pairs are not easy to find,[2] the
length contrasts *i/í, ü/ű, u/ú* appear to be uncontroversial. Some examples are
given in (4).

(4)  *a.* int         'beckon'              ínt         'tendon' (acc.)
         kürt        'horn'                kűrt        'free exercise (in skating)' (acc.)
         zug         'nook'                zúg         'rumble'
     *b.* hidra       'hydra'               hídra       'to (the) bridge'
         büntettek   'they punished'       bűntettek   'crimes'
         szurok      'tar'                 szúrok      'I stab'

For mid rounded vowels, length distinctions always entail minor quality
differences, but these can be abstracted away from as we saw above.[3] Some
minimal pairs are given in (5).

---

[2] This is because, with respect to high vowels, the phonological value of length is rather vague
in colloquial Hungarian (cf. Nádasdy 1985, Kontra 1995, Pintzuk *et al* 1995). In a large number
of lexical items the length of high vowels vacillates, especially if the vowel is not in the last syllable: [i] ~ [iː] *híradó* 'news', *Tibor* <proper name>; [ü] ~ [üː] *hűvös* 'cool', *szüzek* 'virgins'; [u] ~
[uː] *púpos* 'hunchback', *turista* 'tourist' (as the examples show, the spelling is irrelevant here). It
appears that the vacillation concerns 'long' vowels (i.e. it is a case of variable shortening, rather
than variable lengthening) since a number of words have invariable short high vowels (e.g. *liba*
'goose', *üveg* 'glass', *buta* 'stupid') but there are practically none in which a non-final high vowel
(more exactly: one not in the last syllable of the word) would be invariably long.
   Word final high vowels in polysyllabic words tend to be short in colloquial Hungarian (cf.
Nádasdy and Siptár 1998). In compounds, this applies if the last compound member is itself
polysyllabic, hence words like *férc*‖*mű* 'hack work' are exempt. This shortness is not affected
by suffixation: high vowels are not lengthened before a suffix (unlike low vowels, cf. section
3.1.1).
   In monosyllabic words, on the other hand, final high vowels are regularly long. (This is also
true for compounds whose last members are monosyllabic.) There is a handful of exceptions to
this generalization, all of them function words with short [i]: *ki* 'who', *ki* 'out', *mi* 'we', *mi* 'what',
*ti* 'you' (pl.), *ni* 'look!'; cf. section 5.4.1.
[3] Word-finally, mid rounded vowels can only be long. This generalization is exceptionless, and
applies to loanwords and foreign names as well, e.g. *presto* [prestoː], *Cocteau* [koktoː], *pas de
deux* [pɔdödöː]; cf. section 5.4.1.

(5)  *a.*  tör    'break'      tőr    'dagger'
          por    'dust'       pór    'peasant'
     *b.*  növel  'increase'   nővel  'with a woman'
          koma   'friend'     kóma   'coma'

The rest of the vowels—*é, e, á, a*—never contrast in length without differing in quality in non-trivial ways. Phonetically, they do not constitute long/short pairs; the question is if they do so in the phonological system.[4] Pairs like *ken* 'smear'–*kén* 'sulphur', *való* 'real'–*váló* 'divorcing' are proper minimal pairs showing that the highlighted vowels contrast in some feature(s), but they do not tell us if that feature is length or something else. Thus, we have to look beyond distributional facts and consider the phonological behaviour of these segments.

A number of stems exhibit length alternation when certain suffixes are added. Consider some examples in the plural and with the derivational suffix *-izál* '-ize', with high and mid vowels in (6*a*) and *é, e, á, a* in (6*b*).[5]

(6)    sg.   pl.
     *a.*  víz–vizek    'water'    analízis 'analysis'–analizál 'analyse'
          tűz–tüzek    'fire'     miniatűr 'miniature'–miniatürizál 'minia-
                                  turize'
          út–utak      'road'     úr 'gentleman'–urizál 'play the gentleman'
          tő–tövek     'root'     pasztőröz 'pasteurize'–pasztörizál 'id.'
          ló–lovak     'horse'    agónia 'agony'–agonizál 'agonize'
     *b.*  kéz–kezek    'hand'     prémium 'bonus'–premizál 'award a bonus'
          nyár–nyarak  'summer'   kanális 'canal'–kanalizál 'canalize'

In the pairs in (6), the relation between members is the same throughout: the stem vowel 'gets shortened'. This suggests that the *é–e, á–a* relationships are the same as *í–i* and the others, that is, the former constitute long/short pairs as well. It would be a good idea to represent them identically (length apart).

For [ɛ]–[eː] it has been suggested (cf. Nádasdy and Siptár 1998) that [low] should be done away with as a contrastive feature. If we want to do the same with respect to [ɔ]–[aː], we have to subscribe to the generative tradition (going back to Szépe 1969) that takes [ɔ] to be underlyingly non-round, its surface roundness being due to a late adjustment rule. In that case, using the symbols /a/ and /e/ for the phonological segments underlying [ɔ] and [ɛ] respectively, we would have the following system:

---

[4] Orthography suggests that they do; but this is obviously not a relevant piece of evidence (although it reflects the linguistic intuitions of those who first applied Latin script to Hungarian).

[5] Stem vowel shortening will be discussed at greater length in section 3.1.2.

(7)                          [–back]                    [+back]
                      [–round] [+round]          [–round] [+round]
     [+high]            i   i:    ü   ü:                       u   u:
     [–high]            e   e:    ö   ö:          a   a:       o   o:

Alternatively, we could keep [low] for back vowels but avoid using [round] since it is fully predictable (for back vowels). This would give us identical representations for /a/ and /a:/ without claiming that [ɔ] starts out as unrounded. Thus, all lowness specifications for front vowels and all roundness specifications for back vowels could be added in a fill-in (structure building) fashion and no structure changing operations would be involved.

(8)                          [–back]                    [+back]
                      [–round] [+round]           [–low] [+low]
     [+high]            i   i:    ü   ü:          u   u:
     [–high]            e   e:    ö   ö:          o   o:   a   a:

In order to make a principled choice between the two possibilities sketched in (7) and in (8), let us translate both into the Clements/Hume feature system we adopt in this book (cf. Clements and Hume 1995). Using the three unary articulator features LAB, COR, DOR, plus a binary aperture feature [±open], the system in (7) can be represented diagrammatically as (9):

(9)

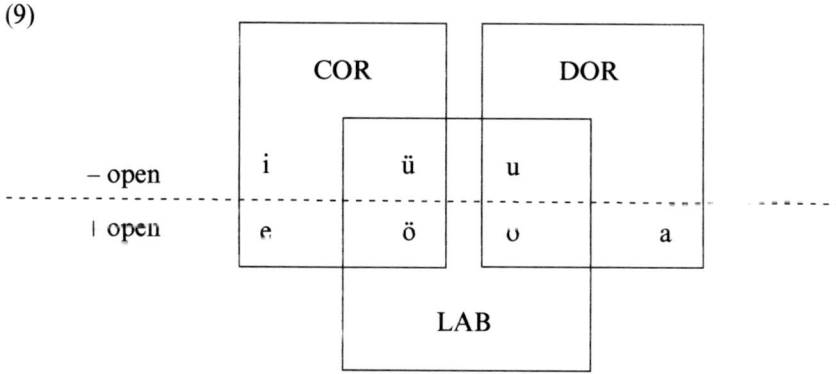

This is phonetically quite accurate: it mirrors the facts that the front rounded vowels involve the COR and LAB articulators, and that the back rounded vowels involve DOR and LAB; furthermore, it suggests correctly that the height distinction between short and long /e/ or the roundness distinction between short and long /a/ is a matter of phonetic implementation (in particular, the system in (9) does not claim that [e:] derives from a low vowel or that the surface roundness of [ɔ] is due to a feature changing operation turning [–round] into [+round]; this last point is one in which (9) is superior to (7), incidentally). However, (9) has the distinct disadvantage that the natural classes /i e/ (transparent vowels in terms of vowel harmony, see 3.2) or /e a/ (low vowels targeted by LVL, see 3.1.1) cannot be directly referred to since the lack (or non-involvement) of LAB is not a phonological object. Therefore, we will replace this system by (10) which is roughly equivalent to (8), except for the use of unary place features:[6]

(10)

| | | COR | LAB | DOR | |
|---|---|---|---|---|---|
| −open$_1$ | | i | ü | u | −open$_2$ |
| | | | ö | o | +open$_2$ |
| +open$_1$ | | e | | a | |

It must be added that the system in (10) does not involve phonological underspecification in the sense that e.g. /ü/ will not be assigned the feature COR, or /u/ the feature LAB, anywhere in the course of the derivation. Rather, each vowel segment has exactly one articulator feature throughout the phonology, and the phonetic implementation module will directly interpret [LAB, −open$_2$] as 'high *front* rounded', [DOR, −open$_2$] as 'high back *rounded*', and so on. Short [ɛ], then, will have the phonological representation [COR, +open$_1$, +open$_2$], translated phonetically as 'front *low* unrounded', whereas long [e:], having the same phonological representation except that it is linked to two X slots, will be interpreted as 'front *mid* unrounded'. Similarly, short [ɔ] will be represented as [DOR, +open$_1$, +open$_2$] and implemented as 'low back *rounded*', whereas long [a:] will share the same phonological features but come out phonetically as 'low(est) central *unrounded*' (cf. (1)). Note further that the feature values [−open$_1$] for /i ü u ö/ and [+open$_2$] for /e a/ are strictly speaking redundant, but will be assumed to be underlyingly specified (in order to avoid underspecification, wherever possible). In sum, the underlying vowel system of Hungarian will be taken to be specified as shown in (11).[7]

---

[6] The rough equivalences can be spelt out as follows: [−open$_2$] = [+high]; [+open$_2$] = [−high]; [−open$_1$] = [−low]; [+ open$_1$] = [+ low]; DOR = [+back]; COR = [−back,−round]; LAB = [−back, +round]. Cf. Clements and Hume (1995: 275–83) for discussion.

[7] However, some harmonically irregular stems will be analysed in Chapter 6 as involving vowels specified for more than one articulator to encode their exceptional behaviour.

(11)

| | i | ü | u | ö | o | e | a |
|---|---|---|---|---|---|---|---|
| COR | ● | | | | | ● | |
| LAB | | ● | ● | | | | |
| DOR | | | ● | ● | | ● | ● |
| open₁ | (–) | (–) | (–) | (–) | – | + | + |
| open₂ | – | – | – | + | + | (+) | (+) |

## 3.1. VOWEL LENGTH ALTERNATIONS

Alternations in the length of vowels are governed by two types of regularities: Low Vowel Lengthening (LVL) and Stem Vowel Shortening (SVS). In the following subsections we will look at these. Although their names suggest a process-oriented approach, LVL and SVS will be considered here as static alternation patterns. A dynamic account will be given in Chapter 6 below.

### 3.1.1. Low vowel lengthening

Unlike the occurrence of word final high and mid vowels,[8] that of word final low ([+open₁]) vowels is not phonologically restricted (even if *á* occurs with certain limitations).[9] But morpheme final low vowels are subject to an important condition: they have to be long before a suffix. This means that short final low vowels get lengthened before suffixes.[10] Examples:

(12)  /a/ → /aː/

| | | | |
|---|---|---|---|
| fa | 'tree' | fát | 'tree' (acc.) |
| alma | 'apple' | almás | 'apple' (adj.) |
| tartja | 'he holds it' | tartják | 'they hold it' |
| háza | 'his house' | házán | 'on his house' |
| létra | 'ladder' | létrám | 'my ladder' |
| marha | 'cattle' | marhái | 'his cattle' (pl.) |
| kutya | 'dog' | kutyául | 'like a dog' |
| delta | 'delta' | deltáig | 'as far as the delta' |
| Varga | <last name> | Vargáné | 'Mrs V.' |
| porta | 'reception' | portára | 'to reception' |
| lusta | 'lazy' | lustább | 'lazier' |

---

[8] See footnotes 2 and 3, respectively. For further details, cf. Nádasdy and Siptár (1998) and section 5.4.1.

[9] In particular, word final *á* occurs in function words, in suffixes (hence in suffixed words), and in interjections quite freely, but as far as nominal stems are concerned, only a handful of examples occur (*burzsoá* 'bourgeois', *hajrá* 'a rush'; *zéhá* 'written examination', *géemká* 'enterprise co-operative', and other acronyms; *fá* 'fa', *lá* 'la' [sol-fa letters]; cf. section 5.4.1).

[10] Cf. Vago (1978*b*, 1980*a*: 3–4), Abondolo (1988: 43), Jensen and Stong-Jensen (1989*a*), Olsson (1992: 75–6), Ritter (1995: 11–12), Nádasdy and Siptár (1994: 67–70, 1998: 157–9).

| /e/ → /eː/ | medve | 'bear' | medvét | 'bear' (acc.) |
|---|---|---|---|---|
| | epe | 'bile' | epés | 'bilious/malicious' |
| | vitte | 'he carried it' | vitték | 'they carried it' |
| | képe | 'his picture' | képén | 'in his picture' |
| | vese | 'kidney' | vesém | 'my kidney' |
| | sörte | 'bristle' | sörtéi | 'his bristles' |
| | mérce | 'measure' | mércéül | 'as a measure' |
| | csempe | 'tile' | csempéig | 'up to the tile' |
| | Bene | \<last name\> | Benéné | 'Mrs B.' |
| | este | 'evening' | estére | 'by evening' |
| | fekete | 'black' | feketébb | 'blacker' |

This alternation is independent of the word class membership of the stem and it does not matter whether the final low vowel is part of the stem (*alma* 'apple') or of some suffix (*tart-ja* 'he holds it'). Although this is obviously a productive phonological process, we will postpone its discussion as such to section 6.2.1. At the moment, let us capture the regularity in terms of a filter (negative condition).[11]

(13)    \*X
        |       /___ ] Y
$[+\text{open}_1]$

where ] = morpheme boundary, Y = the first segment of a suffix

Of course, long final low vowels do not alternate as they are not affected by this condition (in other words, they conform to the required output configuration of the related lengthening process): *kordé* 'cart'–*kordét* (acc.), *burzsoá* 'bourgeois'–*burzsoát* (acc.).

There are apparent counterexamples where something is added to a low-vowel-final lexical item and the vowel remains short; see (14). These cases will be accounted for in section 6.2.1.

| (14) | *a.* | baltanyél | 'hatchet handle' | kefekötő | 'brush-maker' |
|---|---|---|---|---|---|
| | | hazamegy | 'go home' | belelép | 'step into it' |
| | *b.* | kutyaszerű | 'dog-like' | meseszerű | 'like a fairy tale' |
| | | macskaféle | 'felid' | medveféle | 'like a bear' |

---

[11] Jacques Durand (personal communication) points out that a negative filter cannot guarantee that the low vowel in this context will become long rather than get deleted or undergo some other change. This is true if we think in terms of processes; a derivational account of this phenomenon is given in section 6.2.1 below. However, in a declarative framework it is exactly constraints of this kind and their interaction with other constraints that are supposed to account for the observed patterns of alternations.

58     SYSTEMS

c. távoztakor   'on his departure'  megérkeztekor  'on his arrival'
   tortaként     'as a cake'         sörteként      'as bristles'
   példaképp(en) 'for instance'      mérceképp(en)  'as a measure'
   hazai         'domestic'          megyei         'county' (adj.)

### 3.1.2. Stem vowel shortening

In many Hungarian stems, the vowel (or one of the vowels) is underlyingly long but appears as short before certain suffixes: *kéz* 'hand'–*kezek* 'hands', *szintézis* 'synthesis'–*szintetikus* 'synthetic'. We had a preview of the data in (6); now we will take a closer look at them.[12] Compare the behaviour of the following examples:

(15) a. gép      'machine' gépen      'on a machine' gépek      'machines'
        kéz      'hand'    kézen      'on a hand'    kezek      'hands'
     b. akadémia 'academy' akadémiák  'academies'    akadémikus 'academic'
        szintézis 'synthesis' szintézisek 'syntheses' szintetikus 'synthetic'

It appears that this is not a purely phonological alternation since SVS only applies to certain stems (*kéz*, *szintézis*) and not to others (*gép*, *akadémia*); also, certain suffixes may trigger SVS (-*ek*, -*ikus*), others never do (-*en*). The data in (15a–b) also suggest that SVS has two different domains of application: call them Final Stem Vowel Shortening (FSVS, (15a)) and Internal Stem Vowel Shortening (ISVS, (15b)), respectively. The phonological content of both is the same: a long vowel (which is not the last segment of the stem) is replaced by its short counterpart; however, the circumstances of the two types of alternation, i.e. the stems and suffixes concerned, are different.[13]

### 3.1.2.1. Final stem vowel shortening

FSVS exclusively applies in final syllables of mono- and bisyllabic stems. The target vowel is followed by a single consonant (or an empty consonant slot that gets interpreted as [v] precisely when SVS has applied, otherwise it goes uninterpreted: *ló* 'horse'–*lovak* 'horses'). FSVS is primarily attested in nouns; in the conjugation system it is sporadic (e.g. *lő* 'shoot'–*lövök* 'I shoot'); some verb stems shorten before derivational suffixes (e.g. *úszik* 'swim'–*uszoda* 'swimming-pool') but this is also infrequent.

[12] Cf. Vago (1980a: 121–3), Abondolo (1988:46), Jensen and Stong-Jensen (1989a), Olsson (1992: 123–4), Ritter (1995: 9–10), Nádasdy and Siptár (1994: 70–8, 1998: 160–7).
[13] Note that high-vowel examples are mostly vacillating (due to the general vagueness of length in high vowels; see footnote 2 above): *strukturális* 'structural' /u/ ~ /uː/. This complication will be ignored, except for cases where the form suggested by the spelling never occurs (e.g. *vízi* 'water' (adj.)).

In Table 9 all shortening nominal stems are listed (overleaf). Vacillation is not widespread; *nyű, szú,* and *lég* rarely occur suffixed, and native intuitions concerning them are vague.

FSVS primarily affects low (non-high nonround) vowels (*á, é*), less frequently high vowels (*í, ú, ű*), while mid (non-high round) vowels (*ó, ő*) shorten in a few irregular (*v*-inserting) stems only.[14]

Let us note here that all FSVS stems listed in Table 9 are 'lowering stems'.[15] That is, they all require a (linking) vowel before the accusative suffix (*vizet* 'water' (acc.), *egeret* 'mouse' (acc.), even though *\*vízt/\*vizt, \*egért/ \*egert* would be phonotactically well-formed), and the back linking vowel they take is /a/, rather than /o/ (e.g. *nyar-ak* 'summers', *nyul-am* 'my rabbit').

We do not give an exhaustive list of FSVS suffixes here, but a few examples are listed in (16).

(16) -*k*      (plural)       vizek    lovak     kezek     kanalak
    -*t*      (accusative)   vizet    lovat     kezet     kanalat
           (possessive)   vizem    lovad     keze      kanalunk
    -*s*      (adjective)    vizes    lovas     kezes     kanalas
    -*z*      (verb)         vizez    (kövez)   kezez     kanalaz
    -*l*      (verb)         vizel    loval     kezel     (fenekel)
    -*cska*   (diminutive)   vizecske lovacska  kezecske  kanalacska
    -*nként*  (distributive) (utanként) lovanként hetenként kanalanként

Another FSVS distributive suffix is -*nta/nte* (*nyaranta* 'each summer', *hetente* 'each week'). Similar but more or less isolated examples are: *tiz-ed* 'one-tenth', *husz-adik* 'the twentieth', *negy-ven* 'forty', *zsir-adék* 'fats', etc.[16]

Verb stems exhibiting FSVS effects, primarily before derivational suffixes, include the following: *ír* 'write'–*irat* 'document', *szív* 'inhale'–*szivattyú* 'pump', *tűr* 'put up with'–*türelem* 'patience', *bűn* 'crime'–*büntet* 'punish', *szúr* 'stab'–*szurony* 'bayonet', *bújik* 'hide'–*bujkál* 'lie low', *húz* 'pull'–*huzat* 'draught', *rúg* 'kick'–*r[u]gás* 'a kick', *úszik* 'swim'–*uszoda* 'swimming-pool', *óv* 'protect'–[o]*voda* 'nursery school', *sző* 'weave'–*szövet* 'cloth', *vág* 'cut'–*vagdal* 'chop up', etc. Note that most of these cases involve high vowels.

---

[14] Even more irregular types like *tó* 'lake'–*tavak* 'lakes', *hő* 'heat'–*heve* 'its heat' will be ignored here; these show vowel quality alternation in addition to shortening and are too sporadic to be treated phonologically.
    *Lélek* 'soul' is peculiar in that its affected vowel appears not to be in the last syllable. But the *e* that follows is 'unstable' (cf. section 8.1.1) and fails to appear precisely before FSVS suffixes (*lelk-em* 'my soul'). On the other hand, if *lélk-* is taken to be a 'monosyllabic' stem (having the surface alternants *lélek* and *lelk-*), what is peculiar about it as an FSVS stem is that the target vowel is followed by two consonants rather than one.

[15] Cf. Vago (1980*a*: 110–12), Olsson (1992: 116–18), Törkenczy (1992), Kornai (1994: 30–47), Nádasdy and Siptár (1994: 155–9) and section 8.1.3 below.

[16] High vowels exhibit FSVS-like behaviour in a number of other stems but not for all speakers and not with all of the above suffixes: *szín* 'colour'–*sz[i]nek* 'colours', *hús* 'meat'–*h[u]sos* 'meaty'; also with suffixes not normally triggering FSVS: *út* 'road'–[u]*ti* (adj.), *tűz* 'fire'–*t[ü]zön* 'on the fire', *víz* 'water'–*v[i]zi* (adj.).

TABLE 9. *Nominal FSVS stems*

| Vowel | Monosyllabic stems | | Monosyllabic *v*-stems | | Bisyllabic stems | |
|---|---|---|---|---|---|---|
| *í* | *híd* | 'bridge' | – | | – | |
| | *ín* | 'tendon' | | | | |
| | *nyíl* | 'arrow' | | | | |
| | *víz* | 'water' | | | | |
| *ű* | *szűz* | 'virgin' | *fű* | 'grass' | – | |
| | *tűz* | 'fire' | (*nyű* | 'maggot') | | |
| *ú* | *kút* | 'well' | (*szú* | 'woodworm') | – | |
| | *lúd* | 'goose' | | | | |
| | *nyúl* | 'rabbit' | | | | |
| | *rúd* | 'pole' | | | | |
| | *úr* | 'gentleman' | | | | |
| | *út* | 'road' | | | | |
| *ő* | – | | *cső* | 'pipe' | – | |
| | | | *kő* | 'stone' | | |
| | | | *tő* | 'stem' | | |
| *ó* | – | | *ló* | 'horse' | – | |
| *é* | *kéz* | 'hand' | *lé* | 'liquid' | *egér* | 'mouse' |
| | *réz* | 'copper' | | | *szekér* | 'cart' |
| | *mész* | 'lime' | | | *tenyér* | 'palm' |
| | *ész* | 'mind' | | | *kenyér* | 'bread' |
| | *szén* | 'coal' | | | *gyökér* | 'root' |
| | *név* | 'name' | | | *levél* | 'leaf' |
| | *légy* | 'fly' | | | *kötél* | 'rope' |
| | *ég* | 'sky' | | | *fedél* | 'lid' |
| | *jég* | 'ice' | | | *fenék* | 'bottom' |
| | (*lég* | 'air') | | | *kerék* | 'wheel' |
| | *hét* | 'week' | | | *cserép* | 'tile' |
| | *tér* | 'square' | | | *közép* | 'middle' |
| | *dér* | 'frost' | | | *szemét* | 'rubbish' |
| | *ér* | 'vein' | | | *elég* | 'enough' |
| | *bél* | 'bowels' | | | *veréb* | 'sparrow' |
| | *nyél* | 'handle' | | | *nehéz* | 'heavy' |
| | *fél* | 'half' | | | *tehén* | 'cow' |
| | *szél* | 'wind' | | | *fazék* | 'pot' |
| | *dél* | 'noon' | | | *derék* | 'waist' |
| | *tél* | 'winter' | | | | |
| | *lél(e)k* | 'soul' | | | | |

TABLE 9. *Cont'd*

| á | nyár | 'summer' | – | madár | 'bird' |
|---|------|----------|---|-------|--------|
|   | sár  | 'mud'    |   | szamár | 'donkey' |
|   |      |          |   | agár  | 'greyhound' |
|   |      |          |   | bogár | 'beetle' |
|   |      |          |   | kosár | 'basket' |
|   |      |          |   | mocsár | 'marsh' |
|   |      |          |   | mozsár | 'mortar' |
|   |      |          |   | pohár | 'glass' |
|   |      |          |   | sugár | 'ray' |
|   |      |          |   | sudár | 'lash' |
|   |      |          |   | kanál | 'spoon' |
|   |      |          |   | fonál | 'thread' |
|   |      |          |   | darázs | 'wasp' |
|   |      |          |   | parázs | 'embers' |

### 3.1.2.2. Internal stem vowel shortening

The other type of stem vowel shortening, ISVS, may affect any syllable of the stem, and vowels of any tongue height may be equally involved. This type of shortening is only triggered by derivational suffixes, never by inflections.[17] Since there are large numbers of ISVS stems, only examples are given in Table 10.

The major derivational suffixes triggering ISVS are -*ista* '-ist', -*izál* '-ize', -*izmus* '-ism', -*ikus* '-ic', -*atív* '-ative', -*itás* '-ity', -*ális*/-*áris* '-al/-ary', -*ifikál* '-ify', -*ológus* '-ologist', -*íroz* (verb forming suffix); as well as -*ia* '-y', but the latter only if preceded by two consonants (e.g. -*áns* '-ant' vs. -*ancia* '-ance/-ancy'). These suffixes are bisyllabic, vowel-initial, and harmonically non-alternating.[18] As a result, ISVS always applies in the antepenultimate (or earlier) syllable (cf. Trisyllabic Shortening in English). The only exception is the verb-forming suffix -*ál* which, being monosyllabic, affects the penultimate syllable: *filozofál* 'philosophize', *kulturált* 'civilized', *strukturál* 'structure' (verb), *kurzivál* 'italicize'.

---

[17] Both stems and suffixes that are involved here are usually comparatively recent loanwords of Latin origin or behaving in a 'latinate' manner; this is why ö and ü hardly participate in this process.

[18] ISVS sometimes applies before the second member of a compound. The first member usually changes its ending into -o in such cases, e.g. *Hungária* 'Hungary'–*Hungaroring* (a motor racing track near Budapest), *szláv* 'Slavonic'–*szlavofil* 'slavophile', *kémia* 'chemistry'–*kemoterápia* 'chemotherapy', *cézár* 'Caesar'–*cezarománia* 'megalomania'.

TABLE 10.  *Examples of ISVS stems*

| | | | | |
|---|---|---|---|---|
| *i* | *analízis* | 'analysis' | *analitikus* | 'analytical' |
| | *aktív* | 'active' | *aktivitás* | 'activity' |
| | *vízió* | 'hallucination' | *vizionál* | 'hallucinate' |
| | *mítosz* | 'myth' | *mitológia* | 'mythology' |
| | *motívum* | 'motive' | *motivál* | 'motivate' |
| | *stílus* | 'style' | *stiláris* | 'stylistic' |
| *ű* | *miniatűr* | 'miniature' | *miniatürizál* | 'miniaturize' |
| *ú* | *kultúra* | 'culture' | *kulturális* | 'cultural' |
| | *múzeum* | 'museum' | *muzeológus* | 'museologist' |
| | *fúzió* | 'fusion' | *fuzionál* | 'merge' |
| | *úr* | 'gentleman' | *urizál* | 'play the gentleman' |
| *ő* | *pasztőröz* | 'pasteurize' | *pasztörizál* | 'pasteurize' |
| *ó* | *periódus* | 'period' | *periodikus* | 'periodical' |
| | *história* | 'story' | *historizmus* | 'historicism' |
| | *paródia* | 'parody' | *parodizál* | 'take so off' |
| | *filozófia* | 'philosophy' | *filozofál* | 'philosophize' |
| | *kódex* | 'codex' | *kodifikál* | 'codify' |
| *é* | *prémium* | 'bonus' | *premizál* | 'award a bonus' |
| | *téma* | 'topic' | *tematika* | 'set of topics' |
| | *hérosz* | 'hero' | *heroizmus* | 'heroism' |
| | *matéria* | 'matter' | *materiális* | 'material' (adj.) |
| | *szintézis* | 'synthesis' | *szintetikus* | 'synthetic' |
| | *sumér* | 'Sumerian' | *sumerológus* | 'Sumerologist' |
| | *szuverén* | 'sovereign' (adj.) | *szuverenitás* | 'sovereignty' |
| | *analfabéta* | 'illiterate person' | *alfabetikus* | 'alphabetical' |
| | *klérus* | 'clergy' | *klerikális* | 'clerical' |
| *á* | *május* | 'May' | *majális* | 'May Day picnic' |
| | *banális* | 'banal' | *banalitás* | 'banality' |
| | *elegáns* | 'elegant' | *elegancia* | 'elegance' |
| | *náció* | 'nation' | *nacionalizmus* | 'nationalism' |
| | *szláv* | 'Slavonic' | *szlavista* | 'Slavist' |
| | *privát* | 'private' (adj.) | *privatizál* | 'privatize' |
| | *diplomácia* | 'diplomacy' | *diplomatikus* | 'diplomatical' |
| | *kurátor* | 'trustee' | *kuratórium* | 'board of trustees' |
| | *plakát* | 'poster' | *kiplakatíroz* | 'post' |
| | *mágnes* | 'magnet' | *magnetikus* | 'magnetic' |

## 3.2. VOWEL HARMONY

Perhaps the most interesting (but certainly the most widely discussed) phenomenon in Hungarian phonology is vowel harmony.[19] Given the plethora of
accounts and approaches proposed in the literature, the following discussion
will be relatively non-technical and data-oriented. (The reader may wish to
consult the references listed in footnote 19 for technical solutions in various
frameworks.) In section 6.1 below, we will propose our own analysis in terms
of the Clements/Hume feature system.

The domain of vowel harmony is the phonological word in the sense
defined by Vogel (1989), i.e. a single stem plus any number of suffixes.
Compounds (including preverb + verb combinations) contain as many harmonic domains as they have stem morphemes (e.g. *narancs*][*lé* 'orange juice',
*kő*][*por* 'rock flour', *át*][*köt* 'tie up').[20] Within a harmonic domain, all vowels
agree in backness. This statement is too strong as it stands; various qualifica-
tions are necessary, although a very large majority of Hungarian word forms
satisfies the above formulation, too (cf. *perd-ül-és-etek-től* 'from your (pl.)
twirling round' vs. *ford-ul-ás-otok-tól* 'from your (pl.) turning round'). However, in a number of stems front and back vowels co-occur (*bika* 'bull', *kordé*
'cart', *sofőr* 'driver', *nüansz* 'nuance') and a number of suffixes fail to alternate harmonically (*ház-ért* 'for a house', *öt-kor* 'at five (o'clock)', *dressz-íroz*
'train' (verb)).

Notice that most 'mixed' morphemes and word forms contain front
unrounded vowels (/i:/, /i/, /e:/, /e/) along with back vowels, e.g. *liba* 'goose',
*hernyó* 'caterpillar', *papír* 'paper', *tányér* 'plate', *patika* 'pharmacy',
*konkurencia* 'rivalry'; *kulcs-ért* 'for a key', *nyolc-ig* 'until eight', *tan-ít-ó*
'teacher'. It is reasonable to take front unrounded vowels to be 'neutral', i.e.
neither front nor back (as far as harmony is concerned). This is better than
allowing for huge numbers of exceptions; exceptionality is best restricted to
items like *sofőr* 'driver', *nüansz* 'nuance'. There is, however, a stronger reason
for *i, i, é, e* to be taken as neutral: the fact that they let harmony 'pass
through' them (i.e. they are transparent). If the word has another vowel that
is harmonic (non-neutral), suffixes will be harmonized to that vowel: *rövid-en*
'briefly' but *hamis-an* 'falsely'; *örmény-től* 'from an Armenian' but *kastély-tól*

---

[19]   See, among others, Becker Makkai (1970a), Clements (1976), Vago (1976, 1978a, 1980a, b),
Ringen (1977, 1978, 1980, 1982, 1988a, b), Jensen (1978, 1984), Phelps (1978), Zonneveld (1980),
Battistella (1982), Booij (1984), Goldsmith (1985), van der Hulst (1985), Kontra and Ringen
(1986, 1987), Farkas and Beddor (1987), Abondolo (1988), Ringen and Kontra (1989), Kornai
(1987, 1990b, 1994), Kontra, Ringen, and Stemberger (1991), Olsson (1992), Ritter (1995:
190–290), Ringen and Vago (1995, 1998a, b), Polgárdi and Rebrus (1996), Polgárdi (1998).

[20]   Although in 2.4.1 we saw that preverb + verb combinations do not behave syntactically as
compounds but rather as phrases (with the preverb serving as a 'verb modifier'; see 2.4.4, too),
we are going to simplify the following discussion by lumping them together with regular compounds as the distinction is not phonologically relevant. The prosodic constituent involved is
referred to as 'clitic group' by Vogel (1989).

'from a manor house'; *kever-ék-et* 'mixture' (acc.) but *marad-ék-ot* 'remnants'
(acc.). If *hamis, kastély, maradék* had a front (harmonic) vowel in the last syllable, we could never explain why they take back-vowel suffixes.[21]

According to their role in harmony, Hungarian vowels will be classified as
follows:

(27) | *Surface* | *Harmonic status* | |
|---|---|---|
| *backness* | Harmonic vowels | Neutral vowels |
| front | *ö, ő, ü, ű* | *i, í, é, e* |
| back | *a, á, o, ó, u, ú* | |

From now on, 'front-harmonic' will be used to refer to non-neutral front
vowels, whereas all back vowels are also 'back-harmonic'.

### 3.2.1. Suffix harmony

Hungarian vowel harmony is of the 'stem-controlled' kind which means that
it is always the harmonic value of stems that controls that of affixes, never the
other way round. Also, harmony is directional (left-to-right), i.e. only suffixes
are affected.[22]

Suffixes are normally alternating, i.e. their vowel has a front and a back
alternant selected by (agreeing with) the stem vowel(s).[23] The various types of

---

[21] Note that the neutrality of [ɛ] is ambiguous (and controversial), cf. *kódex-ek* 'codices' but
*haver-ok* 'pals', see section 3.2.3.2. It has been argued in several papers (Ringen 1978, 1988*b*,
Ringen and Kontra 1989, Ringen and Vago 1995) that [ɛ] is harmonic whereas [eː] is neutral.
Ringen and Kontra cite empirical evidence for this claim, based on questionnaire studies in
which a group of native speakers were asked to provide suffixed forms of various lexical items
containing front unrounded vowels in their final syllable(s) along with back-harmonic ones in a
preceding syllable. The data suggest that there is some variation with all front unrounded vowels
but the incidence of front-vowel responses is statistically higher in the case of words having [ɛ]
in their last syllable than in the case of the other front unrounded vowels. From this, the authors
conclude that [ɛ] 'is best viewed as front harmonic not neutral'. However, they also point out that
'Hungarian neutral vowels are not equally neutral. The high front unrounded vowels seem most
neutral, the mid front unrounded vowel less neutral and the low front unrounded vowel not neutral at all' (Ringen and Kontra 1989: 190–1). While we agree that there is gradience in the data,
it would be a more faithful summary of the facts if the above quotation ended 'the low front
unrounded vowel is the least neutral of all'. Note that the analysis given in section 6.1 of the present book accounts for the variation as well as the invariable cases without calling [ɛ] a front
harmonic vowel—but this terminological issue loses much of its import in our framework where
long and short /e/ are represented identically (apart from length) and where transparency vs.
opacity is encoded in terms of the difference between linking and spreading rather than in terms
of the featural make-up of the segments involved.

[22] This may be a by-product of the fact that Hungarian has very few prefixes. Verbal prefixes
(preverbs) like *meg-* (perfectivizer), *föl-* 'up', *át-* 'through' are actually compound members as we
said above; the superlative prefix *leg-* 'most' has a transparent vowel or (rather) is outside the harmonic domain; loan prefixes like *poszt-, un-*, etc., behave in a parallel fashion with suffixes of the
*-ista, -ális* type.

[23] Whether it is the first or the last harmonic stem vowel that governs suffix harmony is a
framework-dependent issue; the position taken in this book is that frontness/backness is a property of the whole stem, not of each vowel separately.

alternating suffixes are listed in (28) in terms of the vowel pairs/triplets that alternate in them; the examples on the right stand for larger sets of suffixes exhibiting the same type of alternation.

(28)  ú    ű    (láb-ú '-legged', fej-ű '-headed')
      u    ü    (ház-unk 'our house', kert-ünk 'our garden')
      ó    ő    (vár-ó 'waiting' (adj.), kér-ő 'asking' (adj.))
      o    ö, e (ház-hoz 'to (the) house', föld-höz 'to (the) land', kert-hez
                 'to (the) garden')
      á    é    (vár-ná 'he would wait for it', kér-né 'he would ask for it')
      a    e    (ház-ban 'in (the) house', kert-ben 'in (the) garden')

In the case of o/ö/e, there are three alternants: the one with e occurs with front stems whose last vowel is unrounded (cf. section 3.2.4).

Non-alternating suffixes either exclusively contain neutral (transparent) vowels (cf. (29a–b)) or contain a back-harmonic vowel that fails to harmonize (29c).[24]

(29) *Non-alternating suffixes*

a. Neutral: /i/, /iː/

| | | |
|---|---|---|
| -i | nyári 'summer' (adj.) | |
| -i | lábai 'his legs' | |
| -ni | futni 'to run' | |
| -ig | hatig 'up to six' | |
| -ik | mászik '(it) crawls' | |
| -ik | harmadik '(the) third' | |
| -nyi | maroknyi 'handful' | |
| -int | koppint 'knock' | |
| -is | normális 'normal' | |
| -is | Julis 'Julia' (dim.) | |
| -ci | apuci 'Daddy' | |
| -csi | Karcsi 'Charles' (dim.) | |
| -sdi | katonásdi 'playing at soldiers' | |
| -ít | tanít 'teach' | |

b. Neutral: /eː/

| | |
|---|---|
| -é | lányé 'belonging to a girl' |
| -ért | hazáért 'for (one's) country' |
| -né | Kovácsné 'Mrs Kovács' |
| -nék | adnék 'I would give' |
| -ék | Kovácsék 'the Kovács family' |
| -ék | maradék 'remainder' |
| -lék | adalék 'admixture' |
| -dék | váladék 'discharge' |
| -ték | nyomaték 'emphasis' |
| -ként | kulcsként 'as a key' |
| -nként | hármanként 'three at a time' |
| -képp(en) | voltaképp(en) 'in fact' |

c. Opaque/domain-external

| | | | |
|---|---|---|---|
| -kor | ötkor 'at five' | -iroz | dresszíroz 'train' (verb) |
| -ol | überol 'outdo' | -ista | centrista 'centrist' |
| -us | cicus 'kitten' | -izmus | defetizmus 'defeatism' |

---

[24] This latter type might be analysed as being outside the harmonic domain by definition (e.g. in terms of lexical levels); alternatively, their back vowel is opaque (initiating a harmonic domain of its own). The latter view is supported by the fact that subsequent alternating suffixes (if any) will be back-vowelled.

| | | | |
|---|---|---|---|
| *-u* | Icu 'Helen' (dim.) | *-ifikál* | elektrifikál 'electrify' |
| *-kó* | Ferkó 'Frank' (dim.) | *-ikus* | szeizmikus 'seismic' |
| *-a* | Ila 'Helen' (dim.) | *-izál* | fetisizál 'make a fetish of' |
| *-ológ-* | szexológus 'sexologist' | *-ia* | szexológia 'sexology' |
| *-ál* | recenzál 'review' (verb) | etc. | |

Notice that the vowel *e* does not occur in (29), an argument *against* its being neutral (cf. Ringen 1978, 1988*b*, Ringen and Kontra 1989 for additional discussion; but see footnote 21 again). The list in (29*a–b*) is meant to be exhaustive; (29*c*) may be further extended if further non-alternating endings turn out to be morphemes of Hungarian. E.g. *-us* as in *patikus* 'pharmacist', *-atív* as in *konzultatív* 'consultative', and other similar elements, mostly of Latin origin, normally attached to bound stem morphemes.

### 3.2.2. Stem classes

In this section, we present the data in terms of a preliminary classification that will serve as a point of departure in section 6.1 below. This classification is based on the harmonic vs. neutral character of vowels. If the last vowel is harmonic, we call the stem a harmonic stem (I); if the last vowel is neutral, we call the stem a neutral stem (II). Within both major classes, we distinguish simple (A) and complex (B) instances: the simple harmonic stems (IA) are ones in which all harmonic vowels are of the same sort (all front or all back); they may also contain neutral vowels in any of their non-final syllables; whereas the complex harmonic stems (IB) contain at least two conflicting harmonic vowels (where conflicting means differing in backness). Simple neutral stems (IIA) exclusively contain neutral vowels, whereas complex neutral stems (IIB) have a harmonic vowel in some earlier syllable. In all four classes, front and back suffix selection are both found (indicated by 'f' and 'b', respectively). Stems in which neutral and back-harmonic vowels co-occur (*bika, papír, kódex*, etc.) are traditionally referred to as 'mixed stems'; note that this label covers subsets of various stem classes as defined here.

**Class IA:** Simple harmonic stems
The last vowel of these stems is harmonic; further harmonic vowels, if any, agree in backness with the last one; they may also contain neutral vowels (but not in their last syllable). Suffix harmony is governed by the backness value of the harmonic vowels:

| (30) *a.* IA–f: *TŰZ* type | *dative* | *ablative* |
|---|---|---|
| tűz 'fire' | tűznek | tűztől |
| tükör 'mirror' | tükörnek | tükörtől |
| öröm 'joy' | örömnek | örömtől |

|  | | |
|---|---|---|
| szemölcs 'wart' | szemölcsnek | szemölcstől |
| rézsű 'slope' | rézsűnek | rézsűtől |

*b.* IA–b: *HÁZ* type

| | | |
|---|---|---|
| ház 'house' | háznak | háztól |
| kupa 'goblet' | kupának | kupától |
| koszorú 'wreath' | koszorúnak | koszorútól |
| bika 'bull' | bikának | bikától |
| csíra 'germ' | csírának | csírától |
| példa 'example' | példának | példától |
| hernyó 'caterpillar' | hernyónak | hernyótól |

As can be seen, both classes contain stems in which the last non-neutral vowel is preceded by neutral vowel(s) (*szemölcs* etc., *bika* etc.).[25]

**Class IB:** Complex harmonic stems
The last vowel of such stems is harmonic; (at least one of) the preceding vowel(s) is another harmonic vowel that disagrees in backness with the last vowel; they may contain any number of neutral vowels (but not in the last syllable). These stems are often referred to as 'disharmonic'. Thus, in stems of this class, back vowels co-occur with front rounded vowels.[26] Suffix selection is governed by the backness value of the last harmonic vowel:

(31)   *a.* IB–f: *SOFŐR* type

| | *dative* | *ablative* |
|---|---|---|
| sofőr 'driver' | sofőrnek | sofőrtől |
| allűr 'mannerism' | allűrnek | allűrtől |
| kosztüm 'outfit' | kosztümnek | kosztümtől |

      *b.* IB–b: *NÜANSZ* type

| | | |
|---|---|---|
| nüansz 'nuance' | nüansznak | nüansztól |
| amőba 'amoeba' | amőbának | amőbától |
| pözsó 'Peugeot' | pözsónak | pözsótól |

**Class IIA:** Simple neutral stems
In these stems, all vowels are neutral. Since neutral vowels surface as front, it is no wonder that most stems belonging here select front-vowel suffixes:

(32) IIA–f: *VÍZ* type

| | *dative* | *ablative* |
|---|---|---|
| víz 'water' | víznek | víztől |
| szegény 'poor' | szegénynek | szegénytől |
| kert 'garden' | kertnek | kerttől |

---

[25] In this case the neutrality of these vowels cannot be demonstrated (beyond the co-occurrence facts) given that there are no harmonizing prefixes in Hungarian and that harmony applies left to right.

[26] It is interesting but phonologically irrelevant that the words belonging here are usually recent loanwords.

However, some 60 simple neutral stems select back-vowel suffixes. This class can be referred to as 'antiharmonic'. Most words belonging here have $i$ or $í$, some of them have $é$ (*cél* 'aim', *héj* 'crust', *derék* 'waist').[27]

(33) IIA–b: *HÍD* type

| | | dative | ablative |
|---|---|---|---|
| | híd 'bridge' | hídnak | hídtól |
| | cél 'aim' | célnak | céltól |
| | derék 'waist' | deréknak | deréktól |
| | | *become X* | *made X* |
| | ritka 'rare' | ritkul | ritkított |
| | néma 'dumb' | némul | némított |
| | | *may X* | *X-ing (adj.)* |
| | szid 'scold' | szidhat | szidó |
| | nyílik 'open' (intr.) | nyílhat | nyíló |

**Class IIB:** Complex neutral stems
The last vowel of these stems is neutral but they also contain harmonic vowel(s). Suffix selection is (by definition) not governed by the neutral (transparent) vowel in the last syllable but by the preceding harmonic vowel. In the *üveg* type this is not evident: whether suffix selection is governed by the harmonic or the neutral vowel, the result would be front in either case. One of the reasons why these are better analysed as they are here (rather than collapsed with the *víz* type) is that there is no complex-neutral analogue of the *híd* class. If front harmony in e.g. *rövid* 'short' or *büfé* 'buffet' was due to *i/é*, not to *ö/ü*, the lack of antiharmonic items of the *CöCiC* or *CüCé* type would be an accidental gap in the pattern. On the other hand, it is in the *papír* type that the transparent nature of neutral vowels (i.e. the fact that they let harmony pass through them without interfering with it) is plainly evident.

(34)   *a.*  IIB–f: *ÜVEG* type

| | | dative | ablative |
|---|---|---|---|
| | üveg 'glass' | üvegnek | üvegtől |
| | rövid 'short' | rövidnek | rövidtől |
| | örmény 'Armenian' | örménynek | örménytől |

      *b.*  IIB–b: *PAPÍR* type

| | | | |
|---|---|---|---|
| | papír 'paper' | papírnak | papírtól |
| | taxi 'taxi' | taxinak | taxitól |
| | dózis 'dose' | dózisnak | dózistól |

---

[27] Some stems of this class end in *-a* when unsuffixed but drop that *-a* before certain suffixes (*tiszt-a* 'clean', *ritk-a* 'rare', *sim-a* 'smooth', *bén-a* 'paralysed', *ném-a* 'dumb'); similarly, *fiú* 'boy' selects back suffixes even when its /u/ is dropped: *fi-am* 'my boy/my son'. The noun *férfi* 'man' is quite peculiar: most suffixes vacillate with it (*férfi-nak/-nek*, *-tóll-től*) but alongside *férfias* 'masculine', *férfiak* 'men' there is no *\*férfies*, *\*férfiek*.

| kordé 'cart' | kordénak | kordétól |
| tányér 'plate' | tányérnak | tányértól |
| kávé 'coffee' | kávénak | kávétól |
| haver 'pal' | havernak | havertól |
| balek 'dupe' | baleknak | balektól |
| maszek 'self-employed' | maszeknak | maszektól |

### 3.2.3. The behaviour of neutral vowels

Although they are phonetically front, the vowels *i, í, e, é* are phonologically neutral because (i) they are transparent with respect to harmony; (ii) they often co-occur with back vowels in stems. Within Class IA, neutral vowels that precede back-harmonic ones (*hernyó* 'caterpillar', *csíra* 'germ', *iskola* 'school', *interjú* 'interview', *periódus* 'period') have a peculiar status. They are never relevant for suffixation, as the latter is governed by the harmonic vowel that follows. But then this is also true of the earlier portion of disharmonic words (IB: *sofőr, nüansz*). Hence, from the data alone it is impossible to tell whether words like *hernyó* are just 'mixed' (i.e. neutral + harmonic, IA) or indeed disharmonic (i.e. front-harmonic + back-harmonic, IB). A similar indeterminacy arises with respect to *szemölcs* 'wart', *rézsű* 'slope', etc. Here, there is absolutely no way to tell whether these are mixed (neutral + harmonic) or plain harmonic (front + front).

The behaviour of neutral vowels is more relevant where they follow, rather than precede, harmonic (in particular, back-harmonic) vowels: *papír, analízis, oxigén, kódex*. These are expected to go into the *papír* type (IIB–h) but not all of them do. Two kinds of such unexpected behaviour are described below.

### 3.2.3.1. Mixed vacillating stems: complex neutral or disharmonic?

Some stems look as if they belonged to the *papír* type but vacillate between selecting front and back-vowel suffixes (see Ringen and Kontra 1989 for statistical data on the extent of variation in a number of individual items). The front-harmonic variants can only be accounted for if we assume that the vowel of the final syllable(s) is not neutral but harmonic: an *opaque* segment that takes over the role of harmonic governor for the rest of the word (including suffixes) just like the IIB–f: *sofőr* type. For instance, in *dzsungel* 'jungle', the *e* may be transparent, letting the *u* govern harmony: *dzsung(e)l-ban* 'in the jungle', or else it may be opaque: *dzsun⟧⟦gel-ben* 'id.'. Such stems will be called 'mixed vacillating' stems. Mixed stems having an *e* in the last syllable predominantly belong here (35a), some stems of this type have *é* in their last syllable (35b); and stems ending in several neutral-vowel syllables are mostly vacillating (35c).

(35) *a.* dzsungel-ben/ban 'in the jungle', Ágnes-nek/nak 'for Agnes', ban-
kett-en/on 'at a banquet', zsáner-ről/ról 'about a genre', hotel-ek/ok
'hotels'

    *b.* konkrét-en/an 'concretely', Tihamér-ről/ról 'about T.', affér-ben/ban
'in a quarrel'

    *c.* analízis-sel/sal 'with analysis', aszpirin-től/tól 'from aspirin', agg-
resszív-en/an 'aggressively', klarinét-tel/tal 'with a clarinet', matiné-
re/ra 'to a morning performance', szanitéc-nek/nak 'to a medical
orderly'

In Table 11, these words will appear in both relevant subclasses, IIB–b and
IB–f, with % standing for vacillation.

### 3.2.3.2. *Mixed disharmonic stems*

A classification paradox is presented by mixed stems that look like those of
IIB–b (*papír*) but exclusively take front-vowel suffixes, i.e. behave like IB–f
(e.g. *kódex* 'codex', *november* 'November', *operett* 'operetta', *oxigén* 'oxygen'). Although we would expect them to skip the final vowel(s) and take
back-vowel suffixes (*\*kódex-nak*, *\*november-ban*), they actually behave as
disharmonic stems (like *sofőr*). Stems belonging here are of two kinds: they
either end in several syllables containing neutral vowels (36*b*) or just one but
that must contain *e* (36*a*).

(36) *a.* kódex-ben 'in a codex', József-et 'Joseph' (acc.), október-től 'from
October'

    *b.* oxigén-nel 'with oxygen', operett-ek 'operettas', acetilén-ből 'from
acetylene', november-ben 'in November', varieté-hez 'to a variety
show'

Notice that stems ending in several neutral-vowel syllables tend to belong
here if their last vowel is *e* or *é* but to the vacillating class if it is *i* or *í*. In Table
11, 'mixed disharmonic' stems will be included in class IB–f.

    The behaviour of mixed stems can be summarized as in (37). As can be
seen, not all theoretically possible cases actually exist. If the penult is back
(i.e. there is only one neutral-vowel syllable at the end of the stem), suffixation shifts from neutral to disharmonic in correlation with vowel height
(diagonally in the table): with *i, í* it is always neutral, with *é* it is mostly neutral but sometimes vacillating, whereas with *e* it is predominantly vacillating,
sometimes disharmonic, and in a few cases neutral. (The neutrality of *e* is
most clearly shown by the small and closed set of *haver, maszek*, etc.) On the
other hand, if there are at least two neutral vowels in the final syllables, the
same diagonal distribution is found with a shift to the right: we have

| (37) penult | ult | neutral (IIB–b) | vacillating (IIB–b/IB–f) | disharmonic (IB–f) |
|---|---|---|---|---|
| back | i, í | papír-nak dózis-nak | | |
| neutr. | i, í | | analízis-nak/nek agresszív-nak/nek | |
| back | é | kávé-nak rostély-nak | konkrét-nak/nek Tihamér-nak/nek | |
| neutr. | é | | matiné-nak/nek klarinét-nak/nek | oxigén-nek varieté-nek |
| back | e | haver-nak maszek-nak | dzsungel-nak/nek hotel-nak/nek | kódex-nek október-nek |
| neutr. | e | | | november-nek operett-nek |

vacillation with i, í, vacillation or disharmonic suffixation with é, and only disharmonic behaviour with e.[28]

Vacillating stems appear to exhibit some sensitivity to larger context (cf. Kontra, Ringen, and Stemberger 1991). Thus *ezzel a pulóverrel* 'with this jumper' and *azzal a pulóverral* 'with that jumper' occur more frequently than *ezzel a pulóverral* 'with this jumper' and *azzal a pulóverrel* 'with that jumper'. If this is true in general (i.e. beyond the experimental setting in the paper referred to), the phenomenon is an interesting case of harmony-at-a-distance.

Let us summarize some properties of e whose neutrality is debated in the literature (cf. Ringen 1978, 1988b, Ringen and Kontra 1989).

[28] This can be seen more clearly if the table is repeated in two distinct parts:

| (i) | penult | ult | neutral | vacillating | disharmonic |
|---|---|---|---|---|---|
| | back | i, í | papír-nak dózis-nak | | |
| | back | é | kávé-nak rostély-nak | konkrét-nak/nek Tihamér-nak/nek | |
| | back | e | haver-nak maszek-nak | dzsungel-nak/nek hotel-nak/nek | kódex-nek október-nek |
| (ii) | penult | ult | neutral | vacillating | disharmonic |
| | neutr. | i, í | | analízis-nak/nek agresszív-nak/nek | |
| | neutr. | é | | matiné-nak/nek klarinét-nak/nek | oxigén-nek varieté-nek |
| | neutr. | e | | | november-nek operett-nek |

(38) *a.* there are no non-alternating suffixes containing *e* (cf. (29));

   *b.* there are no antiharmonic stems containing *e* in their last or only syllable (cf. (33));

   *c.* among stems having a back penult and *e* in their last syllable, neutral (*haver*), vacillating (*dzsungel*), and mixed-disharmonic (*kódex*) types are found but firmly neutral instances are few.

It is perhaps best to treat *e* as a primarily neutral vowel: this is supported by its articulatory features (front unrounded) and its neutral behaviour in *haver* etc. Where its neutrality is neither supported nor refuted (*beton, szemölcs*), it can be taken to be neutral for reasons having to do with system economy. It is only where *e* is demonstrably non-transparent (opaque) that we have to take it to be a harmonic vowel (*kódex*); this can be accounted for by positing a (structurally) different underlying representation (see section 6.1).[29]

The classification of stem types proposed in this section (and the previous one) can be summarized as in Table 11.

### 3.2.4. Rounding harmony

As we saw in (28) above, in suffixes like -*hoz/hez/höz* 'to', -*tok/tek/tök* (2pl.), and -*on/en/ön* 'on', there is a three-way suffix alternation: this time, the roundness of front stem vowels is also relevant.

(39)  tűz-höz       'fire'      víz-hez    'water'   ház-hoz    'house'
      szemölcs-höz  'wart'      kötény-hez 'apron'   hernyó-hoz 'caterpillar'
      sofőr-höz     'driver'    kódex-hez  'codex'   nüansz-hoz 'nuance'

Rounding harmony is considerably less complex than backness harmony is: it does not extend to stem-internal distribution (in other words, *szemölcs* or *kötény* are not irregular or 'disharmonic' with respect to rounding in the same way as *sofőr* or *kódex* are with respect to backness); there are no neutral vowels with respect to rounding (i.e. no vowel is skipped by the process, it is always the last stem vowel that governs rounding harmony—provided it is a front vowel); and there are no antiharmonic stems (i.e. no parallel of the *híd* case), at least not with respect to ternary suffixes. One phenomenon that resembles antiharmony is *hölgy-ek* 'ladies' (*\*hölgy-ök*), *tüz-et* 'fire' (acc.) (*\*tüz-öt*) but notice that this is a case of quaternary harmony, a property of 'lowering stems', cf. section 8.1.3; with the ternary suffixes this never happens: *hölgy-höz* (*\*hölgy-hez*), *tűz-ön* (*\*tűz-en*).

Rounding harmony is usually treated in the literature as some minor subsidiary pattern that is not essentially related to backness harmony. It is either

---

[29] In fact, the *e* of *kódex* 'codex' and *haver* 'pal' will be represented identically; the difference will be encoded in whether the back vowel preceding them is lexically linked to DOR or not. See section 6.1 for the details.

TABLE 11. *Stem classes for vowel harmony*

---

**I.** HARMONIC STEMS (their last vowel is harmonic)

---

**IA** SIMPLE HARMONIC STEMS
(containing one [or several compatible] harmonic vowel[s])
**IA–f:** *tűz* 'fire'
*tükör* 'mirror', *kürt* 'horn', *gőz* 'steam', *köszörű* 'grinder',
*szemölcs* 'wart', *ripők* 'cad', *rézsű* 'slope', *esztendő* 'year',
*revü* 'variety show', *kesztyű* 'glove'
**IA–b:** *ház* 'house'
*kupa* 'goblet', *város* 'town', *koszorú* 'wreath', *duplum* 'duplicate
copy', *bika* 'bull', *hernyó* 'caterpillar', *izom* 'muscle', *tégla* 'brick',
*opera* 'opera', *bitó* 'gallows', *patika* 'pharmacy', *stílus* 'style',
*beton* 'concrete', *konkurencia* 'rivalry'

---

**IB** COMPLEX HARMONIC STEMS
(containing non-compatible harmonic vowels)
**IB–f:** *sofőr* 'driver'
*attitűd* 'attitude', *operatőr* 'cameraman', *allűr* 'mannerism',
*kaszkadőr* 'stuntman', *kosztüm* 'outfit'; *kódex* 'codex', *október*
'October', *oxigén* 'oxygen'; %*dzsungel* 'jungle', %*konkrét*
'concrete', %*analízis* 'analysis'
**IB–b:** *nüansz* 'nuance'
*amőba* 'amoeba', *pözsó* 'Peugeot', *bürokrácia* 'bureaucracy',
*cölibátus* 'celibacy'

---

**II.** NEUTRAL STEMS (their last vowel is neutral)

---

**IIA** SIMPLE NEUTRAL STEMS
(containing only neutral vowels)
**IIA–f:** *víz* 'water'
*rét* 'meadow', *szegény* 'poor', *rekettye* 'gorse', *bili* 'potty',
*fillér* 'penny', *menedzser* 'manager', *kemping* 'camping site';
%*férfi* 'man'
**IIA–b:** *híd* 'bridge'
*cél* 'aim', *derék* 'waist'; %*férfi* 'man' [antiharmonic stems]

---

**IIB** COMPLEX NEUTRAL STEMS
(containing harmonic vowels as well)
**IIB–f:** *üveg* 'glass'
*rövid* 'short', *tőzeg* 'peat', *örmény* 'Armenian', *gyülevész*
'riff-raff' (adj.), *keszőce* 'morello soup', *hübrisz* 'hubris'
**IIB–b:** *papír* 'paper'
*kordé* 'cart', *kuvik* 'little owl', *haver* 'pal', *reverzális* 'mutual
concession', *csiricsáré* 'dawdry'; *pönálé* 'forfeit', *föderatív* 'federal';
%*dzsungel* 'jungle', %*konkrét* 'concrete', %*analízis* 'analysis'

---

not analysed explicitly (if mentioned at all) or analysed in a manner that is totally different from the analysis of backness harmony (cf. Polgárdi and Rebrus 1998). In Chapter 6 of the present book, we shall present an analysis in which backness harmony and rounding harmony are intertwined (although the differences noted in the preceding paragraph will not be ignored).

# 4

# THE CONSONANT SYSTEM

The lexical consonant inventory, listed and exemplified in section 2.2.2, will be classified along the lines indicated in (1):

(1)

```
                          Consonants
                         /         \
              Obstruents             Sonorants
             / |      \              /      \
        Stops Fricatives Affricates Nasals  Liquids
```

| | Stops | Fricatives | Affricates | Nasals | Liquids |
|---|---|---|---|---|---|
| Labial | p/b | f/v | — | m | — |
| Dental | t/d | s/z | $t^s$/– | n | l/r |
| Palatal | $t^y$/$d^y$ | š/ž | č/ǰ | $n^y$ | j |
| Velar | k/g | x/– | — | — | — |

The place-of-articulation categories of (1) may appear to be oversimplified from a phonetic point of view but are quite sufficient for a phonological classification. Labials include the bilabial stops /p b/, the bilabial nasal /m/, as well as the labiodental fricatives /f v/.[1] The class of dentals comprises the

---

[1] Note that /v/ will be underlyingly represented as neutral between sonorant (approximant) and obstruent (fricative) status (i.e. unspecified for the feature [son]) to account for its Janus-faced character with respect to voice assimilation and phonotactic constraints. Phonetically, [v] is a fricative in most positions but (except for certain well-defined cases, see section 4.1.1) it is the least fricative-like—the least 'noisy'—of all; hence, although in terms of their behaviour, some /v/'s act as if they were sonorants and others as if they were obstruents, their typical surface realization, just like the underlying representation to be proposed, is actually quite faithfully characterized as something between an approximant and a fricative.

lamino-dental stops /t d/, nasal /n/, lateral /l/, and trill /r/, as well as the lamino-alveolar fricatives /s z/ and affricate /tˢ/. Palatals subdivide into palato-alveolar fricatives /š ž/ and affricates /č ǰ/ on the one hand and dorso-palatal stops /tʸ dʸ/, nasal /nʸ/, and approximant /j/ on the other.[2] Finally, the class of velars includes /k g/ and /x/; the latter segment is realized as a voice-less glottal glide [h] prevocalically and as a voiceless velar fricative [x] else-where. The phonetic implementation rules defining the exact points of articulation (as well as, for palatals, the exact articulator) fall outside the scope of the present study.

This chapter is organized as follows. Section 4.1 discusses the class of obstruents. After justifying the treatment we suggest for /v/ and /x/ in 4.1.1, the subclasses of stops, fricatives, and affricates are characterized in 4.1.2–4, respectively. We turn to the class of sonorants in 4.2; finally, a set of repre-sentations of Hungarian underlying consonants is proposed in 4.3.

## 4.1. OBSTRUENTS

The two most salient common properties of (oral) stops, fricatives, and affricates are that they occur in voiced/voiceless pairs in the inventory (two voiceless obstruents, /tˢ/ and /x/, do not have underlying voiced counterparts but [dᶻ] and [ɦ] both occur as surface segments) and that they participate in voice assimilation, both as targets and as triggers. In this respect, /v/ is half-way between obstruents and sonorants in that it undergoes voice assimila-tion: *tévhit* [fh] 'misbelief', *távkapcsoló* [fk] 'remote control panel' but does not trigger it: *részvét* \*[zv] 'sympathy', *pótválasztás* \*[dv] 'by-election'. On the other hand, /x/ shows the opposite behaviour: it does not undergo voice assimilation but devoices a preceding obstruent (see examples further below). In this section, we consider the status of these two segments first; then we will consider the three subclasses of obstruents one by one.

### 4.1.1. Obstruent clusters, voicing, and the status of /x/ and /v/

In Hungarian, adjacent obstruents must agree in terms of voicing. Word ini-tial consonant clusters that do not contain a sonorant are always voiceless throughout as in (2a); even irregular initial clusters tend to conform to this pattern, see (2b):

---

[2] The latter set will be treated here as non-anterior coronals rather than high non-back dor-sals, for reasons that will become apparent in due course. The approximant /j/ behaves as a con-sonantal sonorant (i.e. a liquid) rather than a glide, under any of the possible interpretations of the latter term. See section 4.2.

(2) *a.* sport [šp] 'sports', stég [št] 'landing-stage', skála [šk] 'scale', szpáhi [sp] 'Turkish cavalryman', sztár [st] 'leading man/lady', sztyeppe [st^y] 'prairie in Russia', szkíta [sk] 'Scythian'

*b.* psziché [ps] 'psyche', xilofon [ks] 'xylophone', szfinx [sf] 'sphinx'

There is a single set of counterexamples that will turn out to be significant as we go on: in words like *tviszt* [tv] 'twist' (the dance), *kvarc* [kv] 'quartz', *szvit* [sv] 'suite', *svung* [šv] 'momentum', we find a voiceless obstruent followed by (what is traditionally classified as) a voiced fricative where, however, the latter is invariably /v/.

Other morpheme-internal (intervocalic or morpheme-final) obstruent clusters are either all-voiceless as in (3*a, b*) or else all-voiced as in (3*c, d*):

(3) *a.* pitypang [t^yp] 'dandelion', puszpáng [sp] 'boxwood', ráspoly [šp] 'file'; szeptember [pt] 'September', bukta [kt] 'sweet roll', kaftán [ft] 'Turkish coat', asztal [st] 'table', este [št] 'evening'; kesztyű [st^y] 'glove', bástya [št^y] 'bastion'; sapka [pk] 'cap', patkó [tk] 'horseshoe', butykos [t^yk] 'pitcher', dafke [fk] 'obstinacy', deszka [sk] 'plank', táska [šk] 'bag', kocka [t^yk] 'cube', bocskor [čk] 'moccasin'; klopfol [pf] 'beat (steak)', bukfenc [kf] 'somersault', aszfalt [sf] 'asphalt', násfa [šf] 'pendant'; kapszula [ps] 'capsule', buksza [ks] 'purse'; tepsi [pš] 'frying-pan', taksál [kš] 'estimate'; nátha [th] 'cold' (noun); kapca [pt^s] 'foot clout', vakcina [kt^s] 'vaccine'; kapcsol [pč] 'link' (verb)

*b.* kopt [pt] 'Coptic', akt [kt] 'nude', szaft [ft] 'gravy', liszt [st] 'flour', test [št] 'body', jacht [xt] 'yacht'; maszk [sk] 'mask', barack [t^yk] 'apricot'; copf [pf] 'plaited hair'; gipsz [ps] 'gypsum', koksz [ks] 'coke'; taps [pš] 'applause', voks [kš] 'vote' (n); also in place-names like Apc, Detk, Batyk, Recsk, Szakcs, Paks, etc.

*c.* rögbi [gb] 'rugby football', azbeszt [zb] 'asbestos'; labda [bd] 'ball', Magda [gd] <a name>, bovden [vd] 'Bowden cable', gazdag [zd] 'rich', rozsda [žd] 'rust'; mezsgye [žd^y] 'ridge'; izgul [zg] 'be excited', pezsgő [žg] 'champagne'; udvar [dv] 'yard', fegyver [d^yv] 'weapon', özvegy [zv] 'widow'; kobzos [bz] 'minstrel', madzag [dz] 'string', lagzi [gz] 'wedding'; habzsol [bž] 'devour'

*d.* smaragd [gd] 'emerald', kezd [zd] 'begin', pünkösd [žd] 'Whitsun', kedv [dv] 'temper', edz [dz] 'train' (verb)

Notice that, again, /v/-final clusters defy this regularity: *pitvar* [tv] 'porch', *akvárium* [kv] 'fishbowl', *köszvény* [sv] 'gout', *posvány* [šv] 'mire'.

Loanwords that originally contained an obstruent cluster of heterogeneous voicing (or happen to have a spelling suggesting one) automatically get adjusted to this pattern (but cf. *Pickwick* [kv], *Ruszwurm* [sv] etc.):

(4) a. abszolút [ps] 'absolute', obstruens [pš] 'obstruent', abcúg [ptˢ] 'down with him!', abház [ph] 'Abkhaz', Buddha [th], joghurt [kh] 'yogurt'

   b. futball [db] 'football', Macbeth [gb], matchbox [ǰb] 'toy car', Updike [bd], anekdota [gd] 'anecdote', afgán [vg] 'Afghan'

In suffixed forms, stem-final voiceless obstruents become voiced if the suffix begins with a voiced obstruent (5a) and vice versa: stem-final voiced obstruents become voiceless if the suffix begins with a voiceless obstruent (5a):

(5) a. kalap-ban [b:] 'in (a) hat', kút-ban [db] 'in (a) well', fütty-ben [dʸb] 'in (a) whistle', zsák-ban [gb] 'in (a) sack', széf-ben [vb] 'in (a) safe', rész-ben [zb] 'in part', lakás-ban [žb] 'in (a) flat', ketrec-ben [dᶻb] 'in (a) cage', Bécs-ben [ǰb] 'in Vienna'

   b. rab-tól [pt] 'from (a) prisoner', kád-tól [t:] 'from (a) bath-tub', ágy-tól [tʸt] 'from (a) bed', meleg-től [kt] 'from the heat', szív-től [ft] 'from (a) heart', víz-től [st] 'from water', garázs-tól [št] 'from (a) garage', bridzs-től [čt] 'from bridge' (the card game)

This assimilation process is regressive and (right-to-left) iterative:

(6)    liszt-ből [stb] → [sdb] → [zdb] 'from flour'
       pünkösd-től [ždt] → [žtt] → [štt] (→ [št]) 'from Whitsun'

It also applies across a compound boundary (rab⟧⟦szolga [ps] 'slave', lit. 'captive-servant'), across a word boundary (nagy kalap [tʸk] 'large hat') and indeed across any higher boundary as long as no pause intervenes; furthermore, as the examples in (4) show, it applies in non-derived environments as well, hence it is postlexical (but obligatory and non-rate-dependent).

Sonorants do not participate in the process: they do not voice a preceding obstruent (7a), nor are they devoiced by a following voiceless obstruent (7b):

(7) a. kalap-nak 'to (a) hat', kút-nak 'to (a) well', fütty-nek 'to (a) whistle' zsák-nak 'to (a) sack', széf-nek 'to (a) safe', rész-nek 'to (a) part', más-nak 'to sth else', léc-nek 'to (a) lath', csúcs-nak 'to (a) peak'

   b. szem-től 'from (an) eye', bűn-től 'from (a) sin', torony-tól 'from (a) tower', fal-tól 'from (a) wall', őr-től 'from a guard', száj-tól 'from (a) mouth'

There are two segments that behave asymmetrically with respect to this process. One is /v/ that undergoes devoicing (szívtől [ft] 'from (a) heart') but does not trigger voicing (hatvan *[dv] 'sixty'). The other one is /x/ that triggers devoicing (adhat [th] 'he may give') but does not undergo voicing before an obstruent. The usual solution for /x/ is to assume that this segment is /h/ at the underlying level and to characterize it as [−cons] (this would be quite appropriate phonetically as long as [+cons] is defined as 'constriction *in the oral cavity* at least equal to that found in fricatives'). Then, in order to exclude /h/ from the class of potential targets, the input to voice assimilation can be restricted to [+cons, −son] segments. However, the glottal realization of this segment does not occur preconsonantally; what does occur is its velar realization [x] (see section 8.2.2). It is this [x] that resists voice assimilation (e.g. *pechből* [xb], *[γb] 'out of bad luck')—but then it cannot be claimed to be [−cons]. Several possibilities suggest themselves at this point, none of them very satisfactory. One would be to order the putative rule /h/ → [x] after voice assimilation, such that this rule, *h*-strengthening, counterfeeds voicing.[3] Another possibility would be not to restrict voice assimilation to [+cons] segments and let /h/ undergo it (in principle, at least).[4] The solution we opt for here is stipulating an *ad hoc* filter to the effect that *[γ] is disallowed in Hungarian surface representations (or representations at any level, for that matter). This will do the job: we can simplify the rule of voice assimilation (by omitting [+cons] which, without rule ordering, and especially if the underlying segment is /x/ rather than /h/, would be useless anyway), yet keep our grammar from generating *[γ].

Turning to the issue of /v/, the classical solution (couched in SPE terms) of Vago (1980a) is to specify /v/ as a sonorant consonant (on a par with liquids, as far as major class features are concerned). This move successfully eliminates /v/ as a potential trigger of the rule of voice assimilation; but makes it disappear from the set of possible targets as well. Therefore, Vago has to state the devoicing of /v/ in a separate rule.[5] Using SPE conventions for collapsing rules (and disregarding the problem of *h* we mentioned above), Vago (1980a: 35) formulates voice assimilation (including the devoicing of /v/) as follows:

(8) $\left\{ \begin{array}{c} [-son] \\ \left[ \begin{array}{c} +cons \\ -cor \\ +cont \end{array} \right] \end{array} \right\} \rightarrow [\alpha \text{ voice}] / \underline{\quad} (\#) \left[ \begin{array}{c} -son \\ \alpha \text{ voice} \end{array} \right]$

(Obstruents and /v/ are assimilated to a following obstruent in voicing.)

---

[3] Both rules being postlexical, this ordering would have to be stipulated.

[4] Zsigri (1994, 1998) suggests to (do that and yet) exempt [x] from undergoing the rule by introducing the notion of 'phonetic quotations'. He points out that voiceless obstruents that are clearly non-Hungarian do not get voiced: *Bath-ba* [θb], *[ðb] 'to Bath', as if they were 'encapsulated' or surrounded by 'quotation marks'. He then claims that all Hungarian [x]-final lexical items are exactly like this example in that they refuse to be affected by Hungarian phonological rules (in particular, voice assimilation). This suggestion would be perfectly all right if [x]-final items were indeed few and clearly non-native. However, as we will see in section 8.2.2, this is far from being the case. We are therefore left with the brute-force solution proposed in the text. (See Szigetvári 1998 for extensive discussion and an alternative proposal.)

[5] He further assumes that, once it is devoiced, this segment automatically switches from sonorant to obstruent status via the redundancy rule that specifies all [−voice] segments as [−son].

A further argument in favour of this solution is that it explains the phono-tactic oddity of /v/ noted above (immediately below (2)): if /v/ is [+son], it is not surprising that it occurs in initial clusters as in *tviszt* 'twist', *kvart* 'fourth' (in music), *szvetter* 'sweater', *svéd* 'Swedish', cf. parallel examples with other non-nasal sonorants like *tréfa* 'joke', *klassz* 'superb', *szleng* 'slang', *srác* 'kid'.

However, the phonotactic evidence is not as unambigous as it might seem at first. In word-final clusters /v/ patterns with obstruents: it occurs after sonorants in such clusters (*ellenszenv* 'dislike', *érv* 'argument', *elv* 'principle', *ölyv* [jv] 'hawk').[6] It also occurs in final *-dv* clusters, e.g. *kedv* 'disposition', *üdv* 'salvation'; this constitutes a violation of Sonority Sequencing in any case but the violation is at least not unprecedented if /v/ is a fricative (an obstruent); whereas if it is a sonorant, cases like this would violate the otherwise excep-tionless generalization that (on the surface) no final cluster can consist of a sequence of obstruent plus sonorant, in that order.

In onsets that are not part of a word initial cluster and in non-branching codas both voiced fricatives and liquids occur practically unrestricted; hence such positions do not offer any evidence as to the status of /v/'s occurring in them (except, crucially, codas followed by a voiceless obstruent where voiced fricatives undergo voice assimilation whereas liquids do not: as we saw above, /v/ patterns with obstruents in this case, too).

In sum, a /v/ occurring in an onset behaves as a sonorant (this is manifest in word initial clusters and in onsets preceded by a voiceless obstruent but remains latent in onsets preceded by a (syllable ending in a) voiced obstruent, a sonorant consonant, or a vowel, or not preceded by anything, cf. the ex-amples in (9*a*), listed in this order), whereas a /v/ occurring in a coda (be it branching or non-branching) behaves as an obstruent (again, this is manifest in branching codas and in codas followed by a voiceless obstruent but remains latent in non-branching codas followed by a voiced obstruent, a sonorant, or nothing, cf. the examples in (9*b*)).[7]

(9) *a.* kvarc 'quartz', pitvar 'porch';
     medve 'bear', olvas 'read', kova 'flint', vér 'blood'
   *b.* terv 'plan', hívsz [fs] 'you call', óvtam [ft] 'I protected';
     révbe 'to port', bóvli 'junk', sav 'acid'

---

[6] This is similar to what we find for other voiced fricatives, cf. *nemz* 'beget', *vonz* 'attract', *torz* 'distorted', *rajz* 'drawing', whereas liquids do not occur after nasals or other liquids (except that *l* marginally occurs—in names and recent loanwords like *fájl* 'file', *görl* 'chorus-girl'—after the other two liquids; on postconsonantal final *j*; cf. section 4.1.3).

[7] Phonetically, the degree of friction seems to correspond nicely to the pattern presented here: forms in (9*b*) tend to exhibit more noisiness than forms in (9*a*); in fact, the first line of (9*a*) may be the least fricatival, the first line of (9*b*) the most fricatival—obviously so for the voice-assim-ilated cases, while the second lines in each group are in between these two extremes. Note that word initial clusters are not analysed in this book as branching onsets (cf. section 5.2.2).

There are various ways to account for this distribution. First, we could claim that there are two distinct underlying segments involved here: an obstruent whose occurrence is restricted to codas and a sonorant whose occurrence is restricted to onsets.[8] Second, we could take all instances of /v/ to be sonorants underlyingly and derive the obstruent where we have to.[9] Conversely, we could take all /v/'s to be obstruents underlyingly and derive the sonorant where we have to.[10] All these solutions involve feature-changing operations and, with their abundancy, exemplify the excessive power of SPE-like frameworks.

The solution we will propose in section 7.3 utilizes a bit of all but is crucially based on underspecification. We will assume that /v/ is underlyingly neither a fricative ([-son]) nor a liquid ([+son]) but neutral—in that it is unspecified for [son] but has a laryngeal node (like voiced obstruents). Voice assimilation will be defined as in (10):

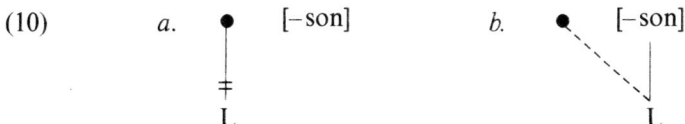

(10)    a.    ● [-son]        b.    ● [-son]

               |                     ⟍
               ‡                      ⟍
               L                       L

Thus, in cases like *pitvar* 'porch', (10b) will be inapplicable since the /v/ is not [-son]. In cases like *révbe* 'to port', either both parts of the rule apply and [vb] will surface sharing the L node of /b/ or neither of the two parts applies and both segments retain their own L nodes (see section 7.3 for discussion). In either case, the cluster surfaces as voiced throughout. Finally, in cases like *óvtam* 'I protected', (10a) severs the link between the root of /v/ and its L

---

[8] However, given that this is a classic case of complementary distribution (with sufficient 'phonetic similarity') and that completely automatic alternation is observable wherever a given /v/ switches from coda to onset status (cf. *tervez* 'plan' (verb), *hívok* 'I call', *óvott* 'he protected', *révén* 'by means of', *savas* 'acidic'), this is not a particularly insightful solution.

[9] Vago's solution alluded to above is an implicit version of this idea. Olsson (1992) offers a more direct (though rather vague) implementation: he also takes /v/ to be a sonorant underlyingly (actually, he classifies it as a glide rather than a liquid, but this is irrelevant here) but posits a rule of 'Structural *v*-strengthening' where the term 'structural' refers to the fact that /v/ is not supposed actually to change into something else (viz., an obstruent) but just behave 'as if' it were [-son] before a consonant or pause. The technical solution is unsatisfactory but the idea is worth pursuing: we could introduce a rule of *v*-obstruentization that changes [+son] into [-son] either where this is strictly necessary (i.e. in C_# and before a voiceless consonant) or in all positions where sonorant status is not essential (i.e. everywhere except in C_V) or else, the golden mean, simply in coda position (cf. Zsigri 1998 for discussion and a quite different solution).

[10] The traditional account (claiming that /v/ is a voiced fricative that exceptionally does not trigger voice assimilation) is an implicit version of this idea, the converse of Vago's solution in a way. Olsson's suggestion (see previous note) could also be tried in reverse: by positing a rule of 'Structural *v*-weakening' that specifies /v/ as 'behaving like' a sonorant before a vowel. Or else a straightforward rule of *v*-sonorization might actually turn a /v/ into a sonorant either in C_V position or, equivalently, in an onset.

node. Thus, the /v/ of this last form will enter phonetic implementation lacking an L node and unspecified for [son].

Given that in the framework assumed here (cf. Lombardi 1995*a*, 1996) it is the job of the phonetic implementation module to interpret obstruents lacking an L node as voiceless and to make sonorants (all of which lack an L node) spontaneously voiced, all we have to do is assume a fill-in rule located in this component of the grammar that turns onset /v/'s into sonorants and coda /v/'s into obstruents by specifying their empty [son] feature as plus and minus, respectively. With this move, all representative examples in (9) will come out exactly as required: all types in (9*a*) will exhibit spontaneously voiced approximants; *terv*, *bóvli*, and *sav* will contain voiced obstruents having their own L nodes and specified as [–son]; *révbe* will have an L node shared between its /v/ and /b/; finally, *hívsz* and *óvtam* will have [–son] labial continuants lacking an L node: i.e. voiceless [f]'s. The details of this analysis will be spelled out more fully in section 7.3.

### 4.1.2. Stops

In this book, the term 'stop' will refer to plosives, i.e. oral stops; nasal stops will simply be referred to as 'nasals'. Thus, stops constitute a subclass of obstruents, rather than cutting across the obstruent/sonorant dichotomy. /p b t d k g/ are uncontroversially members of this class; however, /tʸ/ and /dʸ/ are sometimes characterized in the literature as affricates (see Szende 1992: 119 ff. and references cited there).

Their surface realization may indeed be affricate-like to a variable extent, depending on phonetic context. Before stressed vowels (*tyúk* 'hen', *gyár* 'factory') and word finally (*fütty* 'whistle', *vágy* 'desire') they are quite strongly affricated; before an unstressed vowel much less so (*ketyeg* 'tick', *magyar* 'Hungarian'), and before a stop not at all (*hagyta* [tʸt] 'he left it', *ágyba* 'to bed'). The fricative component is usually absent before /r/ (*bugyrok* 'bundles'); before /l/ lateral release can be observed as in stops (compare *fátylak* 'veils' with *hátlap* 'reverse side'), and only under strong emphasis do we find a fricative component as with true affricates (cf. *vicclap* 'comic journal'). Of the nasals, /m/ may be preceded by slight affrication (*hagyma* 'onion'), but /n/ and /nʸ/ may not (*hagyna* 'he would leave some', *hegynyi* 'as large as a hill'). The degree of affrication depends further on style and rate of speech: in slow, deliberate speech it is much stronger than in fast or casual styles. This wide range of variables and varieties should raise our suspicion that we have basically stops here which, under the appropriate circumstances, are more or less affricated due to well-known physiological factors; notice that true affricates do not exhibit such extensive variability.

All this is quite suggestive but what we would need at this point is some concrete evidence that makes the stop interpretation of /tʸ dʸ/ not only possible but strongly motivated as well. Two such pieces of evidence come to mind.

The first concerns the pre-stop allophones of stops vs. affricates. In this position, stops can be realized by their non-released variants, e.g. *kapta* [kɔp̚tɔ] 'he got it', *rakta* [rɔk̚tɔ] 'he put it', whereas affricates obviously cannot, since they do not have such allophones: *bocskor* [boĉkor] (*[bot̚kor]) 'moccasin', *barack* [bɔrɔt͡sk] (*[bɔrɔt̚k]) 'peach'. Now, /tʸ dʸ/ are usually unreleased in this position, cf. *hegytől* [hɛtʸ̚töl] (*[hɛt͡ʸçtöl]) 'from (a) hill', *hagyd* [hɔdʸ̚d] (*[hɔd͡ʸjd]) 'leave' (imp.); in some cases (before velars?) there may be vacillation: *hetyke* [hɛtʸ̚kɛ] (~ [hɛt͡ʸçkɛ]) 'pert'. This property clearly shows that they pattern with stops.

The other argument is based on the phenomenon that affricates are resistant to OCP-driven fusion across a word boundary.[11] Sequences of identical stops are merged into geminates in any style of speech (*szép pár* [se:p:a:r] 'nice couple', *két tag* [ke:t:ɔg] 'two members', *sok kör* [šok:ör] 'many circles'), whereas affricates remain unmerged in careful speech (*rác cég* [ra:tˢ-tˢe:g] 'Serbian firm', *bölcs csere* [bölĉ-ĉɛrɛ] 'wise change'). In colloquial speech, the first affricate may lenite into a fricative ([ra:stˢe:g], [bölšĉɛrɛ]); it is only in casual speech that the OCP has its way, followed by degemination where appropriate ([ratˢ:e:g], [bölĉ:ɛrɛ] ~ [bölĉɛrɛ]). Now if we look at phrases like *ramaty tyúk* 'decrepit wench', *nagy gyár* 'big factory', we find that the merger applies automatically and obligatorily, i.e. no release is observed in the middle of the cluster—as it is expected for stops, as opposed to true affricates.[12]

---

[11] We assume here that a proper geminate stop/affricate, i.e. a linked structure with two X-slots and a single melody, is phonetically interpreted as a segment with a lengthened closure phase and a single release (burst noise) at the end, whereas a fake geminate, i.e. two X-slots separated by some kind of morphological boundary but dominating (two instances of) identical material, would correspond to a sequence of two phonetic segments, each with its own release. We further hypothesize that this becomes an OCP violation at the point where the relevant bracket is erased and that it is remedied at that point by merging the two melodies and creating a linked structure which is then interpreted as stated above. That repair operation is obligatory for stops but not for affricates.

[12] In over-careful speech, two separate (released) short consonants may occur with a brief pause sandwiched in between: [rɔmɔtʸ-tʸu:k], [nɔdʸ-dʸa:r], but then this is also possible for the other stops. However, 'deaffricated' forms like *[rɔmɔçtʸu:k] or *[nɔjdʸa:r] are totally unacceptable.

Let us note in passing that Olsson (1992) offers a classification of Hungarian consonants that seemingly paves the way for a compromise concerning the stop vs. affricate status of palatal non-continuants. Compare the following with our classification given in (1) above:

| | Labial | Dental | Alveolar | Palato-alv. | Palatal | Velar |
|---|---|---|---|---|---|---|
| Nasals | m | n | | | nʸ | |
| Stops/Affricates | p/b | t/d | tᶻ/dᶻ | ĉ/ǰ | tʸ/dʸ | k/g |
| Fricatives | f | | s/z | š/ž | | x |
| Liquids/Glides | v | l | r | | j | |

In sum: /tʸ dʸ/ are palatal stops in Hungarian; in the appropriate phonetic contexts, under appropriate conditions in terms of stress, speech rate, and speech style, they become affricated, as is to be expected for physiological reasons. However, this does not warrant their classification as affricates.

### 4.1.3. Fricatives

The set of underlying fricatives includes labiodental /f/, velar /x/, as well as four coronal segments: alveolar /s z/ and palato-alveolar /š ž/. The latter four are (redundantly) [+strident] and are traditionally known—together with the affricates /tˢ č ǰ/—as sibilants.[13] In addition, the following non-underlying fricatives occur in surface representations:

– [v] as the realization of the 'semi-sonorant' /v/, primarily in coda positions (cf. section 7.3);

– [j] as the realization of /j/ in the last position of word-final clusters if the preceding consonant is voiced; and

– [ç] as the realization of /j/ in the last position of word-final clusters if the preceding consonant is voiceless and either nothing follows or the following onset is also voiceless.

In this section, the array of /j/ and /x/ allophones will be briefly considered. The actual feature composition of /j/ will be discussed in section 4.2; the various processes affecting /x/ will be described in section 8.2.2.

The segment /j/ was traditionally classified as a fricative. This is, however,

---

Our policy in (1) was to reduce the number of place-of-articulation classes, accounting for predictable place differences in terms of manner of articulation. By contrast, Olsson collapses the manner classes of stops and affricates (he also considers collapsing liquids with glides as in the above table but finally decides to keep them apart) but differentiates six places. (A seventh place would be Glottal for the glide *h* but he takes the underlying segment—as we do here, see section 8.2.2—to be the velar fricative, deconsonantalized prevocalically into a glide by rule.) Given that stops and affricates are united into a single class, one would think that the old debate concerning whether /tʸ/ and /dʸ/ are stops or affricates is resolved in an elegant manner. However, the idea of economy in the use of distinctive features is not one of Olsson's main concerns. He defines and uses twenty different features, at least one third of which are predictable from the others, hence redundant. One of these surplus features—redundant in terms of the above classification—is 'gradual release' (SPE: delayed release). Accordingly, Olsson takes a stance on the affricacy issue with respect to /tʸ/ and /dʸ/, claiming that they are affricates. In a subsequent paper (Olsson 1993) he explains that one should take the variant in the strongest position as basic (where 'strongest' is understood as 'most resistent to lenition'). Given that the oral palatal noncontinuants are phonetically realized as affricates before a stressed vowel and as stops in various weaker (more lenition-prone) positions (this is *his* assessment of the facts; see the second paragraph of 4.1.2 for details), it follows that they are underlyingly affricates. This reasoning is based on a possibly misguided principle and does not tally with Olsson's own treatment of *h* (see above); more substantially, it disregards the fact that genuine affricates are never realized as stops, no matter how weak the position.

[13] For the various postlexical processes affecting sibilants, see section 7.2.2.

not borne out by either its phonetic or phonological properties. Phonetically, the 'elsewhere' allophone of /j/ is an approximant since no noise is generated as it is produced. There is, however, one type of context where its fricative allophones appear: postconsonantal final position (before another consonant or pause). Here, if the preceding consonant is voiceless (and if its effect is not overridden by that of a subsequent voiced obstruent), a voiceless (fortis) palatal fricative ([ç]) is pronounced: *kapj* 'get' (imp.), *rakj* 'put' (imp.), *döfj* 'stab' (imp.); if the preceding consonant is voiced, a lenis palatal fricative ([ʝ]) occurs. This fricative is fully voiced if a consonant-initial word follows (except where the following consonant is a voiceless obstruent: voice assimilation applies to the whole word-final cluster in this case: *vágj ki* [kçk] 'cut out' (imp.)); before pause, [ʝ] loses much of its voicing due to a very general and very late (possibly non-language-specific) process but does not become fortis: *férj* 'husband', *szomj* 'thirst', *dobj* 'throw' (imp.).

Phonologically, /j/ cannot be an obstruent either; if it were, it should participate in voice assimilation. Except for the case just mentioned where /j/ is obstruentized first and then becomes [ç] either through progressive voice assimilation as in *kapj* etc. or through the general rule of (regressive) voice assimilation as in *vágj ki* etc., this segment neither undergoes nor triggers voice assimilation (cf. *ajtó* *[ɔçtoː] 'door'; *fáklya* [faːkjɔ], *[faːgjɔ] 'torch'). In sum, underlying /j/ must be a sonorant. Whether it should be analysed as a glide (as in Vago 1980a, Olsson 1992, etc.) or as a liquid (as in Nádasdy and Siptár 1989: 15–16, Siptár 1993a, 1995) will be considered in section 4.2.

Turning to /x/, its prevocalic alternant [h] could be loosely characterized as a 'glottal fricative' but it will be technically analysed here as an obstruent glide ([–cons, –son]). The major alternation patterns it participates in are as follows. Intervocalically, it is represented by a voiced glottal approximant [ɦ] or is deleted (in casual speech). In postconsonantal onset position it may also be deleted (in fast casual speech) but it is mostly represented by [h] (and always so in a postpausal onset). In coda position, /x/ is realized as [x]. This velar fricative may be phonetically palatalized in a front-vowel context (just like the other velars: [k], [g]), but—contrary to traditional claims—is not neutralized with the voiceless palatal fricative [ç]. Geminate /xx/ is always realized as [xː]. A process-oriented account of all these alternations (with sets of examples) will be given in section 8.2.2.[14]

Let us conclude this section with a comparison of the behaviour of /v/, /j/, and /x/ in various contexts. Representative examples are given in (11).

---

[14] It has been suggested in the literature that both /h/ and /x/ should be posited as underlying segments (Tálos 1988) or else the glottal allophones should be derived by rule from an underlying /x/ (Olsson 1992). These positions are argued against in Siptár (1995). In the present book, we adopt the position advocated by Olsson; see Siptár (1998b) for additional discussion.

| (11) | /v/ | /j/ | /x/ |
|---|---|---|---|
| *Onset* | | | |
| [ __ V | vegyes 'mixed' | jegyes 'fiancé' | hegyes 'pointed' |
| [C __ V | kvarc 'quartz' | — | — |
| V. __ V | páva 'peacock' | pálya 'course' | léha 'frivolous' |
| VC. __ V | rakva 'putting' | rakja 'puts it' | rakhat 'may put' |
| | medve 'bear' | szablya 'sabre' | szabhat 'may cut' |

| *Coda* | | | |
|---|---|---|---|
| V __ ] | sav 'acid' | zaj 'noise' | sah 'Shah' |
| C __ ] | — | lépj 'step!' | — |
| | nedv 'humidity' | dobj 'throw!' | — |
| V __ C] | hívsz 'you call' | sajt 'cheese' | jacht 'yacht' |
| | szívd 'suck it!' | majd 'later' | — |
| V __ .CV | óvtam 'I warned' | ajtó 'door' | pechtől 'from bad luck' |
| | révbe 'to port' | mélybe 'down' | méhbe 'to womb' |

In the first onset context (utterance/phrase initially), /v/ is represented by an approximant/fricative (i.e. a voiced continuant characterized by a variable amount of friction between zero and slight), /j/ is represented by an approximant (frictionless continuant), while /x/ is realized as a voiceless glottal glide [h]. In a word initial cluster, none of our three segments occurs regularly as the first segment.[15] As the second segment, only /v/ occurs and is represented as an approximant (note the lack of voice assimilation). (In the isolated example *fjord* 'id.' /j/ occurs as second segment of a word initial cluster.) Intervocalically, /v/ is an approximant/fricative, /j/ is an approximant, and /x/ is a voiced glottal glide [ɦ] (voicing may be absent foot-initially, i.e. before a stressed vowel: *a hír* 'the news' [h] ~ [ɦ]). In a postconsonantal onset, /v/ is an approximant (note the lack of voice assimilation in *rakva*) with slight optional friction added (after labial stops, this friction is actually quite strong, cf. *lopva* 'stealthily', *dobva* 'throwing'), /j/ is an approximant, while /x/ is a voiceless glottal glide [h] (or disappears with or without lengthening of the preceding consonant).

In coda contexts, /v/ and /x/ are always represented by fricatives, a voiced (or, if assimilated, voiceless) labiodental and a voiceless velar fricative, respectively; /j/, however, is only fricativized as the last segment of a final cluster, into [ç] or [ʝ], depending on surrounding consonants. Note the lack of voice assimilation in *sajt* and *ajtó* (where /j/ is a sonorant) vs. the presence of voice

---

[15] Although sporadic examples like *Wrangler* [vr-] or *Hradzsin* [xr-] 'the castle in Prague', *Hruscsov* [xr-] 'Khrushchev' crop up, these can be safely disregarded as irregular. Note that the first consonant in such clusters would not be in the onset under the analysis presented in Chapter 5 below. Similarly, the last consonants in *nedv*, *lépj*, and *dobj* are not part of the coda but are syllabified as an appendix (see section 5.2.4.3).

assimilation in *hívsz* and *óvtam* (where /v/ is an obstruent/). On the other hand, the /x/ of *méhbe* is an obstruent but it is not voice-assimilated into *[ɣ] (see section 4.1.1).

### 4.1.4. Affricates

There are a number of surface affricates in Hungarian that are not reflexes of a corresponding underlying affricate but are derived from something else. As we saw in section 4.1.2 above, [t͡ç] and [d͡ʝ] are possible surface realizations of the palatal stops. In addition, (long) affricates can arise from the coalescence of stop + fricative (and other) sequences: *látsz* [laːt͡sː] 'you see', *kétség* [keːt͡ʃeːg] 'doubt', see section 7.2.2. The set of underlying affricates, on the other hand, definitely includes /t͡s/ and /t͡ʃ/; the question is whether their voiced counterparts are also part of the underlying phoneme inventory.[16]

The speech sound [d͡z] can come from three sources in Hungarian. It can be a voiced (i.e. voice-assimilated) allophone of /t͡s/ (*lécből* [leːd͡zbøl] 'from lath', *táncba* [taːnd͡zbɔ] 'into (a) dance'), where obviously no underlying /d͡z/ is involved. It can occur in words like *pénz* [peːnd͡z] 'money', *benzin* [bɛnd͡ʲin] 'petrol'; here, however, we have /nz/ clusters where [d] is an (optional) intrusive stop element like [p] in *szomszéd* 'neighbour', [b] in *oromzat* 'gable', [tʲ] in *München* 'Munich', etc. (and could be analysed along the lines of Clements 1987).

Finally, in words like *madzag* 'string', *bodza* 'elder', *pedz* 'nibble', [d͡zː] can

$$\text{be analysed in one of two ways: either as geminate } \underset{d \; z}{\overset{X \; X}{\bigwedge}} \rightarrow [d͡zː], \text{ cf. } \textit{vicces}$$

'funny' $\underset{t \; s}{\overset{X \; X}{\bigwedge}} \rightarrow [t͡sː]$, or as $\underset{d \quad z}{\overset{X \; X}{\diagup}} \rightarrow [d͡zː]$, cf. *látszik* 'seem' $\underset{t \quad s}{\overset{X \; X}{\diagup}} \rightarrow [t͡sː]$.

The first option would involve positing underlying /d͡z/.

But this underlying segment would have a rather skewed distribution: it would not occur word initially or postconsonantally at all; preconsonantally it would merely occur in a handful of suffixed forms; whereas intervocalically and finally (between a vowel and a word boundary) it would only occur doubled (long). This peculiar distribution, not found for any other member of the Hungarian consonant inventory, would be automatically explained—at least in its gross outlines—by the cluster analysis (assuming an independently motivated realization rule converting a cluster of stop + sibilant into a long affricate). Let us consider what objections can be made to such an analysis.

Three types of possible counter-arguments come to mind. (i) The surface contrast between long affricates as in *madzag* 'string' and [d] + [z] clusters as in *vadzab* 'wild oats' appears to show that the former cannot be derived from an underlying cluster. (ii) $C_iC_jC_k$ clusters (e.g. *kardvirág* 'cornflag') do not

---

[16] This section is based on Nádasdy and Siptár (1989: 21–3).

generally get simplified—apart from fast-speech consonant elision, cf. section 9.5—whereas $C_iC_iC_j$ clusters (e.g. *keddre* 'by Tuesday') do. Given that a stem-final (long) *dz* is shortened before a consonant-initial suffix, it follows—the argument goes—that it cannot be a cluster. (iii) Words like *vakaródzik* 'scratch oneself' can have short intervocalic [dᶻ]; this makes the distribution less skewed and the analysis with a unitary /dᶻ/ more plausible. These counter-arguments are, however, untenable:

(i) The phonetic difference between *madzag* 'string' ([dᶻ:]) and *vadzab* 'wild oats' ([d-z]) is totally parallel to that between *metszi* 'he cuts it' ([tˢ:]) and *hátszél* 'tail-wind' ([t-s]); in *vadzab/hátszél* internal word boundary (compound boundary) occurs between the stop and the fricative, and it is that boundary that blocks their coalescence into a single long affricate (at least in non-casual speech). Hence, any counter-argument based on surface contrast of the *madzag/vadzab* type is unfounded (this observation is due to É. Kiss and Papp 1984).

(ii) Next to another consonant, all Hungarian intramorphemic geminates get shortened (*sakktól* [šoktol] 'from chess'); this applies to [dᶻ:] as well (*edzve* [ɛdᶻvɛ] 'being trained'). This, however, only proves that the immediate input to degemination is [dᶻ:] (rather than a cluster); what it does **not** prove is that [dᶻ:] should derive from underlying ⤬ and not ⤙. Hence, this counter-argument fails, too.

(iii) In words like *vakaródzik* 'scratch oneself', there is free variation (for some speakers) between short [dᶻ] and long [dᶻ:] (as well as simple [z]). This seems to refute our claim above, i.e. that there are no intervocalic short [dᶻ]'s. But free variation proves exactly that length is irrelevant in this position: to put it differently, no short : long opposition is possible here. Since in non-vacillating cases (*madzag*) it is always long [dᶻ:] that occurs, it is quite easy to see that in words like *vakaródzik* the segment in question is not short /dᶻ/ but a long [dᶻ:] whose actual length varies (tends to become reduced in long words like this); this [dᶻ:], in turn, may just as well derive from a /d-z/ cluster. Hence, all three potential counter-arguments turned out to be cases that can be easily accounted for in terms of the cluster analysis, too: the fact that at some point in the derivation [dᶻ:] is a true geminate affricate rather than a cluster does not prove it is an affricate underlyingly.

The existence of */dᶻ/ as an underlying segment, therefore, is not supported by any valid argument at all.

The case of [ǰ], however, is different in that arguments for /d-ž/ are more or less balanced by arguments for /ǰ/. Word initial occurrence (as in *dzsámi* <a type of mosque>, *dzseki* 'jacket', *dzsóker* 'Jolly Joker') points toward /ǰ/, whereas the behaviour of word internal [ǰ]'s is practically identical with that of [dᶻ], thus supporting a /d-ž/ analysis. This ambiguity could be resolved, in principle, in three different ways.

First, we could assume that—obviously with the exception of voice assimilated cases like *rácsban* [raːɟbɔn] 'in grating'—[ɟ] always derives from a /d-ž/ cluster. In this case, the scope of degemination should be extended to include word initial position. Since word initial geminates are impossible anyway, such a redundancy rule (or constraint) is needed in any case—it should simply be allowed to operate during a derivation in which an offending representation is created by the coalescence of /d-ž/ into [ɟː].

Second, it would be possible to claim that *dzsámi* 'jami' is /ɟaːmi/ but *hodzsa* 'hodja' is /hodžɔ/; this would explain the ambiguity referred to above but would give /ɟ/ a rather skewed distribution (and it would be impossible to decide whether e.g. *lemberdzsek* 'anorak' is /lemberɟek/ or /lemberdžek/ (→ lemberɟːek → [lɛmbɛrɟɛk]).

Third, we could accept the traditional view that [ɟ] corresponds to /ɟ/ everywhere; but then it is to be explained why its intervocalic (*rádzsa* 'rajah') and final (*bridzs* 'bridge' (card game)) occurrences tend to be long (with a few exceptions like *fridzsider* [-iɟi-] 'refrigerator'). It might be suggested that a kind of loanword gemination is at work here (cf. *dopping* 'doping', *szvetter* 'sweater', *sokk* 'shock', *meccs* '(football) match', see Nádasdy 1989*a*, Törkenczy 1994*a*). This looks quite feasible for items like *menedzser* 'manager' and *bridzs*; the trouble is that the layer of vocabulary including *hodzsa* 'hodja' (Turkish loans) does not in general exhibit this process, cf. *mecset* (*meccset, *mecsett*) 'mosque' (we owe this last point to L. Kálmán, personal communication).

The first solution is technically neat and logically coherent; also, it establishes a parallel between [dᶻ] and [ɟ]; unfortunately, it does not conform to speakers' intuition and is rather abstract. What is more serious, /dž/ as an initial cluster does not fit the overall pattern of permissible initial clusters (cf. section 5.2.2). Although the second and third solutions are less elegant (and open to the objections raised above), it appears that either of them—or, most probably, some kind of combination, e.g. the gradual diffusion of /ɟ/ through the lexicon, to the detriment of earlier /dž/—is more realistic. Hence, the interpretation of /ɟ/ as a unitary underlying segment can be accepted.

In sum, the question we raised at the beginning of this section can be answered as follows. The inventory of Hungarian underlying consonants includes three affricates: /tˢ/ as in *cica* 'kitten', /č/ as in *csúcs* 'peak', and /ɟ/ as in *dzsem* 'jam'.

## 4.2. SONORANTS

Sonorant consonants (like vowels but unlike obstruents) fail to participate in voiced/voiceless oppositions; phonetically, they are all voiced but this is irrelevant throughout the phonology of Hungarian (as in other languages). Hence, sonorants will be underlyingly unspecified for voicing (in fact, they will lack a Laryngeal node); their surface voicing will be left unmentioned until the phonetic implementation module. Two major subclasses will be set up: nasals (/m n nʲ/) and non-nasals; the latter include what are traditionally known as liquids (/l r/), plus the segment /j/ whose exact feature composition has not been conclusively defined yet. This is the topic we turn to now. (With respect to the 'semi-sonorant' /v/, cf. sections 4.1.1, 4.1.3.)

The segment /j/ is definitely not a fricative in Hungarian (see 4.1.3). However, as we saw in section 2.2.1 above, it cannot be part of a branching nucleus, either. But the arguments we listed there against its interpretation as the non-head constituent of a diphthong do not exclude its being a glide in onset/coda position. To recapitulate briefly, the facts that there are no co-occurrence restrictions between a /j/ and a following/preceding vowel or that there are no diphthong/monophthong alternations are quite reconcilable with the view (going back to Vago 1980*a*) that /j/ is a glide. Furthermore, the fact that *j*V-initial words select the 'preconsonantal' allomorph of the definite article does not necessarily entail that /j/ should be consonantal: actually, *h*V-initial words select the same alternant and prevocalic [h] is realized as [−cons]. (According to the analysis presented in section 8.2.2, it is not underlyingly specified as [+cons], either.) Hence, this allomorph of the definite article is more properly called 'pre-onset'. Similarly, forms like *vajjal* 'with butter', although they constitute evidence against a branching nucleus interpretation, have nothing to say about the feature content of /j/ as long as it occupies the coda.

Nevertheless, we wish to maintain the claim that Hungarian /j/ is not a glide ([−cons, +son]) but a liquid ([+cons, +son]). Part of the reason resides in the fricative allophones we encountered in section 4.1.3; these are technically easier to derive if /j/ is [+cons] to begin with. But the claim that /j/ is not simply the vowel melody /i/ occurring in a non-nuclear syllable position (= a glide) can be supported by empirical evidence, too.

Part of this evidence concerns syllabification. On the assumption that syllable structure is assigned in the course of derivation rather than listed in the lexicon, the minimal pairs and near-minimal pairs in (12) cannot be properly syllabified if /i/ and /j/ are melodically identical.

(12)    mágia [maːgiɔ] 'magic' (noun)      vs.    máglya [maːgjɔ] 'stake'
        ion 'id.'                           vs.    jön 'come'
        dió 'walnut'                        vs.    fjord 'id.'

As can be seen from the examples, prevocalic *i/j* can be syllabified either as another nucleus or as an onset: the choice is more or less arbitrary (although it must be admitted that *jön* and *dió* are the expected patterns as opposed to *ion* and *fjord*, the word medial cases are strictly unpredictable). With postvocalic *i/j*, we find a similar degree of arbitrariness (concerning whether it will be a nucleus or a coda):

(13)   laikus 'layman'                   vs.    pajkos 'naughty'
       fáit 'his trees' (acc.)           vs.    fájt 'it hurt' (past)
       női 'feminine'                    vs.    nőj 'grow' (imp.)

There are suffixes consisting of a sole *i* and the imperative marker consists of a sole *j*, cf. *kéri* 'ask' (3sg def.) vs. *kérj* 'ask' (imp.), *fali* 'wall' (adj.) vs. *falj* 'devour' (imp.), *Mari* 'Mary' (dim.) vs. *marj* 'bite' (imp.). Pairs like *síel* [ʃiːɛl] 'ski' (verb) and *milyen* [mijɛn] 'what kind' indicate that an *i/j* associated to two timing slots can be syllabified as either a branching nucleus or a nucleus plus an onset. Finally, the nouns *íj* 'bow', *díj* 'prize', *szíj* 'strap' would contain the common melody of *i/j* associated to *three* timing slots and multiple ambiguity would arise as to how to syllabify them: *íj* could in principle be \**ji*, \**jij*, \**iji*, or *ijj* as well (the last version actually does occur as an alternative pronunciation for *íj* 'bow'). All these complications are avoided if /i/ and /j/ are segmentally represented in two different manners.

Further considerations supporting the conclusion that /j/ is consonantal include processes in which /j/ acts as a (consonantal) trigger, e.g. *l*-palatalization as in *alja* [ɔjːɔ] 'its bottom' (see section 7.1.1), or as a (consonantal) target, e.g. *j*-assimilation as in *moss* (< *mos+j*) 'wash' (imp.) (see section 7.2.1). A final argument can be based on the phenomenon of hiatus filling (see section 9.3) where an adjacent /i/ spreads to fill in an empty onset position, resulting in a [j]-type sound which may be weaker, more transient than the realization of an underlying /j/. Compare *kiáll* 'stand out' with *kijár* 'go out' or *baltái* 'his hatchets' with *altáji* 'Altaic'; the difference is quite noticeable in careful speech, although it may get blurred in a more colloquial rendering. Now, if we assume that /j/ is a liquid (whereas the hiatus filler is obviously a glide), this (potential) phonetic difference is readily explained.

In sum: we have a number of good reasons to claim that the set of Hungarian liquids includes three members: /l r j/ (again, cf. section 4.1.1 with respect to the 'semi-sonorant' /v/).

The other subclass of sonorants, that of nasals, includes three underlying segments: bilabial /m/, dental /n/, and palatal /nʲ/. On the surface, labiodental [ɱ], palato-alveolar [n̠], and velar [ŋ] are also attested. The occurrence of the latter three segments is severely restricted: they exclusively occur before homorganic consonants: [ɱ] only before /f v/ (e.g. *honfi* 'patriot', *honvéd* '(Hungarian) soldier'), [n̠] only before /č ǰ/ (e.g. *kincs* 'treasure', *findzsa* 'cup'), and [ŋ] only before /k g/ (e.g. *fánk* 'doughnut', *ing* 'shirt'). Which underlying nasal should we derive these from? Given that, across morphemes, nasal place assimilation is practically restricted to /n/ (see section 7.4 for discussion), the dental nasal seems to be the obvious choice. But then, notice that the only consonants that can follow [m] morpheme internally are /p b/ (cf. footnote 17 below) and the only consonants that can follow [nʲ] morpheme internally are /tʲ dʲ/. What underlying nasal is there in *komp* 'ferry', *lomb* 'foliage', *ponty* 'carp', and *rongy* 'rag', then? There are three possible answers to this question. If, in order to get maximal mileage out of our rules, we are ready to posit abstract underlying representations, /konp/, /lonb/, /pontʸ/, and /rondʸ/ suggest themselves; the attested surface forms are then derived via nasal place assimilation (with /n/ as input). If we wish to follow the principle that underlying forms of morphemes should be identical with their surface forms unless there is some good reason for them to diverge, and therefore we want our underlying representations to be as close to surface pronunciations as possible (up to allophonic variation), we are led to posit /komp/, /lomb/, /ponʸtʸ/, and /ronʸdʸ/. Finally, if we maintain that predictable information is to be factored out of (omitted from) lexical representations, we end up with an underspecified nasal (with no place features), thus /koNp/, /loNb/, /poNtʸ/, /roNdʸ/. For consistency, then, all preconsonantal nasals that surface as homorganic with the following consonant are to be represented as N: /troNf/ 'trump', /elːeNseNv/ 'antipathy', /poNt/ 'dot', /reNd/ 'order', /kiNč/ 'treasure', /beNjoː/ 'banjo', /baNk/ 'bank', /raNg/ 'rank'. Then, we would need *both* a fill-in rule that spreads the place features of the following consonant onto an underspecified nasal *and* a postlexical rule of nasal place assimilation, restricted to /n/ as input. But notice that an even more radically underspecified solution is also available. Suppose that we posit three underlying nasals as follows: /m/ as in *ma* 'today', *ima* 'prayer', *ám* 'though'; /nʸ/ as in *nyíl* 'arrow', *anya* 'mother', *íny* 'gums'; and /N/ (= [+nas]) as in *nő* 'woman', *ünő* 'heifer', *én* 'I', as well as for all preconsonantal nasals.[17] We assume that N receives place features (by spreading) before labials and palatals in the lexical phonology but remains unspecified for place before other consonants (and in non-preconsonantal positions) until the postlexical component. There, a number of rules apply to /m/ and to /nʸ/ which survived lexical phonology practically unscathed, while /N/ may fully assimilate to liquids (optionally), delete with nasalization of the preceding vowel before continuants, or receive place features from the following consonant; finally, all remaining N's are specified as [n] by default. All these processes will be discussed in section 7.4.


## 4.3. THE UNDERLYING CONSONANT SYSTEM

To summarize the upshot of the foregoing discussion, (14) tabulates the representations of Hungarian underlying consonants. R stands for the root node, L for the laryngeal node, whereas LAB, COR, and DOR are the artic

---

[17] There are a few sporadic exceptions with morpheme-internal preconsonantal /m/ (rather than /N/), including *szomszéd* 'neighbour', *emse* 'sow', *homlok* 'forehead', *kamra* 'larder', *tömjén* 'incense', but on the whole the generalization about the homorganicity of (the overwhelming majority of) morpheme internal nasal + consonant clusters is valid (for nasal + stop/affricate clusters it is exceptionless). Significantly, there are no counterexamples involving [n] before a non-homorganic consonant inside a morpheme.

ulator nodes (under the C-place node, abbreviated as PL in the rules to follow but not appearing in the diagram here). The geometrical arrangement assumed here was discussed in section 1.3 above.

| (14) | Labial | | | | | Dental | | | | | | | | Palatal | | | | | | | | Velar | | |
|---|---|---|---|---|---|---|---|---|---|---|---|---|---|---|---|---|---|---|---|---|---|---|---|---|
| | p | b | f | v | m | t | d | s | z | tˢ | n | l | r | tʸ | dʸ | š | ž | č | ǰ | nʸ | j | k | g | x |
| R | • | • | • | • | • | • | • | • | • | • | • | • | • | • | • | • | • | • | • | • | • | • | • | • |
| [cons] | + | + | + | + | + | + | + | + | + | + | + | + | + | + | + | + | + | + | + | + | + | + | + | |
| [son] | – | – | – | | + | – | – | – | – | – | + | + | + | – | – | – | – | – | – | + | + | – | – | – |
| [nas] | | | | | + | | | | | | + | | | | | | | | | | + | | | |
| [lat] | | | | | | | | | | | | + | | | | | | | | | | | | |
| [cont] | – | – | + | + | | – | – | + | + | ± | | + | + | – | – | + | + | ± | ± | | + | – | – | + |
| L | | • | | • | | | • | | • | | | | | | • | | • | | • | | | | • | |
| LAB | • | • | • | • | • | | | | | | | | | | | | | | | | | | | |
| COR | | | | | | • | • | • | • | • | | • | • | • | • | • | • | • | • | • | • | | | |
| [ant] | | | | | | + | + | + | + | + | | + | + | – | – | – | – | – | – | – | – | | | |
| DOR | | | | | | | | | | | | | | | | | | | | | | • | • | • |

All consonants except /x/ are specified as [+cons]; cf. section 8.2.2 with respect to /x/. The features [nas] and [lat] are either privative (unary) throughout the phonology, or else their negative values are supplied by default rules. Nasals are not underlyingly specified for [cont]; their stop character is predictable by rule:

(15)    *Nasal Stop Spell-out*
        [+nas] → [–cont]

Otherwise, the values for [cont] are specified as in (14). With respect to affricates, the sign ± is meant to suggest the temporally ordered sequence [–cont][+cont]. Stridency is a redundant feature but, given that some rules will need to refer to it, we assume the following rule to introduce it:

(16)    *Stridency Spell-out*

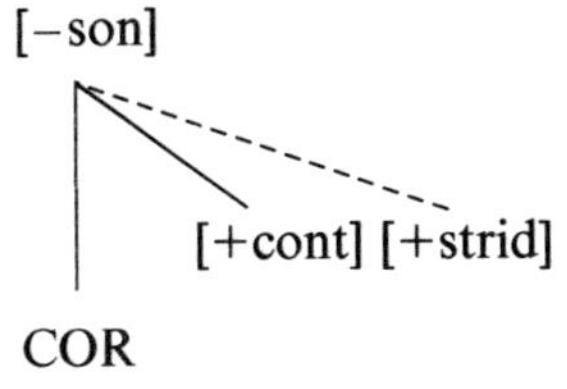

        [–son]
          ⟍ ⟍ ⟍
          |    [+cont] [+strid]
          |
        COR

This formulation is meant to cover affricates as well; the rule scans representations for [–son] roots dominating COR and [+cont], irrespective of whether [–cont] is or is not also dominated by the same root node.

Voicing distinctions are encoded in terms of the laryngeal node (L). Voiced obstruents and /v/ have one (dominating the unary feature [voice], not shown in (14)), whereas regular sonorants and voiceless obstruents do not; with respect to /v/ that has an L node but is unspecified for [son], cf. section 7.3.

Finally, in accordance with the discussion in 4.2 (immediately above the present section), /n/ is assumed to be unspecified for place; although it is listed under the label Dental for convenience, it is strictly speaking a root node merely specified as [+cons, +son, +nas].

# 5

## PHONOTACTICS: SYLLABLE STRUCTURE

### 5.1. INTRODUCTION

The phonological or phonotactic well-formedness of a word can be seen as an interplay of two factors: a prosodic and a non-prosodic one. On the one hand, a phonologically well-formed word must be parsable into (well-formed) prosodic units. It is generally assumed that the prosodic unit that is chiefly responsible for phonotactic well-formedness is the syllable,[1] but there are well-known examples of phonotactic constraints whose domain is a higher level prosodic unit such as the foot or the prosodic word.[2] As the foot does not seem to play an important role in Hungarian, a phonotactically well-formed Hungarian word is a unit which is exhaustively parsable into well-formed syllables. Thus, the phonotactic well-formedness of a word is derivable from well-formedness conditions on syllables (Syllable Structure Constraints (SSCs)). This relation between the well-formedness of words and syllables, however, is not symmetrical: while it holds that a well-formed word consists of a string of well-formed syllables, it is not true that any string of well-formed syllables constitutes a well-formed (potential[3]) word: there are transsyllabic constraints that obtain between syllables, or more precisely, between adjacent subconstituents of different syllables. These constraints do not refer to a prosodic unit higher than the syllable, but impose restrictions on the bonding of syllable edges (interconstituent clusters). In addition, as we have pointed out above, a language may have constraints on prosodic structure that directly refer to prosodic units higher than the syllable (e.g. conditions on word minimality, etc.).

There is evidence (Kaye 1974, Kenstowicz and Kisseberth 1977, Booij 1995, 1999, Hammond 1997) that the phonotactic well-formedness of words also depends on constraints independent of prosodic structure. The relevant constraints are Morpheme Structure Conditions (MSCs) and sequence con-

---

[1] Naturally, this is only true of frameworks that refer to the syllable (compare SPE, Government Phonology).

[2] For instance, the distribution of /h/ in English; cf. Anderson and Ewen (1987), Harris (1994).

[3] Naturally, well-formedness conditions of any kind must define units that are *potentially* well-formed in the given language, in other words, they must not treat accidental gaps as ill-formed (cf. e.g. Chomsky 1964, Halle 1962, Vogt 1954).

straints. MSCs define possible morpheme shapes and may refer to categorial information (word classes). Thus, they can impose constraints on what is a possible morpheme, noun, verb, etc. in a given language. These constraints are different from classical generative MSRs/MSCs (e.g. Chomsky and Halle 1968). They only complement SSCs if there are phonotactic regularities in a given language that are only expressible with reference to the morpheme as a domain.[4] A particular language may also have well-formedness conditions that constrain the combination of segments irrespective of their affiliation with prosodic or morphological units. These sequence constraints may state that a given (sequential) combination of segments (or features) XY is ill-formed regardless of whether it is wholly contained within or cuts across structural units such as syllables or morphemes within the word (which is the largest domain within which phonotactic regularities apply in Hungarian).

## 5.2. SYLLABLE STRUCTURE: SSCs

In this section we discuss the constraints that apply within (the constituents of) the syllable and define the syllable template in Hungarian.

### 5.2.1. The Hungarian syllable template: the basic syllable types

If we disregard the possible complexity of the onset, the nucleus and the coda, Hungarian has the following syllable types:[5]

(1)

| | | *word-initial* | *word-medial* | *word-final* |
|---|---|---|---|---|
| | CV | **pa**.tak 'creek' | fe.**ke**.te 'black' | sem.**mi** 'nothing' |
| | V | **i**.on 'ion' | da.**u**.er 'perm' | te.**a** 'tea' |
| | VC | **em**.ber 'man' | a.**or**.ta 'aorta' | ri.**ad** 'get frightened' |
| | CVC | **tom**.pa 'dull' | ke.**men**.ce 'oven' | be.**teg** 'sick' |

(1) exemplifies the basic syllable types and shows that they are free to occur in any position (initial, medial, and final) in the word.

Blevins (1995) proposes the following binary (YES/NO) parameters to account for language-particular variation in syllable typology: Obligatory

---

[4] They may apply to (near) surface representations, derived representations, or underlying representations (or all three). There are recent arguments that underlying morpheme structure constraints are necessary (Booij 1995, 1999, Hammond 1997).

[5] There is a surprising degree of agreement about this: authors of very different theoretical backgrounds agree that (disregarding constituent complexity) these are the basic syllable types in Hungarian, cf. Deme (1961), Kaye and Lowenstamm (1981), Kornai (1994), though see Kassai (1981) who also permits syllables consisting of consonants only. Naturally, authors whose framework excludes one (or more) of these structures come to different conclusions (e.g. Kaye, Lowenstamm, and Vergnaud 1990).

Onset, Coda, Complex Nucleus, Complex Onset, Complex Coda[6] (note that there is no Onset parameter, i.e. languages cannot choose to have no onsets). (2) shows the parameter settings for Hungarian[7] (disregarding the last three parameters which refer to constituent complexity):

(2)     Obligatory Onset    NO
          Coda               YES

As complex nuclei occur in Hungarian (i.e. the Complex Nucleus parameter is set to 'YES'), the syllable inventory in (1) can be extended:

(3)

| | *word-initial* | *word-medial* | *word-final* |
|---|---|---|---|
| CVV | **só**.vár 'desirous' | sza.**mó**.ca 'wild strawberry' | sző.**lő** 'grape' |
| VV | **í**.ró 'buttermilk' | i.di.**ó**.ta 'idiot' | rá.di.**ó** 'radio' |
| VVC | **ér**.ték 'value' | ki.**ál**.tás 'shout' | le.**ány** 'girl' |
| CVVC | **sár**.kány 'dragon' | ka.**szár**.nya 'barracks' | ta.**lán** 'perhaps' |

A comparison of (1) and (3) reveals that the distribution of long and short vowelled syllables is the same within the word. Furthermore, neither closed nor open syllables are restricted to word-final position and onsetless syllables may occur in positions other than word-initial. Note that long vowels are equally permitted in open and closed syllables (on the distribution of long vowels before consonant clusters, see section 5.4.2).

The three parameter settings discussed so far are fairly uncontroversial. What is more problematic is the setting of the remaining two 'complexity' parameters Complex Onset and Complex Coda. Hungarian words *can* begin and/or end with consonant clusters (e.g. *prém* 'pelt', *ptózis* 'ptosis', *part* 'shore', *akt* 'nude', etc.) but this is not necessarily evidence that these clusters are true onsets or codas.[8] It is a well-known fact that word edges (or certain morphological domain-edges) license special syllable structures. Specifically, there may be consonants or consonant sequences at the edges of these domains that are not incorporated into the onset or the coda of the syllable whose phonetically realized nucleus is the first or the last one in the word, respectively. So the question is whether the consonant clusters that occur word-initially and word-finally in Hungarian are true complex onsets and codas (respectively) or they are 'edge clusters', i.e. clusters occurring at

---

[6] Blevins (1995) has a sixth parameter for Edge Effect that we discuss later.

[7] There are alternative ways of expressing (more or less) the same typological distinctions as these parameters—cf. e.g. Kaye and Lowenstamm (1981), Clements and Keyser (1983), Prince and Smolensky (1993).

[8] Some authors do consider it as evidence (e.g. Kahn 1980), but currently there seems to be an agreement among phonologists that the assumption that word-initial/final consonant clusters are necessarily complex onsets/codas is false, see, e.g. Kenstowicz (1994), Kaye (1992a), Kaye and Lowenstamm (1981), Kaye, Lowenstamm, and Vergnaud (1990), Harris (1994), Steriade (1982). Rubach and Booij (1990), Davis (1990), Törkenczy and Siptár (1999).

domain edges whose initial or final member(s) are licensed by some special mechanism limited to the edges of domains and not by an onset or a coda constituent dominating them. At this point we are not primarily concerned with the actual licensing mechanism, which will be discussed later, but the analysis of onsets and word-initial clusters (section 5.2.2) and codas and word-final clusters (section 5.2.4).

### 5.2.2. Onsets—word-initial clusters

As we have seen already, it is not compulsory for a Hungarian syllable to have a (filled) onset. Thus, both vowel-initial and consonant-initial syllables are possible. In principle, any consonant may be syllabified into a simplex onset.[9] Word-initial two-member and three-member consonant clusters occur—they are shown in Tables 12 and 13. The question is whether these clusters realize branching onsets or not.

TABLE 12. *Word-initial CC clusters*

|    | p | t | tʸ | k | b | d | dʸ | g | tˢ | č | ǰ | f | s | š | v | z | ž | m | n | nʸ | l | r | j | x |
|----|---|---|----|---|---|---|----|---|----|---|---|---|---|---|---|---|---|---|---|----|---|---|---|---|
| p  | + |   |    |   |   |   |    |   |    |   |   |   | + |   |   |   |   | + |   |    | + | + |   |   |
| t  |   |   |    |   |   |   |    |   |    |   |   |   | + |   |   |   |   |   |   |    |   | + |   |   |
| tʸ |   |   |    |   |   |   |    |   |    |   |   |   |   |   |   |   |   |   |   |    |   |   |   |   |
| k  |   |   |    |   |   |   |    |   |    |   |   |   | + | + |   |   |   | + |   |    | + | + |   |   |
| b  |   |   |    |   |   |   |    |   |    |   |   |   |   |   |   |   |   |   |   |    | + | + |   |   |
| d  |   |   |    |   |   |   |    |   |    |   |   |   |   |   |   |   | + |   |   |    |   | + |   |   |
| dʸ |   |   |    |   |   |   |    |   |    |   |   |   |   |   |   |   |   |   |   |    |   |   |   |   |
| g  |   |   |    |   |   |   |    |   |    |   |   |   | + |   |   |   |   | + |   |    | + | + |   |   |
| tˢ |   |   |    |   |   |   |    |   |    |   |   |   | + |   |   |   |   |   |   |    |   |   |   |   |
| č  |   |   |    |   |   |   |    |   |    |   |   |   |   |   |   |   |   |   |   |    |   |   |   |   |
| ǰ  |   |   |    |   |   |   |    |   |    |   |   |   |   |   |   |   |   |   |   |    |   |   |   |   |
| f  | + |   |    |   |   |   |    |   |    |   |   |   |   |   |   |   |   |   |   |    | + | + | + |   |
| s  | + | + | +  | + |   |   |    |   | +  |   | + |   |   |   | + | + | + | + |   |    |   |   |   |   |
| š  | + | + |    | + |   |   |    |   |    | + |   |   |   |   | + | + | + |   |   |    | + | + |   |   |
| v  |   |   |    |   |   |   |    |   |    |   |   |   |   |   |   |   |   | + |   |    |   |   |   |   |
| z  |   |   |    |   |   |   |    |   |    |   |   |   |   |   |   |   |   |   |   |    | + | + |   |   |
| ž  |   |   |    |   |   |   |    |   |    |   |   |   |   |   |   |   |   |   |   |    |   |   |   |   |
| m  |   |   |    |   |   |   |    |   |    |   |   |   |   |   |   |   |   |   | + |    |   |   |   |   |
| n  |   |   |    |   |   |   |    | + |    |   |   |   |   |   |   |   |   |   |   |    |   |   |   |   |
| nʸ |   |   |    |   |   |   |    |   |    |   |   |   |   |   |   |   |   |   |   |    |   |   |   |   |
| l  |   |   |    |   |   |   |    |   |    |   |   |   |   |   |   |   |   |   |   |    |   |   |   |   |
| r  |   |   |    |   |   |   |    |   |    |   |   |   |   |   |   |   |   |   |   |    |   |   |   |   |
| j  |   |   |    |   |   |   |    |   |    |   |   |   |   |   |   |   |   |   |   |    |   |   |   |   |
| x  |   |   |    |   |   |   |    |   |    |   |   |   |   |   |   |   |   |   |   |    |   | + |   |   |

[9] Word-initially, palatal /tʸ/ only occurs in a single morpheme *tyúk* 'hen', but we consider this accidental.

TABLE 12. *Cont'd*

Examples: *ptózis* 'ptosis', *pszichológus* 'psychologist', *pneumatikus* 'pneumatic', *plakát* 'poster', *prém* 'fur', *tviszt* 'twist', *tréfa* 'joke', *xilofon* 'xylophone', *kvarc* 'quartz', *knédli* 'dumpling', *klór* 'chlorine', *krém* 'cream', *blúz* 'blouse', *bronz* 'bronze', *dzéta* 'zeta', *drukkol* 'cheer' (pres. 3rd sg indef.)', *gvárdián* 'Father Superior (of Franciscan monastery)', *gnóm* 'gnome', *gladiátor* 'gladiator', *gróf* 'count', *cvekedli* <type of pasta>, *ftálsav* 'phthalic acid', *flóra* 'vegetation', *friss* 'fresh', *fjord* 'id.', *szpícs* 'speech', *sztár* 'star', *sztyepp* 'steppe', *szkíta* 'Scythian', *szcéna* 'scene', *szféra* 'sphere', *szvetter* 'sweater', *szmog* 'smog', *sznob* 'snob', *szláv* 'Slav', *sport* 'id.', *stáb* 'staff', *skorpió* 'scorpion', *scsí* <Russian soup>, *svéd* 'Swedish', *smink* 'makeup', *snassz* 'passé', *slussz* 'finished', *sróf* 'screw', *vlach* 'Walachian', *zlotyi* <Polish currency>, *zrí* 'trouble', *mnemonika* 'mnemonics', *nganaszán* 'Nganasan', *Hradzsin* <proper noun>

TABLE 13. *Word-initial CCC clusters*

|   | pr | tr | kr | kl |
|---|----|----|----|----|
| s |    | +  |    | +  |
| š | +  | +  | +  |    |

Examples: *sztrájk* 'strike', *szklerózis* 'sclerosis', *spriccel* 'spray' (3 sg pres. indef.), *strázsa* 'guard', *skrupulus* 'scruple'

All the words which begin with consonant clusters are loanwords, but this fact does not in itself say anything about the status of the initial clusters: it is perfectly possible that the words in question are phonotactically just as 'normal' as any 'native' item in the lexicon. Indeed, Hungarian speakers can detect no difference between the well-formedness of a word such as *prém* 'fur' and *rém* 'monster'. While there are few words beginning with three consonants, words beginning with two consonants cannot be said to be infrequent (though there are many more consonant-initial words that begin with a single consonant). Also, there appear to be phonotactic restrictions holding between the consonants making up the word-initial clusters. While some of these restrictions are unrelated to syllable structure (e.g. the ban on adjacent obstruents differing in voicing), others seem specific to this position and may be interpreted as holding between the members of a branching onset (e.g. the non-occurrence of geminates).[10] These constraints may be construed as evidence for the well-formedness of branching onsets. Nevertheless, we want to suggest that the setting for the Complex Onset parameter is in fact 'NO' in Hungarian, and all the clusters that occur word-initially are 'edge clusters'. We assume that the non-final consonants in these clusters are licensed by a special mechanism restricted to domain edges, notably, they are syllabified

---

[10] Accordingly, (some of) these clusters have been analysed as branching onsets. For a detailed analysis and specific constraints cf. Törkenczy (1994*a*).

into a subsyllabic constituent called 'appendix'. Thus, they are represented as (4*a*) rather than (4*b*):

(4)     *a.*

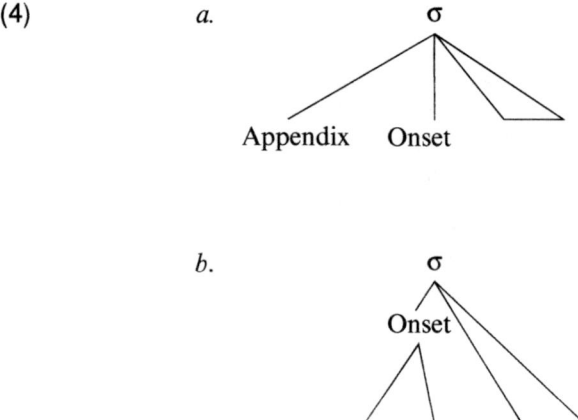

Appendix    Onset

*b.*

Evidence for or against this position may be drawn from alternations/processes that are sensitive to syllable structure and phonotactic patterns. Syllable-structure conditioned alternations (to be discussed in detail in Chapter 8) do not present conclusive evidence since there is no alternation in Hungarian that would require that the clusters in question should *not* be represented as branching onsets. However, it must be pointed out that the relevant alternations/processes (such as vowel ~ zero alternations (cf. Chapter 8) and Fast Cluster Simplification (cf. Chapter 9)) never make a branching onset interpretation *necessary*—i.e. an analysis of these syllable structure sensitive processes is always perfectly compatible with an edge cluster (appendix) interpretation of these consonant sequences.

Phonotactic patterns are another possible source of evidence: if we could show that the 'need' to analyse consonant clusters or substrings of consonant clusters as putatively branching onsets only arises at domain edges, then we could see this as an argument against their branching onset status, as domain edges may license 'special' edge clusters. Given that the word-initial position is suspect (since the clusters occurring there may be edge-licensed as appendix + onset), the most promising place to look for such evidence is medial. In principle—since they could be syllabified in two ways: (i) VC.CV or (ii) V.CCV—two-member medial clusters (CC clusters) can give us a clue if some alternation or distributional fact distinguishes (i) from (ii). For instance, restrictions on the length of vowels in closed vs. open syllables (closed syllable shortening effects as in English, Turkish, Yawelmani, etc.) could distinguish the two syllabifications. Unfortunately, no such fact or phenomenon is available in Hungarian.[11] Medial clusters containing more than two consonants are interesting, however: in such a cluster, syllabification will result in a complex syllabic constituent, either a coda or an onset (-CC.C- or -C.CC-); or vice versa, in a language where complex onsets (or codas) are well-formed, *regular* medial clusters of three consonants (or more) are expected to occur. Therefore, the lack/scarcity/irregularity of medial -CCC- clusters can be taken to suggest that branching onsets are ill-formed.[12]

At first sight, Hungarian seems to abound in word-medial -CCC- clusters. However, the main source of such clusters is analytic suffixation (e.g. [[ [ *kard* ] *ból* ] 'from (the) sword', [[ [ *vers* ] *ról* ] 'about (the) poem', [[ [ *elv* ] *telen* ] 'without principles', etc.) and compounding (e.g. [[ [ *vers* ] [ *láb* ] ] '(metrical) foot', [[ [ *elv* ] [ *társ* ] ] 'comrade', etc.). We have pointed out above that clusters straddling the edge of an analytic domain do not say anything about the phonotactics of the language, they are 'accidental' in the sense that no phonotactic restrictions apply across analytic domain edges—the relevant consonants are just juxtaposed without any restrictions. Thus, 'real data' are monomorphemic items, or words with synthetic suffixation containing medial -CCC-. Interestingly, there are no examples in Hungarian of synthetic suffixation creating -CCC- clusters.[13] There *are* monomorphemic words with -CCC- clusters in the language, but, significantly, their number is rather low, about 300 items in our database. Again, all the relevant words are loans, but, naturally, this does not in itself say anything about their well-formedness in Hungarian (examples: *bisztró* 'bistro', *centrum* 'centre', *komplex* 'complex', *export* 'id', *improvizál* 'improvise (3rd sg. indefinite)', *instancia* 'instance', *ostrom* 'siege', etc.). Furthermore, there are 95 types of clusters altogether that the approximately 300 tokens exemplify, but, typically, the number of tokens in a given type is extremely low (cf. 5.3.2.2 for a full list of the relevant clusters). There are only 7 types with 10 or more tokens and the majority of types (n=48) only have one token. This suggests that medial -CCC- clusters are special/irregular in Hungarian.

Although (monomorphemic) medial -CCC- clusters do display certain regularities (e.g. in a medial -CCC- cluster $C_\alpha C_\beta C_\gamma$, $C_\beta$ is never a sonorant[14]), we claim that these regularities are accidental in Hungarian in that they only reflect some of the regularities of the source languages the relevant words

---

[11] Long vowels other than *é/á* are not permitted before consonant clusters, but this constraint has nothing to do with syllable structure: all consonant clusters (whether they are potentially well-formed as branching onsets or not) behave in the same way: */íːkta/ and */íːkla/ are equally ill-formed. For details, see section 5.4.2.

[12] For a discussion of complex codas, see sections 5.2.4, 8.1.4.4, and 8.1.4.5.

[13] Multiply suffixed past forms of cluster-final stems are the only exception (e.g. [fiŋktɔk] 'fart' (3rd pl. past indef.)). Even these clusters are often broken up ([fiŋgotːɔk], cf. section 8.1.4. See also the behaviour of cluster-initial (-$C_i C_j$V ...) suffixes in section 8.1.2.2.

[14] For a discussion of medial -CCC- clusters, see section 5.3.2.2.

were borrowed from.[15] More precisely, if a constraint obtaining between medial $C_\alpha C_\beta C_\gamma$ is non-accidental, then we have to do with either of the following two situations: (i) it is identical with a constraint obtaining between the consonants of a corresponding two-member medial cluster $C_\alpha C_\beta$ and thus reduces to a constraint applying between a syllable-final consonant and the following syllable-initial one, i.e. it is an interconstituent constraint (e.g. there are no words with medial -*tpC*- in Hungarian, but there are no words containing medial -*tp*- either); (ii) it is an MSC or a sequence constraint and thus it has nothing to do with syllable structure at all (e.g. adjacent obstruents have to agree in voicing in Hungarian). Otherwise, all apparent medial -CCC-specific constraints are accidental, just 'debris' of the constraints that exist in the languages the particular words containing them were borrowed from.

Another argument for the special character of medial -CCC- clusters involves a comparison of medial -CC- clusters and -CCC- clusters. In a language that permits branching onsets we expect to find -$C_\alpha C_\beta C_\gamma$- clusters where $C_\beta C_\gamma$ is a well-formed branching onset and -$C_\alpha C_\beta$- is a permitted interconstituent cluster (-$C_\alpha.C_\beta C_\gamma$-). And vice versa, in general, for every -$C_\alpha C_\beta C_\gamma$-cluster we should find a matching -$C_\alpha C_\beta$- cluster if the latter is a permitted interconstituent cluster. Of course, accidental gaps may exist, but this should be the general tendency. It is interesting to compare English and Hungarian since in the literature English is generally taken to be a language that has branching onsets. As can be seen in (5), English is well-behaved with respect to the generalization above.

(5) | English | $VC_\alpha C_\beta V$ | $VC_\alpha C_\beta C_\gamma V$ |
|---|---|---|
| -kt- | vector | electronic |
| -pt- | chapter | dioptry |
| *-tk- | — | — |
| *-pk | — | — |
| *-tp- | — | — |
| *-kp- | — | — |

Hungarian, on the other hand, is very different: some -$C_\alpha C_\beta C_\gamma$- clusters corresponding to well-formed -$C_\alpha C_\beta$- are curiously missing:

(6) | Hungarian | $VC_\alpha C_\beta V$ | $VC_\alpha C_\beta C_\gamma V$ |
|---|---|---|
| -kt- | akta 'file' | spektrum 'spectrum' |
| -pt- | kapta '(boot) last' | dioptria 'dioptre' |
| -tk- | atka 'mite' | — |
| -pk- | lepke 'butterfly' | — |
| *-tp- | — | — |
| *-kp- | — | — |

[15] Of course, this does not mean that an explanation of why these clusters are not repaired is not in order, see a possible explanation in section 8.1.4.5.

We can either say that the missing clusters are accidental gaps, or the other explanation is that complex onsets are ill-formed. We suggest that the latter interpretation is correct.

It follows from the irregular status of medial -$C_\alpha C_\beta C_\gamma$- clusters that it is never necessary to syllabify two (or more) consonants into an onset in medial position. Therefore we claim that the consonant clusters that occur in word-initial position (the only position where consonant clusters arguably look like complex onsets) do not form onsets, but are edge clusters. Thus, the setting of the Complex Onset parameter is 'NO' in Hungarian. In our interpretation the phonotactic restrictions word-initial clusters display[16] are just (fragmentary) reflections of the constraints that apply in the source languages the relevant words come from.[17]

The fact that branching onsets are not permitted does not in itself explain the scarcity/irregularity of -CCC- clusters. The reason is that a -CCC- cluster can in principle be parsed exhaustively even if it does not contain a branching onset: it could consist of a complex coda and a following non-branching onset: -CC.C-. This raises the question whether complex codas are well-formed in Hungarian.[18] If the answer is negative, it follows that medial -CCC- are ill-formed (assuming that complex onsets are also ill-formed word-initially and word-medially). There are words ending in more than one consonant, but this does not in itself ascertain the status of these final clusters as complex syllabic constituents. We will return to this problem in section 5.2.4.

### 5.2.3. Rhymes

The rhyme may be branching or non-branching in Hungarian. Thus, the following rhyme templates are well-formed:

(7)

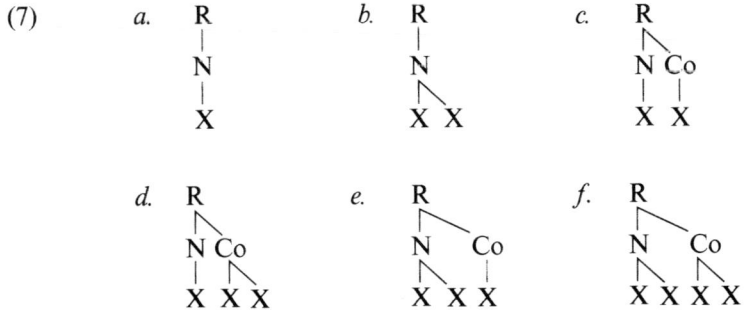

---

[16] Such as the absence of initial /sr/ (compare attested /šr/ *sróf* 'screw'), cf. Törkenczy (1994*a*) and Törkenczy and Siptár (1999).

[17] Chiefly English and German (cf. Siptár 1980).

[18] GP has a 'built-in' negative answer to this question since the theory does not permit complex codas (it does not even have a coda constituent, cf. Kaye, Lowenstamm, and Vergnaud (1990), Harris (1994)).

In general, there is no restriction on nuclei in branching or non-branching rhymes in Hungarian: any vowel can occur in a closed or an open syllable:

(8)     *closed* σ                    *open* σ

| | | | | |
|---|---|---|---|---|
| i | **rit**.mus | 'rhythm' | **szi**.ta | 'sieve' |
| ü | **ül** | 'sit' | **hü**.lye | 'stupid' |
| u | szi.**rup** | 'syrup' | **bu**.ta | 'dumb' |
| e | **em**.lő | 'breast' | te | 'you' |
| ö | **öt**.let | 'idea' | **ö**.reg | 'old' |
| o | o.**rom** | 'peak' | **ro**.ham | 'attack' |
| a | a.**lak** | 'shape' | **pa**.ta | 'hoof' |
| | | | | |
| iː | **sír** | 'grave' | **sí** | 'ski' |
| üː | **űr** | 'space' | **tű** | 'needle' |
| uː | **púp** | 'hump' | **bú**.tor | 'furniture' |
| eː | **sért** | 'hurt' | **mé**.ter | 'metre' |
| öː | **őr** | 'guard' | **ső**.lő | 'grape' |
| oː | **ól** | 'pigsty' | **hó** | 'snow' |
| aː | **fánk** | 'doughnut' | a.**lá** | 'under' |

It is apparent in (8) above that (*a*) any vowel quality is possible and (*b*) long and short vowels equally occur both in branching and in non-branching rhymes, i.e. no rhyme-specific phonotactic statement is necessary. We shall see in later sections that this is an oversimplification because (i) the distribution of vowels in stem/word-final open syllables is different from that in medial open syllables (see section 5.4.1) and (ii) only a very limited set of long vowels can occur in word-medial closed syllables and word-final syllables closed by more than one consonant when these syllables are undivided by a morpheme boundary (see section 5.4.2). The constraints that (i) and (ii) are due to are not SSCs strictly speaking because they refer to the phonological word or apply within the morpheme. There is one phonotactic restriction, however, which seems specific to the rhyme. This constraint concerns the distribution of surface roundedness/labiality within the rhyme. Vowels preceding the nasal + stop clusters /mp, mb/ must be rounded *if the vowel and the entire consonant cluster are within the rhyme*. Accordingly, while there are many words like *lump* 'drunkard', *komp* 'ferry', *tömb* 'block', *domb* 'hill', words like hypothetical \**limp* or \**semb* whose vowels are not rounded are unattested.[19] By contrast, there are many words like *ember* 'human being', *bimbó* 'bud', *lámpa* 'lamp', *némber* 'hag' in which the second member of the nasal + stop clusters is not within the rhyme (*em.ber*, *bim.bó*, *lám.pa*, *ném.ber*) and thus the vowel is not required to be rounded. This constraint can be seen as evidence for the rhyme node in Hungarian. Note that it is 'directional'. It is a constraint on vowels preceding

/mp, mb/ and cannot be seen as a requirement that labiality/rounding has to be shared within a V + nasal + stop rhyme since any vowel quality is possible in rhymes containing non-labial nasal + stop clusters: e.g. /nt/ *bánt* 'hurt', *csont* 'bone', *ment* 'save', *dönt* 'decide', *hint* 'sprinkle'; /ng/ *ing* 'shirt', *ráng* 'jerk', *zeng* 'resound', *döng* 'buzz', *korong* 'disk'. It has to be pointed out that the status of this constraint is unclear. It is (almost) exceptionless, but it does not play an active role in the phonology. There are no alternations that it would condition, and no evidence is available concerning native speakers' intuitions about the well-formedness of strings violating it.

### 5.2.4. Codas—word-final consonant clusters

In Hungarian the coda differs from the onset in that the former may branch. Thus, the setting of the Complex Coda parameter (Blevins 1995) is 'YES'. The coda is maximally binary branching. Furthermore, complex codas may be morphologically complex (i.e. there are suffixes solely consisting of consonants syllabified into the coda).

#### 5.2.4.1. Non-branching codas

Any underlying consonant may systematically occur in a non-branching coda.[20]

#### 5.2.4.2. Branching codas

In Hungarian the surface form of words may end in at most three consonants (*hat* 'six', *part* 'coast', *szfinx* [sfiŋks] 'sphinx'). Nevertheless, we claim that the coda is maximally binary branching, and that the more complex clusters at the ends of words are not (exhaustively) syllabified into a single coda. Furthermore, not all word-final two-term clusters realize branching codas. Let us examine word-final two-term clusters first. The notation used in Table 14 is as follows: a blank space in an intersection of a row and a column means that the relevant cluster is unattested. Numbers occur at intersections when a given cluster is attested: 1 = a cluster that only occurs undivided by a morpheme boundary (analytic or synthetic); 2 = a cluster that only occurs when divided by a morpheme boundary (analytic or synthetic); 3 = a cluster that occurs both monomorphemically and when divided by a morpheme boundary (analytic or synthetic). A box is struck out by dashes to indicate that the relevant cluster(s) is/are subject to (eliminated on the surface by) assimilations.[21]

---

[19] The word *galamb* 'dove' is problematic/exceptional if the constraint is taken to apply to the UR because the vowel of the final syllable is only rounded at the surface /galamb/.

[20] The surface realization of underlying consonants may be determined by syllabic constituency; see the behaviour of /x/ (section 8.2.2).

[21] None of these assimilations are related to syllable structure (i.e. they operate regardless of the syllabification of the cluster to which they apply) and some of them are postlexical (see Chapter 7). Note that Table 14 shows the inventory of word-final clusters *after* these assimilations have applied, i.e. clusters that are subject to assimilations appear in it in the assimilated form.

TABLE 14. *Word-final CC clusters*

| | p | t | tʸ | k | b | d | dʸ | g | tˢ | č | ǰ | f | s | š | v | z | ž | m | n | nʸ | l | r | j | x |
|---|---|---|---|---|---|---|---|---|---|---|---|---|---|---|---|---|---|---|---|---|---|---|---|---|
| **p** | 1 | 1 | | | | — | — | — | | | − | 1 | 3 | 1 | | — | — | | | | | | 2 | |
| **t** | | 3 | — | 1 | | — | — | — | | | | | — | | | — | | | | − | | | | − |
| **tʸ** | | 1 | 1 | | | — | — | — | | | − | 2 | | | | — | | | | − | | | | |
| **k** | 1 | | | 1 | | — | — | — | | 1 | − | | 3 | 1 | | — | | | | | | | 2 | |
| **b** | | | | | 1 | 2 | | | | — | — | | | — | | — | | | | | | | 2 | − |
| **d** | | | | | | 3 | — | | | — | — | | | — | 1 | 1 | | | − | | | | | |
| **dʸ** | | | | | | 2 | 3 | | | — | | | | — | | — | | | | | | | | |
| **g** | | | | | | 3 | | 1 | | — | — | | | — | | — | | | | | | | 2 | − |
| **tˢ** | | | | 1 | | — | — | — | 3 | − | | | | — | | | | | | | | | | |
| **č** | | | | 1 | | — | — | — | | 3 | − | | | — | | | | | | | | | | |
| **ǰ** | | | | | | 2 | | | | — | 1 | | | — | | | | | | | | | | − |
| **f** | | 1 | | | | — | — | — | | | − | 1 | 2 | | | — | | | | | | | 2 | |
| **s** | | 3 | 1 | | | — | — | — | | | − | | 3 | | | — | | | | | | | | − |
| **š** | | 3 | | | | — | — | — | | | − | | | 3 | | | | | | | | | | − |
| **v** | | | | | | 2 | | | | — | — | | | — | | | | | | | | | 2 | − |
| **z** | | | | | | 3 | 2 | 1 | | — | — | | | — | | 2 | | | | | | | | − |
| **ž** | | | | | 2 | 3 | 2 | | | — | — | | | — | | | | | | | | | | − |
| **m** | 1 | 2 | | | 1 | 2 | | | 2 | | | 1 | 2 | | 1 | 1 | 1 | 1 | | | | | 3 | |
| **n** | − | 3 | − | 1 | − | 3 | − | 1 | 1 | 3 | | 3 | 1 | 1 | 1 | | | − | 1 | − | 1 | | | − |
| **nʸ** | 2 | 1 | | | | 2 | 3 | | | | | | 2 | 1 | | | | | | | | 3 | | − 1 |
| **l** | 1 | 3 | | 1 | | 3 | 3 | 1 | 1 | 3 | | 1 | 2 | 1 | 1 | | | 1 | | | | 1 | | |
| **r** | 1 | 3 | 1 | 1 | 1 | 3 | 3 | 1 | 1 | 3 | 1 | 3 | 1 | 1 | 1 | 1 | 1 | 1 | 1 | 1 | 1 | | 3 | |
| **j** | 1 | 3 | | 1 | | 3 | | 1 | 1 | 3 | 1 | 1 | 1 | 1 | 1 | 1 | 1 | 1 | | | 1 | | 3 | 1 |
| **x** | 1 | | | | | | | | | | | | | | | | | | | | | | | 1 |

Examples: (two examples are cited if a given cluster occurs divided and undivided by a morpheme boundary): *csepp* 'drop'; *recept* 'receipt'; *copf* 'plait'; *bicepsz* 'biceps', *kapsz* 'get' (2sg pres. indef.); *taps* 'applause'; *lopj* 'steal' (sg. imp. indef.); *ott* 'there', *olvadt* 'molten'; *Detk* <place name>; *pötty* 'dot'; *Batyk* <place name>; *vágysz* 'desire' (2sg pres. indef.); *akt* 'nude' (noun); *sakk* 'chess'; *Szakcs* <place name>; *szex* 'sex', *raksz* 'put' (2sg pres. indef.); *voks* 'vote' (noun); *rakj* 'put' (sg. imp. indef.); *több* 'more'; *dobd* 'throw' (sg. imp. def.); *dobj* 'throw' (sg. imp. indef.); *kedd* 'Tuesday', *vidd* 'carry' (sg. imp. def.); *nedv* 'juice'; *edz* 'train' (3sg pres. indef.); *hagyd* 'allow' (sg. imp. def.); *meggy* 'sour cherry', *adj* 'give' (sg. imp. indef.); *smaragd* 'emerald', *fogd* 'hold' (sg. imp. def.); *agg* 'old'; *fogj* 'hold' (sg. imp. indef.); *barack* 'peach'; *vicc* 'joke' (noun), *látsz* 'see' (2sg pres. indef.); *Recsk* <place name>; *giccs* 'kitch', *táts* 'open wide' (sg. imp. indef.); *tátsd* 'open wide' (sg. imp. def.); *bridzs* 'bridge (game)', *szaft* 'juice'; *treff* 'clubs'; *döfsz* 'stab' (2sg pres. indef.); *döfj* 'stab' (sg. imp. indef.); *paraszt* 'peasant', *löszt* 'yellow soil' (acc.); *groteszk* 'grotesque'; *klassz* 'great', *eressz* 'let go' (2sg pres. indef.); *est* 'evening', *kost* 'ram'; *friss* 'fresh', *hass* 'effect' (sg. imp. indef.); *hívd* 'call' (sg. imp. def.); *hívj* 'call' (sg. imp. indef.); *gerezd* 'slice' (noun), *nézd* 'watch' (sg. imp. def.); *küzdj* 'fight' (sg. imp. indef.); *rezg* 'vibrate' (3sg pres. indef.); *nézz* 'watch' (sg. imp. indef.), *idősb* 'senior'; *pünkösd* 'Whitsun', *vésd* 'etch' (sg. imp. def.); *esdj* 'beg' (sg. imp. indef.); *kolomp* 'bell'; *teremt* 'create' (3sg pres. indef.); *lomb* 'foliage of a tree'; *nyomd* 'push' (sg. imp. def.); *teremts* 'create' (sg. imp. indef.); *tromf* 'trump'; *nyomsz* 'push' (2sg pres. indef.); *hamv* 'ash'; *nemz* 'beget' (3sg pres. indef.); *tömzs* 'lode'; *stramm* 'healthy and strong'; *szomj* 'thirst', *nyomj* 'push' (sg. imp. indef.); *ront* 'mess up' (3sg pres. indef.), *sünt* 'hedgehog' (acc.); *fánk* 'doughnut'; *rend* 'order' (noun), *bánd* 'feel sorry for' (sg. imp. def.); *ring* 'sway' (3sg pres. indef.); *lánc* 'chain'; *kincs* 'treasure', *bánts* 'hurt' (sg. imp. indef.); *fajansz* 'faience', *kensz* 'smear' (2sg pres. indef.); *pátens* 'letter'; *ellenszenv* 'antipathy'; *vonz* 'attract' (3sg pres. indef.); *kinn* 'outside'; *ajánl* 'recommend' (3sg pres. indef.); *lányt* 'girl' (acc.); *konty* 'bun'; *hányd* 'throw' (sg. imp. def.); *rongy* 'rag', *mondj* 'say' (sg. imp. indef.); *hánysz* 'throw' (2sg pres. indef.); *enyv* 'glue'; *könny* 'tear' (noun), *menj* 'go' (sg. imp. indef.); *enyh* 'relief'; *talp* 'sole'; *folt* 'patch', *élt* 'live' (3sg past indef.); *halk* 'quiet'; *küld* 'send' (3sg pres. indef.), *öld* 'kill' (sg. imp. def.); *hölgy* 'lady', *áldj* 'bless' (sg. imp. indef.); *rivalg* 'blare' (3sg pres. indef.); *polc* 'shelf'; *kulcs* 'key', *ölts* 'wear' (sg. imp. indef.); *golf* 'id.'; *élsz* 'live' (2sg pres. indef.); *fals* 'out of tune'; *nyelv* 'language'; *film* 'id.'; *hall* 'hear' (3sg pres. indef.); *szörp* 'soft drink'; *tart* 'hold' (3sg pres. indef.), *várt* 'wait' (3sg past indef.); *korty* 'swig'; *park* 'id.'; *szerb* 'Serb'; *kard* 'sword', *várd* 'wait' (sg. imp. def.); *tárgy* 'object' (noun), *hordj* 'carry' (sg. imp. indef.); *dramaturg* 'director's assistant'; *harc* 'fight' (noun); *korcs* 'mongrel', *tarts* 'hold' (sg. imp. indef.); *turf* 'id.'; *kommersz* 'cheap', *versz* 'beat' (2sg pres. indef.); *sors* 'fate'; *érv* 'argument'; *borz* 'badger'; *törzs* 'tribe'; *reform* 'id.'; *konszern* 'concern'; *árny* 'shadow'; *görl* 'girl in chorus line'; *orr* 'nose'; *fürj* 'quail', *várj* 'wait' (sg. imp. indef.); *lajt* 'water-barrow', *fájt* 'hurt' (3sg past); *hüvelyk* 'thumb'; *majd* 'later', *fújd* 'blow' (sg. imp. def.); *cajg* 'cheap cloth'; *Svájc* 'Switzerland'; *nefelejcs* 'forget-me-not', *felejts* 'forget' (sg. imp. indef.); *dölyf* 'arrogance'; *fédervejsz* 'talcum powder', *fújsz* 'blow' (2sg pres. indef.); *Majs* <place name>; *ölyv* 'hawk'; *rajz* 'drawing'; *pajzs* 'shield' (noun); *slejm* 'phlegm'; *kombájn* 'combine harvester'; *fájl* 'file'; *ujj* 'finger', *falj* 'devour' (sg. imp. indef.); *bolyh* 'fluff'; *jacht* 'yacht'; *pech* 'bad luck'.

In Table 14 (i) not all the attested clusters are systematic occurrences (i.e. some of them are exceptional/irregular); (ii) not all the systematic (well-formed) clusters represent branching codas (some of the clusters are 'marginal', cf. e.g. Steriade 1982, Kenstowicz 1994, Blevins 1995); and (iii) the morphological complexity of a given cluster may be a result of analytic suffixation (which is a barrier to syllabification) or synthetic suffixation (which is not).

An examination of Table 14 reveals that most of the attested word-final clusters conform to the Sonority Sequencing Principle (SSP) (cf. e.g. Selkirk 1982, Steriade 1982, Clements 1990, Zec 1988, Kenstowicz 1994, Blevins 1995) which requires that sonority has to increase towards the centre of the syllable.[22] In terms of government as defined in section 1.3, this means that within a branching (two-term) coda the second consonant must govern the first. However, some of the clusters in Table 14 seem to violate this requirement. Some of these 'exceptions' are systematic. First, given that government is assumed to apply at the skeletal tier (cf. section 1.3), it cannot hold between the timing slots dominating the root node of a geminate since the melodic content of the two slots is the same and thus they are equally sonorous. Nevertheless, geminate codas are well-formed. The licensing of final clusters

---

[22] There is a weaker version of the principle that tolerates sonority plateaus (cf. e.g. Törkenczy 1994*a*, Blevins 1995). Here we adopt the stronger version.

consisting of geminates may be attributed to (i) root-binding (cf. section 1.3) or (ii) some special licensing mechanism. For expository reasons, here we shall simply assume that the relevant licensing mechanism is root-binding and defer the argumentation until later in this section. So let us state the following (partly universal, partly language specific) constraint for branching codas in Hungarian:

(9) *Hungarian Branching Coda Constraint*
    Branching codas must be licensed either by government or root-binding.

Given the assumption that the direction of government is universally right to left in a coda, (9) upholds the SSP and permits geminate codas because they contain a shared root node (the first X is bound).

Some of the attested clusters in Table 14 do not conform to (9). These are the following:

(10) *a.*   ps, tʸs, ks, fs,
             pj, kj, bj, gj, fj, vj, mj, rj,
             bd, dʸd, gd, ǰd
     *b.*   pt, kt,
             tk, tʸk, tˢk, čk,
             pf,
             pš, kš,
             kč, dv, dz, nl

There is an important subdivision within the set of clusters in (10). While the clusters in (10*b*) are always monomorphemic, those in (10*a*) are predominantly polymorphemic.[23] Specifically, the latter type of clusters are the result of suffixation by definite imperative -*d*, indefinite imperative -*j*, or 2sg present indefinite -*sz*. We shall discuss these suffixes more fully below. They are peculiar in that they may be added to consonant-final stems freely, i.e. without regard to phonotactic constraints. The clusters in (10*b*), in contrast to those in (10*a*), are 'lexically restricted' in the sense that they only occur in a handful of words (all of which tend to be loans or place names). The complete list is shown in (11):

(11)   pt      recept 'receipt', korrupt 'corrupt'
       kt      absztrakt 'abstract', akt 'nude', defekt 'puncture', direkt 'on purpose', egzakt 'exact', indirekt 'indirect', intakt 'intact', kompakt 'compact', korrekt 'unbiased', perfekt 'perfect', verdikt 'judgement', viadukt 'viaduct'

---

[23] Note that (rarely) /gd, ps, ks, mj, rj/ also occur monomorphemically.

| tk | Detk <place name> |
|---|---|
| tʸk | Batyk <place name> |
| tˢk | barack 'peach', palack 'bottle', tarack 'howitzer' |
| čk | Recsk <place name> |
| pf | copf 'plait' |
| pš | arabs 'Arab horse', taps 'applause' |
| kš | voks 'vote' |
| kč | Szakcs <place name> |
| dv | kedv 'mood', nedv 'fluid', üdv 'salvation' |
| dz | edz 'train' (verb), pedz 'begin to understand' |
| nl | ajánl 'recommend'[24] |

The monomorphemic occurrences of some of the clusters in (10a) mentioned above are also only attested in very few stems:

(12)    gd    smaragd 'emerald'

        mj    szomj 'thirst'

        rj    férj 'husband', fürj 'quail', sarj 'offspring'

        ks    bilux '(dual beam) headlights', bórax 'sodium borate', bokafix 'ankle-length socks', boksz 'boxing', exlex 'lawless', fix 'certain', főnix 'phoenix', -impex <suffix occurring finally in the name of foreign trade companies: e.g. Medimpex (company specializing in the import and export of pharmaceuticals)>, index 'id.', keksz 'biscuit', kódex 'codex', koksz 'coke', komplex 'complex', konvex 'convex', krikszkraksz 'unintelligible markings', lasztex 'lastex', nikotex 'denicotinized', ónix 'onyx', ortodox 'orthodox', paradox 'id.', reflex 'id.', suviksz 'shoeshine', turmix 'milkshake'

        ps    bicepsz 'biceps', gipsz 'gypsum', mumpsz 'mumps', ripsz 'repp', ripsz-ropsz 'at once', snapsz 'schnapps', zsupsz 'crash!'

We claim that—as opposed to the polymorphemic clusters in (10a)[25]—those in (11) and (the monomorphemic occurrences in) (12) are irregular, thus the stems listed in (11) and (12) are accidental occurrences and do not characterize the phonotactic structure of Hungarian.[26]

Thus, we have seen that the clusters that violate (9) either contain the suffixes -d, -sz, -j or are irregular. This, however, does not mean that those that conform to (9) are necessarily *all* permitted/well-formed.

First, it must be noted that of the fricative + stop/affricate clusters (i.e. the only type of obstruent cluster that is not already excluded by the requirement

---

[24] This word is often pronounced with a final geminate or non-geminate *l*: [ɔjaːlː] or [ɔjaːl].

[25] The licensing of these clusters is discussed later in this section.

[26] Presumably, native speakers are able to identify these items as 'foreign/strange/non-Hungarian'. Unfortunately, no experimental evidence is available to test this prediction.

of right-to-left government within a branching coda), only those are well-formed where *both* consonants are coronal. There are very few exceptions to this requirement, all of which (we claim) are phonotactically irregular: [ft, sk, zg, žb]. The following is a full list of the stems containing these clusters:

(13)   ft     lift 'elevator', kuncsaft 'customer', seft 'illegal deal', szaft 'gravy', taft 'taffeta'

       sk     arabeszk 'arabesque', baszk 'Basque', burleszk 'slapstick', etruszk 'Etruscan', groteszk 'grotesque', humoreszk 'humorous piece of writing', kioszk 'news-stand', maszk 'mask', obeliszk 'obelisk', odaliszk 'odalisk', pittoreszk 'picturesque'

       zg     rezg 'vibrate'[27]

       žb     idősb 'elder', kevesb 'fewer', nemesb 'nobler'[28]

The irregularity of non-coronal obstruent clusters in the coda is confirmed by the fact that they are broken up by epenthesis if (synthetic) suffixation should create such a cluster while coronal clusters are not: compare /žira:f-t/ 'giraffe' (acc.) [žira:fot] and /koš-t/ 'ram' (acc.) [košt]. Note also that affricates are disallowed in coda obstruent clusters regardless of the place of articulation of the other consonant.

Coda clusters containing sonorants are not constrained by the above requirements: e.g. *halk* 'quiet' *lomb* 'foliage', *perc* 'minute', *lánc* 'chain'. Thus, obstruent clusters in branching codas have to obey stricter constraints than other clusters. This suggests that a minimum sonority distance requirement is at play here. Let us assume that there is a minimum of sonority distance that is normally required for government to license clusters as branching codas.[29] Suppose that the sonority distance settings for Hungarian consonants are the following (where < is a smaller sonority distance than <<):

(14)   *Sonority Hierarchy: Hungarian*[30]
       stops, affricates < fricatives << nasals << liquids

Furthermore, let us assume that (15) constrains government in Hungarian:

---

[27] This item is only included because it is usually cited in the literature on Hungarian phonotactics. Actually, it only appears as a bound form before vowel-initial suffixes (as an allomorph of the 'epenthetic' stem *rezeg*, cf. section 3.3.2 and Chapter 8) and thus it is not an exception. *Rezg* as a free morpheme is obsolete and/or poetic.

[28] These obsolete forms are actually polymorphemic. They all contain a no longer productive suffix *-b* and are not used in ECH. The corresponding regular (attested) forms are *idősebb*, *kevesebb*, *nemesebb*.

[29] Recall that, while individual languages may not reverse sonority relationships in the Sonority Hierarchy, they can have different sonority distance settings between segment classes in the hierarchy. Cf. section 1.3.

(15) *a.* Government can apply if the sonority distance between the segments in a governing relationship is at least $S_{min}$

   *b.* $S_{min} = $ << or >>
   (where $S_{min}$ is the minimum sonority distance)

(14) and (15) together leave all obstruent clusters unlicensed. Let us further assume that clusters with *subminimal distance*, i.e. clusters whose members are not equally sonorous but do not conform to (15), may be well-formed if licensed by some special provision in the grammar. Note that this special licensing may not derive from binding because the consonants the clusters discussed consist of do not necessarily share a COR node since they may differ in the value for [anterior]: e.g. *most* [mošt] 'now'. Furthermore, coda clusters with subminimal sonority distance are not licensed by virtue of their simply being COR. It is evident that they display the same directionality effects as the clusters that conform to the minimal sonority distance requirement. For instance, /tš/ is a COR cluster, but is not a well-formed branching coda because, albeit minimally, /š/ *is* more sonorous than /t/. This suggests that it is *government* that licenses these clusters, but if the distance is subminimal between the members of a cluster, then government is subject to the following constraint in a coda:

(16)  *Subminimal government*
   In a coda cluster C1C2, government can apply in a configuration C1>C2 iff C1 and C2 are both COR

This gives the right result for fricative + stop clusters, but also (incorrectly) allows fricative + affricate codas. Note, however, that the relevant clusters (stˢ, sč, štˢ, šč) do not have to be excluded by a constraint specific to the coda, but are unpermitted irrespective of their syllabic constituency and/or affiliation (i.e. they are excluded by a sequence constraint). These clusters only occur in compounds when divided by an analytic domain boundary: [[[*húsz*]][*centis*]]] '20-centimetre long', [[[*tenyész*]][*csődör*]]] 'stud', [[[*hős*]][*cincér*]]] 'oak cerambix', [[[*has*]][*csikarás*]]] 'stomach-ache'.[31] Let us now examine word-final sonorant clusters. Some of them (notably liquid + nasal clusters) are licensed by government. Not all of them occur, but we consider these gaps accidental.[32] In the remaining types (nasal + nasal,

---

[30] Glides are omitted because Hungarian has no glides in our analysis.

[31] There are ten exceptional monomorphemic items containing these clusters: *szcéna* 'scene', *scsí* 'Russian soup', *diszciplina* 'discipline', *proszcénium* 'proscenium', *reminiszcencia* 'reminiscence', *oszcillo-* 'oscillo-', *o[p]szcén* 'obscene', *excentrikus* 'eccentric', *transzcendens* 'transcendent', *excellenciás* 'excellency'.

[32] With one possible exception: /jnʸ/. See the discussion below.

nasal + liquid and liquid + liquid clusters) the second consonant cannot govern the first one. This correctly excludes nasal + nasal, nasal + liquid codas,[33] but incorrectly renders *all* liquid + liquid codas inadmissible. Although the evidence is somewhat meagre (as the relevant clusters only occur marginally in a few stems), we claim that government can apply in some of the liquid + liquid clusters (notably /rl, jl, jr/).[34] This is accounted for if we assume a fine-tuned sonority scale in which different liquids represent different degrees of sonority along the lines described in Hooper (1978), Clements (1990). For instance:

(17)   l << r << j

Again, the sonorant coda clusters that appear to violate the directionality requirement (i.e. the SSP) are all *j*-final.

All sonorant + obstruent codas are licensed by government. Nevertheless, some of these clusters are ill-formed. Let us examine nasal + obstruent clusters first. The problem with these clusters is that in addition to the homorganic ones /mp, mb, nt, nd, nʸtʸ, nʸdʸ/[35] (e.g. *kolomp* 'bell', *lomb* 'foliage', *ront* 'destroy', *rend* 'order', *ponty* 'carp', *gyöngy* 'pearl'), non-homorganic /nʸt, nʸd/ also occur (*lány-t* 'girl' (acc.), *hány-d* 'throw' (sg. imp. def.)). /nʸt, nʸd/ only occur polymorphemically. We have pointed out above that definite imperative *-d* can be added to stems without regard to any phonotactic restriction. Thus, (as we shall see later) the polymorphemic occurrences of /nʸd/ are not licensed by being syllabified into the coda. One might want to argue that the same state of affairs applies to /nʸt/ as well. This is not the case, however. /nʸt/ is always the result of suffixation by the accusative suffix or the past tense suffix. These suffixes are unlike the imperative *-d* in that a 'linking' vowel appears before them to prevent certain consonant clusters from being derived.[36] The behaviour of these (types of) suffixes with respect to nasal-final stems is shown in (18):[37]

---

[33] We have already seen that the single violation *ajánl* is irregular (and is normally repaired by the time it surfaces as [ɔjaːl(ː)]). The other apparent violations are *j*-final and will be discussed later in this chapter.

[34] As pointed out above, the relevant clusters are rare. /rl/ and /jl/ only occur in *görl* 'showgirl', *fájl* 'file', *geil* [gɛjl] 'nauseatingly sweet'; /jr/ does not occur at all. They might be considered exceptional, but there is evidence (independent of coda phonotactics) for the sonority relations in (17); see section 5.3.2.

[35] /mt, nk, ng/ also appear in Table (14). Of these /mt/ only occurs in a single exceptional stem *teremt* 'create'. The latter two are not problematic because they actually surface as homorganic clusters: e.g. *link* [liŋk] 'untrustworthy', *rang* [rɔŋg] 'rank'. [ŋk] and [ŋg] are not even non-homorganic underlyingly since /n/, from which all surface reflexes of [ŋ] derive, is unspecified for place (cf. section 4.3).

[36] See the details in Chapter 8.

[37] In fact there is a third type of suffix behaviour. Suffixes like the 2sg definite are always vowel-initial unless added to a vowel-final stem: compare *nyom-od, bán-od, hány-od* (Type A suffixes, cf. section 8.1.2.2).

(18)          *imperative -d*              *accusative -t*
    m     nyom-d 'push'           szem-et 'eye'
    n     bán-d  'repent'          ón-t    'tin'
    nʸ    hány-d 'throw'           lány-t  'girl'

We claim that this difference is attributable to the fact that accusative *-t* is syl-labified into the coda while imperative *-d* is licensed in a different way. Given this assumption and because the well-formedness of nasal + stop clusters obviously does not depend on the voicing of stops, /nʸt, nʸd/ have to be con-sidered licensed codas. Since nasal + stop clusters are licensed by govern-ment, we have to assume that they also have to meet an additional requirement which filters out those which are ill-formed. This constraint has to disallow non-homorganic nasal + stop clusters, but permit /nʸt, nʸd/. (19) achieves this result:[38]

(19)    In a nasal + stop coda cluster C1C2, C1 must be place-bound unless both of them are COR.

It is not obvious whether the same constraint holds for nasal + affricate clusters or not. Hungarian only has COR affricates, therefore /m/ + affricate codas are predicted to be ill-formed. This prediction appears to be true: there is a single exception /mč/, which only occurs in the morphologically complex form *tere*[mč] 'create' (sg. imp. def.).[39] Note, however, that the constraint extended to nasal + affricates would permit /nʸtˢ, nʸč, nʸǰ, nǰ/ as coda clusters, but they do not occur. Of these, /nǰ/ is permitted since /nč/ occurs (e.g. *mancs* 'paw') and it is unlikely that the voicing difference should entail a difference in well-formedness. The others do not occur because an MSC (which bans preconsonantal nasals with an independent place specification) excludes them within the morpheme, and they may not result from suffixation because there are no suffixes that consist of an affricate. For this reason, permitting them causes no problems and therefore we extend (19) to cover nasal + affricate codas too:

(20)        In a coda cluster  C1    C2

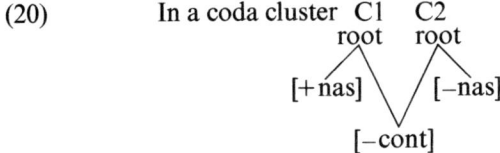

C1 must be place-bound unless both of them are COR.

---

[38] (19) would in fact allow the unattested codas /ntʸ/ and /ndʸ/, but this causes no problems because /n/ is underlyingly placeless and thus these clusters would surface as [nʸtʸ] and [nʸdʸ] as a result of nasal place assimilation (cf. section 7.4.).
[39] On the syllable structure of this form, see section 5.2.4.3.

The distribution of final nasal + fricative clusters is governed by a slightly different constraint. Disregarding /v/ for the moment (whose sonority ranking, as we shall see, is problematic), only those nasal + fricative clusters whose first member is /n/ and whose second member is a coronal seem to be well-formed. Although /mf, ms, mz, mž, nʸs/ are attested, they only occur (i) in polymorphemic clusters whose second consonant is 2nd sg. present indefinite -sz, which always behaves in a special way (/ms/ nyom-sz 'push'; /nʸs/ hány-sz 'throw'), or (ii) in a few irregular stems. The following is an exhaustive list of these stems:

(21)   mf    tromf 'trump'
       mz    nemz[40] 'beget (pres 3rd sg. indefinite)'
       mž    tömzs 'lode'

Given the MSC referred to above, (22) accounts for the observed distribution:

(22)   In a coda cluster C1C2 where C1 is [+nasal] and C2 is [−son, +cont], C2 must be COR.

In the above discussion of branching codas we have disregarded /v/. As argued in section 4.1.1, /v/ is a 'two-faced' consonant: it behaves as a sonorant in onsets, but as an obstruent in codas (this is encoded by its unspecification for the feature [son]). We propose that despite its asymmetrical behaviour in onsets and codas (cf. section 4.1.1) /v/ always has the sonority ranking of a fricative.[41] Given the constraints discussed in the present section and the above assumption about its ranking in the sonority hierarchy, its distribution in final clusters is the expected one. Thus, government does not allow it to co-occur with obstruents as the first or the second element in coda clusters since the sonority distance between fricatives and other obstruents is subminimal.[42] The few exceptional stems in which it does co-occur with obstruents in final position are listed in (11). Because of the directionality of government /v/ + sonorant codas are ill-formed. In codas /v/ is permitted following a sonorant if the sonorant is a liquid or a nasal since government can apply (e.g. érv 'argument', elv 'principle', ölyv 'hawk') and (22) does not constrain nasal + /v/ coda clusters as /v/ is unspecified for [son] (e.g. hamv 'ash', ellenszenv 'antipathy', könyv 'book').[43]

---

[40] This item is obsolete in ECH as a free form.

[41] It would be undesirable to allow the sonority ranking of a segment to vary depending on the position it occurs in. The sonority ranking of /v/ as a fricative in onsets generally does not cause problems because we argue that branching onsets do not occur in Hungarian. /v/, however, does behave in a special way in onsets in interconstituent clusters; see section 5.3.2.1.

[42] Subminimal government is excluded because /v/ is not COR.

[43] /v/ is extremely rare in final clusters containing sonorants other than /r/. The following is a complete list: elv, nyelv 'language', ölyv, hamv (obsolete), -szenv 'feeling' <bound morpheme>, enyv 'glue', könyv. In the present analysis this is an accident.

We have not examined the distribution of preconsonantal /j/ yet. Tör-kenczy (1994a) claims that there are two constraints that apply to pre-consonantal /j/ in branching codas; one requires that obstruents following /j/ must be coronal and the other excludes palatal consonants after /j/. In our view, these constraints are untenable because they make unmotivated and unnecessary distinctions between equally well-formed /jC/ clusters. For instance, /jn/ and /jm/ conform to the above constraints and thus are judged well-formed as opposed to /jnʸ/, which violates them, and is therefore sup-posed to be ill-formed. This seems to make the right prediction since /jn/ and /jm/ are attested, but /jnʸ/ is not. However, this difference is not really signifi-cant since the only stems in which the first two clusters are attested are *kom-bájn* 'combine-harvester' and *slejm* 'phlegm'. In fact, preconsonantal /j/ is rare in final (non-geminate) clusters other than /jt/.[44] Furthermore, /jt/ is the only cluster whose coda status can be tested: as the accusative and the past suffix (which can be realized as [t]) attach to /j/-final stems without an epenthetic vowel (*sóhaj-t* [šoːhɔjt] 'sigh' (acc.), *búj-t* [buːjt] 'hide' (past)), /jt/ must be a possible coda (cf. Section 8.1.4). In order to avoid making untestable well-formedness distinctions within the set of final /jC/ clusters we claim here that all of them are well-formed and no constraint applies specif-ically in this environment.

### 5.2.4.3. Appendices

In the discussion of final clusters in the previous section we have disregarded the final clusters that contain the consonants /d, j, s/ when they realize the definite imperative, the indefinite imperative, and the 2nd sg. present indefi-nite suffix respectively. Final clusters containing these suffixes often violate government (e.g. *lopsz* 'steal' (2nd sg. pres. indef.)) and/or other constraints applying within the coda (*nyomd* 'push' (sg. imp. def.)). In general, there are no phonotactic constraints applying between these suffixes and the final con-sonant of the stem they are attached to. This is completely true of definite imperative -*d* and indefinite imperative -*j*. These suffixes may be added to any stem. The gaps in /j/ or /d/ final clusters in Table 14 are not due to phonotac-tics: they are either (i) accidental (there are no verb stems ending in affricates or /x/, so /tʲj, čj, ǰj/[45] and /xd/, /xj/ do not occur), or (ii) the result of assimila-tions (e.g. [šd] does not appear at the surface, but underlyingly it does, and is

---

[44] /j/ frequently occurs in the final *polymorphemic* clusters /jd, js, jč/ but they all contain ana-lytic suffixes (definite imperative -*d*, indefinite imperative -*j*, or 2sg present indicative -*sz*). The fol-lowing is a complete list of *stems* with final jC clusters other than /jt/: /jp/ *selyp* 'lisper'; /jk/ *sejk* 'sheik', *sztrájk* 'strike', -*ajk* 'lip' <bound form>', *hüvelyk* 'thumb'; /jd/ *fajd* 'grouse', *gajd* 'hubbub', *majd* 'later', *ofszájd* 'offside'; /jg/ *cajg* 'cheap cloth'; /jt'/ *Svájc* 'Switzerland'; /jf/ *dölyf* 'arrogance'; /js/ *fédervejsz* 'talcum powder', *hajsz* <interjection>; /jš/ *Majs* <place name>; /jv/ *ölyv* 'hawk'; /jz/ *csuszpájz* <a kind of vegetable dish>, *rajz* 'drawing', *spájz* 'pantry'; /jž/ *pajzs* 'shield'; /jm/ *slejm* 'phlegm'; /jn/ *kombájn* 'combine harvester'; /jl/ *fájl* 'file'; /jh/ *bolyh* 'fluff'. The items *selyp* and *bolyh* are only included because they are cited in the literature—in ECH they are bound forms that only occur before vowel-initial suffixes; as free forms they are obsolete.

later eliminated by voicing assimilation). To sum up, definite imperative -*d* and indefinite imperative -*j* are phonotactically completely independent of the stems they are added to. This phonotactic independence also manifests itself in the fact that final clusters that contain these suffixes may violate sonority sequencing (i.e. government).[46] The suffix -*sz* behaves similarly, albeit to a somewhat limited extent. Final clusters containing it may violate sonority sequencing (e.g. *kapsz* 'get' (2nd sg. pres. indef.), *vágysz* 'desire' (2nd sg. pres. indef.), *raksz* 'put' (2nd sg. pres. indef.), but there is a phonotactic(ally motivated) phenomenon that concerns -*sz*: it cannot be attached to a stem that ends in a [+strident] consonant; instead the allomorph -*Vl* is selected (e.g. *tesz-el* 'put' (2nd sg. pres. indef.), *néz-el* 'look' (2nd sg. pres. indef.), *keres-el* 'search' (2nd sg. pres. indef.) and not \**te*[sː], \**né*[sː], \**kere*[šs][47]). The allomorphy is certainly phonotactically motivated, but we assume that it is not related to syllable structure.

Another aspect of the independence of these suffixes can be seen if we examine word-final clusters that consist of more than two consonants. There are extremely few monomorphemic words that end in more than two consonants. (23) lists all the relevant items:

(23)   mps   mumpsz 'mumps'
      nks   szfinx 'sphinx', szkunksz 'skunk'
      nst   dunszt 'steam', kunszt 'trick'
      rst   karszt 'karst', verszt 'verst'
      kst   szext 'sixte'
      ršt   vurst 'sausage'
      ršč   borscs <Russian soup>
      jst   lejszt 'hard work'
      jšt   mihelyst 'as soon as'

We consider all the words in (23) exceptional/irregular. Disregarding these words, the polymorphemic final three-term clusters that occur are the ones listed in Table 15:

---

[45] Even if there were such underlying forms, they would be eliminated by assimilation (cf. section 7.2).

[46] In striking contrast, there are just a handful of irregular monomorphemic items (listed in (12)) and none containing suffixes other than -*d*, -*sz*, and -*j* that are in violation of sonority sequencing.

[47] [šs] final forms like *kere*[šs] do rarely occur, but they are obsolete/unusual in ECH.

TABLE 15. *Polymorphemic word-final CCC clusters*

|     | t | d | s | j |
|-----|---|---|---|---|
| dz  | — | + | ——— |  |
| mǰ  | — | + | — |  |
| ŋk  |   | — | + |  |
| ŋg  | — | + | — | + |
| ntˢ | + | — |   |  |
| nǰ  | — | + | — |  |
| ns  | + | — |   | — |
| nš  | + | — |   | — |
| nl  |   | + | + | — |
| lǰ  | — | + | — |  |
| rn  | + |   |   | — |
| rǰ  | — | + | — |  |
| rl  | + |   |   | — |
| ǰ̆  | — | + | — |  |
| js  | + | — |   | — |
| jš  | + | — |   | — |
| jn  | + |   |   | — |

Examples: *edzd* 'train' (sg. imp. def.), *teremtsd* 'create' (sg. imp. def.), *lengsz* 'swing' (2sg pres. indef.), *zengd* 'resound' (sg. imp. def.), *pénzt* 'money' (acc.), *bontsd* 'open' (sg. imp. def.), *ENSZ-t* 'UN' (acc.), *brilliánst* 'brilliant' (acc.), *ajánld* 'recommend' (sg. imp. def.), *ajánlsz* 'recommend' (2sg pres. indef.), *töltsd* 'pour' (sg. imp. def.), *konszernt* 'concern' (acc.), *tartsd* 'hold' (sg. imp. def.), *görlt* 'girl' (acc.), *hajtsd* 'bend' (sg. imp. def.), *fédervejszt* 'talcum powder' (acc.), *mihelyst* 'as soon as', *pajzst* 'shield' (acc.), *kombájnt* 'combine harvester' (acc.), *zengj* 'resound' (sg. imp. def.).

Table 15 shows the final polymorphemic CCC clusters that appear at the surface (the notation is the usual one where the boxes struck out by dashes denote clusters that are/would be eliminated by assimilations). The number of underlying clusters would be higher because there are processes that simplify consonant clusters (e.g. degemination (cf. Chapter 9) turns the underlying triliteral cluster /rrs/ into [rs] in *varrsz* [vɔrs] 'sew' (2nd sg. pres. indef.). The attested clusters either have one of the three suffixes -*d*, -*j*, -*sz* discussed above, or the accusative -*t* as their final element. Let us set aside the accusative for the moment and concentrate on the other three. The fact that they can create final CCC clusters by attaching to stems ending in branching codas further attests to their phonotactic independence. We can account for this independence by claiming that they are in fact not syllabified into the coda, but belong to a special subsyllabic constituent, the appendix. Thus, in Hungarian the extended syllable can have an appendix not only initially, but finally as well:

(24)                                    σ

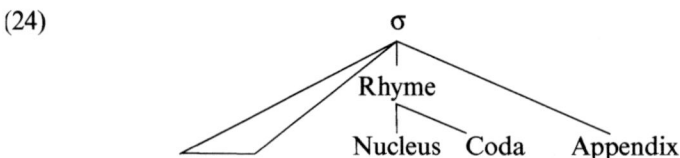

The extended syllable shown in (24) is restricted to the right edge of analytic domains, i.e. the appendix must be peripheral. Furthermore, it may only occur after a coda. Consonant clusters that are (partially) in the appendix are unconstrained by coda restrictions (e.g. may violate sonority sequencing and may consist of more than two consonants). Regularly, on the coda side only analytic suffixes may be in the appendix. -*d* and -*j* are clearly syllabified into the appendix. We have seen that they can be added to any stem-final consonant and they can occur as the last consonant in final CCC clusters, i.e. they can attach to any stem ending in a branching coda. The surface non-occurrence of some final C1C2C3 clusters where C1C2 is a possible coda and C3 is -*d* or -*j* is due to assimilations (e.g. [ltj] does not occur because /tj/ becomes [č] in imperatives (cf. section 7.2.1)); or is unrelated to syllable structure (e.g. /mbd/ and /mbj/ do not occur because the morphology does not generate these combinations (there are no verb stems ending in /mb/, and -*d* and -*j* are verbal suffixes)).

The appendix status of -*sz* is more problematic. We have seen that, *modulo* -*Vl* allomorphy, there are practically no restrictions between the stem final consonant and -*sz*. This is what we expect of an appendix. However, an inspection of Table 15 reveals that very few -*sz*-final CCC clusters are attested (only [ŋks] and [nls]). This is unexpected even if one takes it into consideration that *sz* can only be added to verbs. The reason why there are few -*sz*-final CCC clusters is that, *typically*, -*sz* attaches to stems that end in more than one consonant with a linking vowel (e.g. *látsz*, *\*látasz* 'see' vs. *\*osztsz*, *osztasz* 'distribute'; *adsz*, *\*adasz* 'give' vs. *\*kezdsz*, *kezdesz* 'begin'). Some stems allow forms with and without the linking vowel (e.g. *fingasz*, *fingsz* 'fart'); others only have forms without it (e.g. *varrsz* but *\*varrasz* 'sew').

The occurrence of accusative -*t* in final CCC clusters is a further complication. We have pointed out above that accusative -*t* syllabifies into the coda, and if it cannot, an epenthetic vowel occurs before it (compare *leves-t* 'soup' (acc.) and *zsiráf-ot* 'giraffe' (acc.)), i.e. there is phonotactic interaction between it and the stem-final consonant (see the details in Chapter 8). Since three-term codas are not allowed, we would expect it to be preceded by an epenthetic vowel after stems ending in consonant clusters. Table 15 shows that this is not the case; it can appear as the last consonant in a final CCC cluster. It does not behave like an appendix, however: it never occurs without an epenthetic vowel after stem final clusters C1C2 if C2+[t] is not a possible branching coda (compare *farm-ot* 'farm' (acc.) and *konszern-t*), i.e. there is a phonotactic interaction between the *-t* and the stem-final consonant. On the other hand, the epenthetic vowel is not *always* missing if C2 of the stem-final cluster plus [t] make a possible branching coda. Some of these cases have independent explanations (e.g. all /rnʸ/-final nouns are 'lowering stems', which in itself requires a linking vowel to be present before the suffix even though /nʸt/ is a possible coda: *árny-at* 'shadow' (acc.), *szárny-at* 'wing' (acc.), *szörny-et* 'monster' (acc.), and similarly with other combinations, e.g. *fürj-et* 'quail' (acc.), *törzs-et* 'trunk' (acc.), cf. section 8.1.3. on lowering stems). Others, however, have no independent explanation: /rš/-final stems always have a linking vowel before the accusative, even if the stem is not lowering: e.g. *bors-ot* 'pepper' (acc.) although /rš/ and /št/ are well-formed codas.

Thus the general problem is that the phonotactic restrictions on the melodic content of final clusters and those on the complexity of final clusters seem to suggest conflicting classifications for 2nd sg. present indefinite *-sz* and accusative *-t*.[48] The former suffix is completely insensitive to the melodic content of the stem-final consonant (a typical appendix-like behaviour: *csap-sz* 'hit' (2nd sg. pres. indef.), but *usually* cannot be attached to cluster-final stems without a linking vowel (*oszt-asz* 'distribute' (2nd sg. pres. indef.)). The latter, on the other hand, *is* sensitive to the stem-final consonantal melody (a typical coda-like behaviour: *ón-t* 'tin' (acc.) vs. *nyom-ot* 'trace' (acc.)), but *if the final consonantal melody is right*, can be *sometimes* added to stems ending in a consonant cluster without a linking vowel (*konszern-t*). The question is how to explain the non-appendix-like behaviour of *-sz* and the non-coda-like behaviour of *-t*.

First, let us try to answer the first part of the question. The problem is that—contrary to our expectations—a linking vowel appears after cluster-final stems before the suffix *-sz* (which is assumed to syllabify as an appendix). It is significant that not only the presence, but the quality of the linking vowel is also unexpected. The normal linking vowels are mid *e/ö/o* (i.e. [−open₁])[49] and not low *a/e* (i.e. [+open₁]). The latter quality is the one that we get after lowering stems and suffixes: e.g. *fog-at* 'tooth' (acc.), *szög-et* 'nail' (acc.), *tök-ök-et* 'pumpkins' (acc.) (compare non-lowering *bog-ot* 'knot' (acc.), *rög-öt* 'clod' (acc.)). Here, however, the lowered quality is not due to the stem but to the suffix itself. Verb stems are never lowering, and the quality of the linking vowel before *-sz* is always (unexpectedly) low after all the stems that it can follow (compare *mond-ok* 'say' (1st sg. pres. indef.) and *mond-asz* 'say' (2nd sg. pres. indef.)). Also, a low linking vowel normally does not alternate with zero: it is present after lowering stems even after stems whose final consonant could form a well-formed branching coda with the suffix: e.g. *fal-at* 'wall' (acc.), *has-at* 'stomach' (acc.), *vár-at* 'castle' (acc.)

---

[48] Cf. Rebrus (2000). The past suffix also behaves similarly to the accusative (cf. section 8.1.2.2).

[49] Underlying mid *e* is eventually phonetically implemented in ECH as low [ɛ] (cf. section 6.1).

(compare non-lowering *hivatal-t* 'office' (acc.), *kas-t* 'hive' (acc.), *vér-t* 'blood'
(acc.)).[50] By contrast, the low linking vowel of *-sz* is not stable: it is not present after some stems (compare *hat-sz* 'influence' (2nd sg. pres. indef.) and
*tart-asz* 'hold' (2nd sg. pres. indef.)) We suggest that considering the *-Vsz*
variant of *-sz* as an instance of allomorphy is in harmony with these facts.
Then the *-Vsz* variant appears in the lexicon along with *-sz* and thus the unexpectedness of the vowel quality is then just a lexical fact. The low initial vowel
of the *-Vsz* variant behaves just like any low linking vowel, i.e. it is stable and
does not alternate with zero. The fact that the *-Vsz* allomorph typically[51]
appears after cluster-final stems, i.e. the allomorphy is phonotactically
conditioned, is on a par with the behaviour of *-Vl*, which is a variant of *-sz*
after [+strident] stems. Both can be seen as cases where morphology is
dependent on phonological information. This interpretation makes it
possible to maintain that *-sz* syllabifies as an appendix.

A possible answer to the second part of the question (which concerns the
behaviour of *-t* after cluster final stems) is to claim that the reason why *-t*
attaches without a linking vowel to stems like *konszern* is that in these words
the first consonant of the final cluster is not in the coda but in the nucleus as
shown in (25) (where only the relevant structure is displayed):

(25)

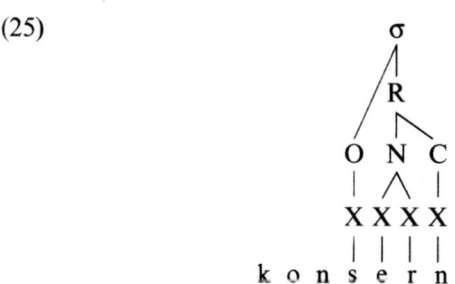

This syllabification would allow *-t* to attach to *konszern* without a linking
vowel (i.e. into the coda of the stem-final syllable) in spite of the fact that
codas are maximally binary branching. In other words, it would explain why
*-t* appears to be insensitive to the number of stem-final consonants (in some
stems) although it must be syllabified as a coda. There are, however, several
problems, which make this explanation untenable.

First of all, a very complicated statement would be needed to specify the
conditions in which a consonant can syllabify into the nucleus. The class of
consonants that could syllabify in this way is not difficult to identify. They
must be [+sonorant], since *-t* always attaches with a linking vowel to C1C2-

---

[50] See section 8.1.3 on lowering stems.
[51] The *-Vsz* allomorph also appears after a more-or-less arbitrary set of stems ending in *t* preceded by a long vowel. All the stems ending in the verb-forming suffix *-it* belong here. E.g. *alakít-
asz* 'form', *vét-esz* 'err', *fűt-esz* 'heat' (compare *lát-sz* 'see').

final stems if C1 is an obstruent, even if C2+*t* is a well-formed coda: *taps-òt* not *\*taps-t* 'applause' (acc.) (compare *kas-t* 'hive' (acc.), *most* 'now'), *sze*[ks]-*et* not *\*sze*[ks]-*t* 'sex' (acc.) (compare *szesz-t* 'alcohol' (acc.), *liszt* 'flour'). Furthermore, [+sonorant] segments could only be in the nucleus of a closed syllable otherwise we would predict that (i) any consonant can follow a sonorant within the same syllable and (ii) no interconstituent constraints refer to sonorant final syllables (these predictions are untrue; cf 5.2.4.2 and 5.3.2). It is a significant fact that additional conditions would also have to be stipulated since /r/ behaves differently from other sonorants when it precedes stem-final obstruents. As pointed out above, -*t* attaches to /rš, rs, rz/-final stems with a linking vowel (*bors-ot* 'pepper' (acc.), *mersz-et* 'courage' (acc.) *borz-ot* 'badger' (acc.)).[52] The problem is that (i) /r/ does not behave in this way before other stem-final consonants (compare e.g. *konszern-t*), and (ii) other sonorants do not behave in this way before stem-final /š, s, z/ (compare e.g. *konstans-t* 'constant' (acc.), *fajansz-t* 'faience' (acc.), *pénz-t* 'money' (acc.)). Thus, the conditions on the hypothesized syllabification of sonorants into the nucleus can hardly be formulated with a sufficient degree of generality.

Furthermore, there is some degree of unpredictable variation. Some stem-final sonorant + /s, š, z, ž, n, nʸ, l, r, j/ clusters show more than one kind of behaviour: in some stems they always take the suffix -*t* without a linking vowel, in others they always require a linking vowel, and there are stems that allow both variants:

| (26)[53] | | -*t* | -*Vt* | -*t*/-*Vt* |
|---|---|---|---|---|
| | nš | protestáns-t | revans-ot | briliáns-t/briliáns-ot |
| | ns | fajansz-t | sansz-ot | reneszánsz-t/reneszánsz-ot |
| | jz | csuszpájz-t | rajz-ot | — |
| | nz | pénz-t | bronz-ot | csimpánz-t/csimpánz-ot |
| | jž | — | — | pajzs-t/pajzs-ot |

A similar behaviour can be attested if we examine stems that end in geminates whose melodic content is such that after the corresponding stem-final non-geminate segments no linking vowel appears. These are the stems ending in /ss, šš, zz, nn, nʸnʸ, ll, rr, jj/.[54] Typically, no linking vowel appears after these geminates before -*t*: *idill-t* 'idyll' (acc.), *finn-t* 'Finnish' (acc.), *plüss-t* 'plush' (acc.), *dzsessz-t* 'jazz' (acc.), etc.[55] Most of the examples with a linking vowel

---

[52] Note that these are not lowering stems, thus the presence of the linking vowel cannot be attributed to a factor independent of the clusters examined.

[53] This chart is based on Papp (1975). The /z, ž/-final clusters surface with /s, š/ respectively, before -*t* because of Voicing Assimilation (cf. section 7.3). Glosses: *briliáns* 'brilliant', *bronz* 'bronze', *csimpánz* 'chimpanzee', *csuszpájz* 'vegetable dish', *pajzs* 'shield', *rajz* 'drawing', *reneszánsz* 'Renaissance', *revans* 'return match', *sansz* 'chance'.

[54] There are no stems ending in /žž/.

[55] These three-term clusters actually surface as two-term as a result of Degemination (cf. section 9.4).

are lowering stems and thus they are irrelevant to the issue at hand: e.g. *ujj-at* 'finger' (acc.), *toll-at* 'feather' (acc.), *gally-at* 'twig' (acc.), etc. (cf. section 8.1.3 on lowering stems). Nevertheless, here too there *is* some idiosyncratic variation: *genny-t/genny-et* 'pus' (acc.), *orr-t/orr-ot* 'nose' (acc.), *bross-t/bross-ot* 'brooch', etc.[56] These facts suggest that idiosyncratic restrictions would have to be imposed on the incorporation of sonorants into the nucleus. Some stems would have to be marked as not allowing it and others as optionally allowing it.

Even if the complexities/difficulties pointed out above were disregarded, the most serious problem with the hypothesis is that, after the relevant clusters, -*t* behaves in the same way even if the vowel preceding the cluster is long: *kombájn-t*, *protestáns-t* 'protestant' (acc.), *pénz-t* 'money' (acc.) ([nst] or [nt*t]), *fájl-t* 'file' (acc.), etc. Therefore, if we maintained that in these stems the postvocalic consonant of the final cluster is in the nucleus, then we would have to allow ternary branching nuclei. In fact, in trying to avoid ternary codas we would end up creating ternary nuclei.

Because of the problems discussed above we consider the syllabification shown in (25) untenable and suggest that the behaviour of -*t* after cluster-final stems like *konszern* is due to the fact that the relevant stems are lexically marked so that they *exceptionally* allow the syllabification of -*t* into the appendix. These stems will then have no linking vowel before -*t*. The stems that show variation (e.g. *briliáns*) appear twice (marked and unmarked) in the lexicon of speakers who use both variants.[57] The lexically unmarked cluster-final stems will always have a linking vowel before -*t*. Thus, all word-final clusters containing the suffixes -*d*, -*j*, -*sz* have the structure coda+appendix and accusative -*t* can also syllabify as an appendix after some cluster-final stems. This makes it possible to maintain that the coda constituent is maximally binary branching in Hungarian, although word-final ternary clusters do occur. In addition to the restriction on its melodic/morphological content, the occurrence of the appendix is subject to the following general condition (which is a version of the Peripherality Condition, cf. Hayes 1995):

(27)     The appendix (i) must be peripheral in an analytic domain and (ii) must not be adjacent to the nucleus.

---

[56] There are no (non-lowering) noun stems that select the variant with the linking vowel only, unless we include examples like *mell-et* 'breast' (acc.), *szenny-et* 'dirt' (acc.) which cannot be identified as lowering stems on the basis of the quality of the linking vowel in ECH (though other dialects show that they are lowering stems, cf. section 8.1.3). There are some comparable verb stems, however, which always require a linking vowel before the past tense suffix -*(t)t* (which behaves similarly to the accusative, cf. section 8.1.4.4): *hall-ott* 'hear' (3sg past indef.), *hull-ott* 'fall' (3sg past indef.), *kell-ett* 'have to' (3sg past indef.), *vall-ott* 'confess' (3sg past indef.). Note that *hull-t* 'fall' (3sg past indef.) is a possible alternative form along with *hull-ott*.

[57] This suggests that for some speakers even these stems can be non-variable, a prediction that appears to be true.

## 5.3. TRANSSYLLABIC CONSTRAINTS

Transsyllabic constraints are constraints applying between adjacent segments belonging to different syllables. Logically, transsyllabic constraints could refer to segment clusters of the following kinds:

(28)  *a.* V.V
      *b.* C.C
      *c.* V.C
      *d.* C.V

(28*a*) shows two adjacent nuclei (hiatus), (28*b*) is a coda followed by an onset (interconstituent cluster), (28*c*) is a nucleus followed by an onset, and (28*d*) is a coda consonant followed by a nucleus. Out of these four possibilities (28*d*) appears to be universally excluded by the Maximal Onset Principle (cf. Blevins 1995) (or any equivalent mechanism designed to capture the fact that a prevocalic consonant syllabifies universally as an onset rather than a coda[58]). There are no transsyllabic constraints applying between a vowel and a following non-tautosyllabic consonant in Hungarian (28*c*). Let us examine the constraints applying in contexts (28*a, b*).

### 5.3.1. Hiatus

Nuclei can be adjacent (hiatus may occur) with the following restrictions: the initial vowel of vowel-initial synthetic suffixes deletes when they are attached to vowel-final stems (compare *ház-on* 'on (the) house' and *kapu-n* 'on (the) gate', cf. section 8.1.4.2). Some of the remaining vowel clusters are broken up by a (postlexical) process of hiatus filling (e.g. *fáig* [faːjig] 'up to the tree', cf. section 9.3). The rest of the hiatuses surface (e.g. *kakaó* [kɔkɔɔː] 'cocoa', *csataordítás* [t͡ʃɔtɔɔrdiːtaːʃ] 'battle cry').

Table 16 shows clusters of two vowels (vowels in hiatus) that occur in Hungarian. Table 16 abstracts away from hiatus filling, but is near surface in the sense that it shows vowel clusters that survive after the deletion of the initial vowels of vowel-initial synthetic suffixes in hiatus. Blanks indicate that the combination in question is not attested, stars mark vowel clusters that only occur when separated by an analytic boundary, and a given vowel cluster is spelt out if there is at least one monomorphemic stem in which it occurs. In most cases the difference between stars and blanks in Table 16 is phonologically accidental. The reason is that morphologically complex hiatuses only survive (i.e. escape deletion) if the two nuclei become juxtaposed as a result of (i) compounding (*disznóölés* 'pig killing'), (ii) prefixation (by preverbs, e.g.

---

[58] Barra Gaelic and Kunjen are sometimes cited as possible counterexamples, cf. Clements (1986), Sommer (1981), and Blevins (1995).

TABLE 16. *VV clusters*

| | i | ü | e | ö | u | o | a | i: | ü: | e: | ö: | u: | o: | a: |
|---|---|---|---|---|---|---|---|---|---|---|---|---|---|---|
| i | * | * | ie | * | iu | io | ia | * | * | ie: | iö: | iu: | io: | ia: |
| ü | * | * | üe | | | * | üa | | | üe: | | | | |
| e | ei | * | * | * | eu | eo | ea | * | * | * | * | * | eo: | ea: |
| ö | | | | | öu | | | | | | | | | |
| u | ui | * | ue | | uu | uo | ua | ui: | | * | * | | uo: | ua: |
| o | oi | | oe | * | | oo | oa | | | oe: | | | | oa: |
| a | ai | * | ae | * | au | ao | * | ai: | * | * | * | * | ao: | * |
| i: | * | * | * | * | | | | | | * | | | | |
| ü: | * | * | * | * | * | | * | * | | * | | * | | * |
| e: | * | * | * | | * | * | * | * | | * | | | | * |
| ö: | * | * | * | * | * | * | * | * | | * | * | * | * | * |
| u: | * | * | * | * | * | * | * | | | * | | | | * |
| o: | * | * | * | * | * | * | * | * | * | * | * | * | * | * |
| a: | a:i | * | * | * | * | a:o | * | * | | * | | * | * | * |

Examples:[59] *kies* 'picturesque', *július* 'July', *liliom* 'lily', *riadt* 'frightened', *diéta* 'diet', *miliő* 'milieu', *fiú* 'boy', *dió* 'walnut', *kiált* 'shout', *menüett* 'minuet', *nüansz* 'nuance', *habitüé* 'regular visitor', *koffein* 'caffeine', *múzeum* 'museum', *neon* 'id.', *tea* 'id.', *sztereó* 'stereo', *leány* 'girl', *Szöul* 'Seoul', *jezsuita* 'Jesuit', *influenza* 'flu', *vákuum* 'vacuum', *fluoreszkál* 'fluoresce', *pápua* 'Papuan', *intuíció* 'intuition', *duó* 'duo', *január* 'January', *sztoikus* 'stoic', *poentíroz* 'embellish with jokes', *kooperál* 'cooperate', *boa* 'id.', *poén* 'punchline', *oázis* 'oasis', *mozaik* 'mosaic', *Izrael* 'Israel', *autó* 'car', *aorta* 'id.', *naív* 'naive', *kakaó* 'cocoa', *Káin* 'Cain', *káosz* 'chaos'

*beleönt* 'pour into') or (iii) analytic suffixation (e.g. *ollóért* 'for scissors'), i.e. when they are separated by an analytic domain edge, a boundary across which no phonotactic or syllable structure constraints hold (cf. section 1.3). In a few cases, blanks are due to a regularity which is unrelated to hiatus. Thus, the lack of stars in the rows for /o, ö/ (the non-occurrence of polymorphemic /o, ö/-initial vowel clusters) is due to the fact that /o, ö/ are not permitted at the end of an analytic domain in general (cf. section 5.4.). Vowel Harmony (cf. sections 3.2 and 6.1) accounts for the absence of monomorphemic vowel clusters containing front rounded vowels and back vowels.[60]

Bearing in mind the above observations, the following regularities specific to hiatus can be observed in Table 16. In hiatus

---

[59] Only monomorphemic examples are given.
[60] *Nüansz* 'nuance' and *Szöul* 'Seoul' are the only exceptions. Note also the exceptional behaviour of domain-final /o/ in 'foreign compounds' discussed below.

(29)  *a.* identical segments cannot occur.
      *b.* a long vowel cannot be prevocalic.
      *c.* the vowels /ö, ö:, ü, ü:/ cannot occur.

No separate statement needs to be included in the grammar of Hungarian to account for (29*a*) since it can be explained with reference to a general constraint on the form of phonological representations, the Obligatory Contour Principle (OCP), which bans adjacent identical elements on the same tier.[61] Vowel clusters containing identical segments within an analytic domain[62] are impossible to represent since (30*a*) below is the representation of long vowels and (30*b*) is excluded by the OCP.

(30)          *a.* X   X        *b.* X   X
                   \ /               |   |
                   V$\alpha$              V$\alpha$  V$\alpha$

Words like *kiirt* 'exterminate', *rakétaautó* 'rocket car', *biliig* 'up to the chamber pot', etc. are only apparent counterexamples since in them the vowel segments making up the relevant clusters are in different analytic domains, they are 'fake' geminate vowels (and can be represented as (30*b*) without violating the OCP): [[[ki][irt]], [[rakéta][autó]], [[bili]ig]].[63] The same applies to vowel clusters consisting of identical long vowels as in [[[lé]ért]] 'for juice'. In Table 16 the monomorphemic occurrence of /oo/ and /uu/ (e.g. *vákuum* 'vacuum', *zoológia* 'zoology') seem truly problematic as they appear to be in violation of (29*a*) (i.e. the OCP). However, the few words that contain them[64] are either pronounced with single long (rarely short) vowels ([va:kum/va:ku:m, zo:lo:giɔ]) or, alternatively, with fake geminates comparable with the cases described above ([va:kuum, zoolo:giɔ]). We assume that in the latter pronunciation these words have been reanalysed as containing more than one internal analytic domain: [[[váku][um]], [[zo][ológia]] to avoid violating the OCP (cf. 8.1.4.5). Note that the latter form is problematic in another respect: it violates the language-specific constraint (58) against domain-final mid vowels (cf. 5.4.1). If the form is pronounced [zoolo:giɔ], the OCP and (58) are in conflict: both of them cannot be upheld at the same time. Apparently, the otherwise

---

[61] For different formulations of and problems with the OCP cf. Leben (1973), McCarthy (1986), Odden (1986, 1988).

[62] We assume that the OCP does not hold across the edge of an analytic domain.

[63] In fact, the fakeness of these geminates can even be heard in Hungarian as the two vowels are pronounced with distinct pulses, compare *kiirt* [kiirt] vs. *sírt* [ši:rt] 'cried'. The difference between fake and true geminate vowels is even more apparent in the case of *á* and *é*: *odaad* [odɔɔd] 'give over to' vs. *kád* [ka:d] 'tub', *leesik* [lɛɛšik] 'fall down' vs. *késik* [ke:šik] 'be late' (cf. section 8.1.4.5).

[64] The following is a complete list of the relevant items in our database: *kooperál* 'co-operate', *koordináta* 'co-ordinate', *zoológia*, *individuum* 'individual', *vákuum*.

inviolable constraint (58) can be suppressed in this case. This behaviour can also be observed in the pronunciation of 'foreign compounds'. These are complex structures whose first member is a bound morpheme of foreign origin such as *para-* 'id', *kvázi-* 'quasi-', *pszicho-* 'psycho-', etc. Although these items may be part of words which are phonologically indistinguishable from monomorphemic ones (e.g. *paralel* 'parallel'), they can be productively used to form compounds whose second member may be a native or a non-native word (e.g. *kvázi-vörös* 'quasi-red', *paraszimpatikus* 'parasympathetic'). These are often phonologically identifiable as compounds: e.g. *kvázi-vörös* (note the lack of vowel harmony), *paraanyag* [pɔrɔɔnʲɔg] 'suberin' (note the fake geminate and the lack of Low Vowel Lengthening in *para-*), *paraszövet* 'phellem' (note the lack of vowel harmony and that of Low Vowel Lengthening in *para-*). Curiously, in ECH short /o/ can occur finally in the first member of these structures even when they are transparently compound-like and other phonological phenomena (such as the lack of vowel harmony) mark them as compounds: e.g. *pszicho-biológia* [psihobiolo:gia, *psiho:biolo:gia] 'psycho-biology', *pszeudo-főnév* [psɛudofö:ne:v, *psɛudo:fö:ne:v] 'pseudo-noun'. It is unclear why (58) (an otherwise very active constraint which even loans have to conform to) can be violated in just these cases.

We have assumed that in Hungarian the members of all long–short vowel pairs are melodically identical and their only difference is that the root of a given feature tree is associated to one timing slot in the short member and two timing slots in the long one. Given this assumption (in addition to clusters of identical vowels) the OCP should also exclude vowel clusters in which a short vowel and its long counterpart combine in any order (e.g. *ií*, *íi*). This prediction is borne out: hiatuses of this type do not occur, either (when undivided by an analytic domain edge).[65] Naturally, the OCP does not exclude these clusters if the vowels they consist of belong to different analytic domains: [[[ki]] [[ír]]] 'write out', [[[le]] [[ég]]] 'burn down', [[[sí]ig]] 'up to the ski', [[[rá]] [[ad]]] 'put on', etc.

(29a) is clearly a systematic (i.e. non-accidental) OCP-based constraint. By contrast, the interpretation of (29b) and (29c) is less obvious. The reason is that although these constraints are almost exceptionless,[66] it is difficult to say if they account for accidental or systematic gaps. (29b) is a better candidate for a systematic regularity, because there are sporadic examples in which an original prevocalic long vowel shortens in loan words adopted into Hungarian

---

[65] The behaviour of *a–á* and *e–é* is exactly like that of the other short–long pairs, which confirms that (despite the phonetic difference in quality between them) the members of these respective pairs should have identical underlying feature melodies, i.e. should be represented on a par with other short–long pairs.

[66] The full list of exceptions is *káosz* in the case of (29b) and *entellektüel* 'intellectual', *menüett*, *müezzin* 'muezzin', *habitüé*, *nüansz*, *enteriőr* 'interior', *exteriőr* 'exterior', *miliő* in the case of (29c).

(e.g. [buik] and not *[buːik] (from English *Buick* [bjuːik])).[67] (29c) may well be an accident due to several (unrelated) factors: Vowel Harmony (see above and sections 3.2 and 6.1), the relative infrequency of front rounded vowels, and accidents of borrowing (most monomorphemic items with hiatus are loans).

(i) One possible analysis is that we take hiatus to be well-formed in general and consider (29b, c) to be systematic regularities. Then they can be expressed as (31a, b):

(31)

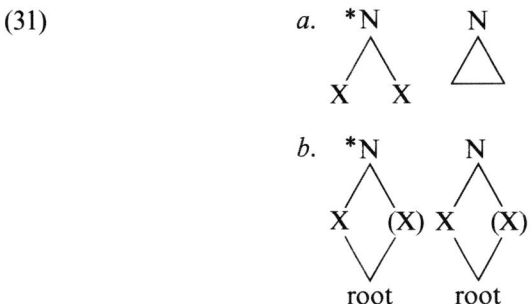

Condition: the segment associated with either (or both) roots is LAB.

As there are no monomorphemic examples of vowel clusters consisting of more than two nuclei,[68] in analysis (i) a separate constraint is needed to exclude *VVV:

(32)

$$\overset{*}{N} \quad N \quad N$$

(ii) It is notable, however, that the number of monomorphemic items actually surfacing with hiatus (i.e. unrepaired by hiatus filling) is low and that most of the relevant items are loans. The number of monomorphemic items with underlying hiatus found in the database used is n=1311. Most of these items, however, contain /i/ or /iː/ in hiatus (n=1075) and surface with compulsory hiatus filling (see section 9.3 for details on hiatus filling). Thus, there are only 236 monomorphemic items actually surfacing with hiatus. This may suggest an alternative analysis in which hiatus is ill-formed:

(33)

$$\overset{*}{N} \quad N$$

---

[67] Similarly, *Zooey* (from J. D. Salinger's novel *Franny and Zooey*) is pronounced [zui] rather than *[zuːi] (note that this example obviously is not a spelling pronunciation).

[68] *Paranoia* 'id' [pɔrɔnojjɔ] does not violate (32). *Dauer* 'perm' ([dɔuɛr] in ECH) is the single exception, but in substandard Hungarian even this word is pronounced [dɔjjɛr].

In this analysis, (31*a*, *b*) are accidental and not part of Hungarian phonology at all and (32) is redundant. Note that there is another way to express the fact that hiatus is ill-formed: we could change the obligatory onset parameter (2) to 'YES'. The ill-formedness of hiatus would follow from this parameter setting. Well-formed onsetless syllables would still exist, but they would be limited to initial position in an analytic domain: e.g. *itt* 'here', *ár* 'price', etc. This can be interpreted as an extension or generalization of the Peripherality Condition (Hayes 1980): exceptional syllable structure is permitted at the edge of a domain, not only in the sense that extra material can be added to the basic syllable template (peripheral extrasyllabic consonants may occur or peripheral consonants may be syllabified into a special constituent (the appendix)), but also in the sense that subminimal syllables may be licensed in peripheral position. Under analysis (ii) what remains to be explained is why violations of (33) are not repaired in items to which Hiatus Filling (see section 9.3) does not apply.

The choice between the two analyses[69] is an empirical one: it depends on native speakers' reactions to words containing hiatus. As our own intuitions are somewhat ambiguous and experimental data pertaining to this problem are not available, we leave this question open.

### 5.3.2. Intervocalic consonant clusters

In this section we discuss the constraints that apply to transsyllabic consonant clusters.

#### 5.3.2.1. Two-member clusters

We have seen that any consonant can occur in a simplex onset or coda in Hungarian. Nevertheless, not all possible combinations of a non-branching coda and a non-branching onset can occur within a word: there are transsyllabic constraints specific to this context. Table 17 shows the attested intervocalic

---

[69] One might suggest a third, sonority-based alternative to account for the predominance of hiatuses containing an /i/ melody. One could say that only vowel clusters containing a [-open$_2$] (high) melody are allowed (hiatuses permitted by this constraint may or may not be seen as subject to (31*a*, *b*) so this analysis does not say anything about the status of (31*a*, *b*)). Hiatus is otherwise disallowed. This can be given a sonority interpretation: as [-open$_2$] vowels are usually assumed to be less sonorous than [+open$_2$] (non-high) vowels (e.g. Goldsmith 1990, Laver 1994) we can say that hiatus is only allowed in Hungarian if there is a sonority difference between the members of the vowel cluster (this presupposes that there is no sonority difference between [+open$_1$] (low) vowels and [-open$_1$] (non-low) vowels). It does not matter which of the two vowels is more sonorous: e.g. *kiabál* 'shout', *mozaik* 'mosaic' (note that even these hiatuses would be subject to postlexical Hiatus Filling [kijɔbaːl, mozɔjik]). The constraint then would be:

(i)  Hiatus is asymmetric: one of the vowels must be governed.

Unfortunately this analysis makes wrong predictions about the well-formedness of clusters both of whose vowels are high. These would be ruled out by (i) because there is no sonority difference between the vowels. This is wrong since hiatuses in which both vowels are [-open$_2$] are well-represented within the set of hiatuses one of whose vowels have a [-open$_2$] specification: e.g. *fiú*, *július*, *jezsuita*, etc.

TABLE 17. *Intervocalic CC clusters*

| | p | t | tʸ | k | b | d | dʸ | g | tˢ | č | ǰ | f | s | š | v | z | ž | m | n | nʸ | l | r | j | x |
|---|---|---|---|---|---|---|---|---|---|---|---|---|---|---|---|---|---|---|---|---|---|---|---|---|
| **p** | pp | pt | * | 10 | — | — | — | — | 15 | 5 | — | 2 | ps | 8 | * | — | — | * | 3 | * | pl | pr | 2 | * |
| **t** | * | tt | — | tk | — | — | — | — | — | — | — | 3 | — | — | tv | — | — | 6 | 3 | — | tl | tr | — | 5 |
| **tʸ** | 2 | * | tʸtʸ | 11 | — | — | — | — | * | — | — | 2 | * | * | 1 | — | — | 4 | * | 4 | 1 | * | — | 1 |
| **k** | * | kt | — | kk | — | — | — | — | ktˢ | 3 | — | 8 | ks | 11 | kv | — | — | km | 8 | 2 | kl | kr | 6 | 6 |
| **b** | — | — | — | — | bb | 7 | * | * | — | — | — | — | 1 | 7 | 3 | * | — | 1 | * | — | bl | br | 4 | — |
| **d** | — | — | — | — | * | 9 | — | * | — | — | — | — | — | — | dv | dz | * | dm | 3 | — | dl | dr | — | * |
| **dʸ** | — | — | — | — | * | * | 7 | * | — | * | — | — | 3 | 2 | 3 | * | 1 | — | — | — | 5 | * | — | — |
| **g** | — | — | — | — | * | * | * | gg | — | — | — | — | — | — | 3 | gz | * | gm | gn | * | gl | gr | 6 | — |
| **tˢ** | * | * | — | tˢk | — | — | — | — | tˢtˢ | — | — | * | * | * | * | — | — | 1 | 3 | * | 5 | * | — | * |
| **č** | * | * | — | čk | — | — | — | — | * | 10 | — | * | * | * | — | — | — | 6 | 3 | * | 1 | * | — | * |
| **ǰ** | — | — | — | * | * | — | * | — | — | 8 | — | — | — | — | — | — | — | * | — | — | * | — | * | *- |
| **f** | * | 8 | 3 | — | — | — | — | — | * | — | — | ff | 1 | * | * | — | — | * | 1 | * | 9 | fr | 1 | * |
| **s** | 15 | st | 2 | sk | — | — | — | — | 3 | * | — | 5 | ss | * | 3 | — | — | sm | 4 | 3 | sl | * | 1 | * |
| **š** | šp | št | 6 | šk | — | — | — | — | * | * | — | 1 | * | šš | 6 | — | — | šm | 2 | 3 | šl | * | *- | 4 |
| **v** | — | — | — | — | * | * | * | * | * | * | — | — | * | * | * | — | — | * | * | * | 5 | 2 | 1 | *- |
| **z** | — | — | — | — | 1 | zd | * | zg | — | — | — | — | — | — | 2 | zz | * | zm | 6 | * | zl | 3 | — | * |
| **ž** | — | — | — | — | * | 2 | 2 | 4 | — | — | — | — | — | — | * | * | * | 6 | 4 | * | 2 | — | — | — |
| **m** | mp | 1 | — | 1 | mb | 2 | * | * | — | 2 | — | 15 | 1 | 1 | 4 | 3 | 5 | mm | 8 | 3 | ml | 3 | 3 | 3 |
| **n** | — | nt | — | nk | — | nd | — | ng | ntˢ | nč | 9 | nf | ns | nš | nv | nz | 4 | — | nn | — | nl | * | — | 3 |
| **nʸ** | * | * | nʸtʸ | * | * | nʸdʸ | * | * | * | * | — | 1 | 1 | 6 | * | * | * | * | nʸnʸ | 1 | * | — | 6 | — |
| **l** | 9 | lt | 1 | lk | 6 | ld | 3 | lg | ltˢ | lč | — | lf | 2 | lš | lv | 2 | 1 | lm | ln | * | ll | *- | — | 9 |
| **r** | rp | rt | rtʸ | rk | rb | rd | 7 | rg | rtˢ | rč | — | rf | rs | rš | rv | rz | rž | rm | rn | rnʸ | rl | rr | rj | rx |
| **j** | 1 | jt | — | jk | 4 | jd | * | 4 | 2 | 4 | — | 2 | js | 1 | 4 | 9 | * | 6 | 9 | 4 | jl | 4 | 11 | 11 |
| **x** | * | * | — | * | 1 | — | — | * | * | * | * | * | * | * | * | — | — | * | 1 | — | 1 | * | * | 1 |

Examples (a question mark marks items in which the relevant cluster may contain an analytic domain boundary): *szappan* 'soap', *kapta* '(boot) last', *lepke* 'butterfly', *kapca* 'foot-cloth', *lépcső* 'stairs', *copfos* 'pigtailed', *apszis* 'apse', *ipse* 'fellow', *srapnel* 'shrapnel', *paplan* 'quilt', *apró* 'tiny', *kopja* 'spear', *suttog* 'whisper', *patkó* 'horseshoe', *hétfő?* 'Monday', *ótvar* 'eczema', *ritmus* 'rhythm', *etnikum* 'ethnic group', *katlan* 'cauldron', *matrac* 'mattress', *nátha* 'flue', *pitypang* 'dandelion', *hattyú* 'swan', *pletyka* 'rumour', *fityfiritty* 'imp', *kotyvaszt* 'concoct', *trutymó* 'suspicious substance', *sa[tʸ]nya* 'stunted', *fátylas* 'veiled', *petyhüdt* 'limp', *akta* 'document', *csökken* 'decrease', *akció* 'action', *bakcsó* 'night heron', *bakfis* 'young girl', *buksza* 'purse', *taksa* 'price', *ekvivalens* 'equivalent', *lakmusz* 'litmus', *akna* 'mine', *szoknya* 'skirt', *lakli* 'tall youngster', *bokréta* 'bunch of flowers', *csuklya* 'hood', *nyikhaj* 'worthless person', *zsibbad* 'go numb', *labda* 'ball', *szubvenció* 'subvention', *dobzoska* 'pangolin', *habzsol* 'devour', *abnormis* 'abnormal', *ablak* 'window', *abrak* 'fodder', *gereblye* 'rake', *addig* 'until then', *dudva* 'weed', *madzag* 'string', *ködmön* 'sheepskin waistcoat', *bodnár* 'cooper', *nudli* 'noodle', *nadrág* 'trousers', *poggyász* 'luggage', *fegyver* 'arms', *jegyző* 'town clerk', *hagyma* 'onion', *naro[dʸ]nyik* 'Narodnik', *kagyló* 'shell', *aggódik* 'worry', *dágvány* 'wallow', *lagzi* 'wedding', *magma* 'id.', *bognár* 'cartwright', *nyegle* 'arrogant', *egres* 'gooseberry', *máglya* 'bonfire', *mackó* 'bear', *icce* <old liquid measure (0.88 litre)>, *cicfarok?* 'achillea', *kecmereg* 'crawl', *fecni* 'slip of paper', *spicli* 'informer', *tacskó* 'dachshund', *gleccser* 'glacier', *kocsma* 'pub', *plecsni* 'stain', *becslés?* 'estimate', *kaftán* 'caftan', *cafka* 'whore', *affér* 'affair', *ofszájd* 'offside', *sufni* 'shed', *kifli* 'roll', *cifra* 'ornamented', *ifjú* 'youth', *aszpik* 'jelly', *asztal* 'table', *kesztyű* 'glove', *deszka* 'plank', *diszciplína* 'discipline', *aszfalt* 'asphalt', *asszony* 'woman', *köszvény* 'gout', *eszme* 'idea', *disznó* 'pig', *tarisznya* 'satchel', *maszlag* 'lie', *csoroszlya* 'old hag', *ispán* 'land-steward', *ostoba* 'stupid', *ostya* 'wafer', *iskola* 'school', *násfa?* 'pendant', *lassú* 'slow', *fösvény* 'miser', *ismer* 'know', *masni* 'ribbon', *rusnya* 'ugly', *pislog* 'blink', *kushad?* 'crouch', *bóvli* 'junk', *sevró* 'kidskin', *szovjet* 'Soviet', *üzbég* 'Uzbek', *gazda* 'owner', *mézga* 'resin', *özvegy* 'widow', *bezzeg* 'by contrast', *csizma* 'boot', *parázna* 'lecherous', *üzlet* 'shop', *ezred* 'thousandth' (fraction), *rozsda* 'rust', *uzsgyi* 'let's go', *vizsga?* 'examination', *zsolozsma* 'chant', *alamizsna* 'alms', *vizsla* 'beagle', *tompa* 'blunt', *tamtam* 'tomtom', *tömkeleg* 'abundance', *tombol* 'rave', *dumdum* 'id.', *csámcsog* 'eat noisily', *kámfor* 'camphor', *szomszéd* 'neighbour', *emse* 'sow', *nyamvadt* 'lousy', *vamzer* 'informer', *tömzsi* 'stocky', *amnesztia* 'amnesty', *cammog* 'trudge', *nyámnyila* 'weakling', *sámli* 'stool', *kamra* 'chamber', *tömjén* 'incense', *lomha* 'slow', *minta* 'pattern', *lankad* 'get tired', *bendő* 'stomach', *angol* 'English', *kanca* 'mare', *szerencse* 'mare', *halandzsa* 'nonsense', *fanfár* 'fanfare', *vánszorog* 'crawl', *közönség?* 'audience', *szenved* 'suffer', *cenzor* 'censor', *avanzsál* 'advance', *dunna* 'quilt', *jelenleg?* 'now', *inhalál* 'inhale', *kulipi[nʸ]tyó* 'small house', *a[nʸ]gyal* 'angel', *kényszer?* 'coercion', *manysi* 'Vogul', *ponyva* 'canvas', *tényleg?* 'really', *dinnye* 'melon', *enyhe* 'slight', *alpári* 'vulgar', *balta* 'hatchet', *kopoltyú* 'gill', *alku* 'deal', *silbak* 'guard', *oldal* 'side', *tölgyes* 'oak-forest', *balga* 'foolish', *délceg* 'dashing', *olcsó* 'cheap', *csalfa* 'deceitful', *alszik* 'sleep', *válság?* 'crisis', *tolvaj* 'thief', *emulzió* 'emulsion', *balzsam* 'ointment', *alma* 'apple', *elnök* 'chairman', *csillag* 'star', *málha* 'pack', *törpe* 'dwarf', *párta* 'girl's headdress', *gyertya* 'candle', *szarka* 'magpie', *borbély* 'hairdresser', *erdő* 'forest', *bárgyú* 'feeble-minded', *márga* 'marl', *herceg* 'prince', *furcsa* 'strange', *férfi* 'man', *erszény* 'purse', *harsány* 'loud', *árva* 'orphan', *borzalom* 'horror', *perzsel* 'scald', *lárma* 'noise', *párna* 'pillow', *ernyő* 'umbrella', *gerle* 'dove', *virrad* 'dawn', *varjú* 'crow', *marha* 'cattle', *selypít* 'lisp', *bojtár* 'young herdsman', *bojkott* 'boycott', *lajbi* 'vest', *vajda* 'voivode', *tajga* 'taiga', *krajcár* 'farthing', *hajcsár* 'drover', *tájfun* 'typhoon', *majszol* 'munch', *hajsókál* 'nurse', *csajvadék* 'riff-raff', *gejzír* 'geyser', *bajmol* 'take trouble', *ajnároz* 'worship', *ejnye* 'Shame on you!', *kajla* 'scatterbrained', *mujré* 'fright', *zsöllye* 'stalls', *kályha* 'stove', *barkohba* <word-game>, *technika* 'technique', *ihlet* 'inspiration', *kehhent* 'cough'.

two-member consonant clusters.[70] Table 17 abstracts away from allophonic differences (hence the lack of [ŋ, m̩] for instance). The notation used is the usual one: a blank space in an intersection of a row and a column means that the relevant cluster is unattested; a star (*) in a box indicates that the relevant cluster only occurs when the two consonants are separated by an analytic morphological domain boundary; a cluster is spelt out if it is attested in monomorphemic items and the number of such items in the database is n>15; numbers have been used to indicate the number of monomorphemic items in the database when the cluster in question is attested in monomorphemic items and the number of such items is n≤15. A box containing a spelt

---

[70] Table 17 is the result of a computer search in the database used (cf. Chapter 1) with some additions since the database does not contain inflected items.

out cluster, a star, or nothing is struck out by dashes to indicate that the relevant cluster(s) is/are subject to (eliminated on the surface by) assimilations.[71]

First it must be pointed out that clusters that straddle analytic boundaries do not reveal the constraints governing interconstituent sequences. Analytic affixation and compounding often create clusters that are not permitted in monomorphemic items; e.g. /nm, kp/ are not permitted monomorphemic interconstituent clusters, nevertheless [[[kan][muri]]] 'stag party' and [[[kerék][pár]]] 'bicycle' are well-formed because the consonants in the relevant clusters belong to different analytical domains. Virtually any cluster can be the result of an analytic morphological operation.[72] The point is that clusters whose member consonants belong to different analytic domains are not relevant to the phonotactic pattern of interconstituent clusters. Therefore, we shall ignore interconstituent clusters that only occur when separated by an analytic domain boundary.

First, let us examine the sonority relationship between the members of an interconstituent cluster. It is often assumed in the literature that interconstituent clusters obey the Syllable Contact Law (SCL) according to which the first consonant in an interconstituent cluster must be more sonorous than the second one (cf. e.g. Vennemann 1988, Clements 1988, 1990, Kaye, Lowenstamm, and Vergnaud 1990, Rice 1992, Harris 1994). In terms of government (as is used in this book, cf. 1.3), this means that in a well-formed interconstituent cluster the coda consonant should be governed by the following onset consonant. (34) shows, however, that the Syllable Contact Law is inoperative in Hungarian: (i) interconstituent clusters may consist of segments of identical sonority (34a), and (ii) often the same segments or segment classes occur in both possible orders (34b) (recall that there are no branching onsets in Hungarian):[73]

(34) a. lepke 'butterfly'
        kapca 'foot cloth'
     b. ak.ta 'document'    at.ka 'mite'
        desz.ka 'plank'     buk.sza 'purse'
        is.ko.la 'school'   tak.sa 'price'
        ron.da 'ugly'       bod.nár 'cooper'

---

[71] None of these assimilations are related to syllable structure (i.e. they all operate regardless of the syllabification of the cluster to which they apply) and some of them are postlexical (see Chapter 7).

[72] Of course, some of these clusters are subject to assimilations, consequently, there are completely unattested clusters at the surface, e.g. /dp/ is subject to regressive voicing assimilation and thus *[dp] is unattested: szabadpiac 'free market' [sɔbɔtpijɔt̚].

[73] The lack of branching onsets is not the only reason why the SCL does not hold in Hungarian: there are interconstituent clusters with rising sonority that would not be well-formed onsets even if branching onsets were permitted, e.g. /tl/ katlan 'cauldron', /knʲ/ szoknya 'skirt', /km/ lakmusz 'litmus', /zn/ vézna 'thin', /nʲl/ tényleg 'really', etc.

al.ma 'apple'          em.lő 'breast'
bal.ta 'hatchet'       kat.lan 'cauldron'
Már.ta 'Martha'        Mát.ra <place name>
maj.ré 'fright'        var.jú 'crow'

However, the fact that the SCL does not hold does not mean that any two consonants can form an intervocalic cluster: systematic gaps do occur.

Let us examine Table 17 and disregard clusters containing /x/ and /v/ for the moment (we shall discuss their behaviour separately at the end of this section). Then, it can be seen in the table that the greatest variety of monomorphemic intersyllabic clusters are of the type in which there is a sonority difference between the two consonants making up the cluster. In general, hardly any special restrictions (pertaining to place of articulation, for instance[74]) apply to clusters of this type. Therefore, we assume that the primary source of licensing for interconstituent clusters is government:

(35)     An interconstituent cluster whose member consonants are in a governing relationship (right-to-left or left-to-right) is well-formed.

As Table 17 shows, however, (i) not all interconstituent clusters whose member consonants are equally sonorous are ill-formed; and (ii) not all interconstituent clusters whose member consonants have different sonority are well-formed. (i) suggests that government is not the only way in which interconstituent clusters can be licensed: geminates and some stop + stop clusters, for instance, are well-formed and thus must be licensed by some other special means of licensing (note that (35) does not imply that clusters whose member consonants are not in a governing relationship are necessarily ill-formed). There are two types of interconstituent clusters that statement (ii) holds true of: clusters consisting of fricatives and stops (in either order) and nasal + stop clusters. Let us disregard the latter type for the moment and focus our attention on the former: some intervocalic fricative + stop and stop + fricative clusters are not well formed. A possible way to handle this problem is to say that *in Hungarian* the sonority difference between fricatives and stops is not great enough for government to apply. Recall that the sonority distance settings for Hungarian consonants and the minimum sonority distance requirement are as follows (cf. section 5.2.4.2):

(36) *a. Sonority Hierarchy: Hungarian*
         stops, affricates < fricatives << nasals << liquids
     *b.* $S_{min}$ = << or >>

---

[74] Nasal + obstruent clusters are an obvious counterexample, cf. the discussion below.

Thus, no governing relationship can obtain between the consonants in fricative + stop and stop + fricative clusters in general, consequently special provisions must be made to license those clusters of this type that *are* well-formed. This would explain why—like other clusters whose members are equally sonorous (e.g. stop + stop or nasal + nasal clusters)—only *some* fricative + stop and stop + fricative clusters are well formed.

Let us now examine the non-analytic interconstituent clusters that are unlicensed by government. We have noted above that some of these clusters are well-formed. Three types of behaviour may be distinguished.

First, all intervocalic geminates are well-formed (the lack of monomorphemic /vː/ and /žː/ is an accidental gap). Geminates obviously cannot be licensed by government if we assume that government applies between timing slots (cf. 1.3) because the two adjacent timing slots have the same segmental content and thus are equally sonorous. Following Rice (1992) we assume that the licensing of geminates is due to the fact that they have shared structure (specifically, a single root node) and thus can be attributed to binding:[75]

(37)    *Interconstituent Binding*
        An interconstituent cluster C1C2 where C1 (the coda consonant) is root-bound is licensed.

The second type of well-formed clusters unlicensed by government consists of liquid + liquid combinations. All possible combinations of the three segments involved (/l, r, j/) seem to be permitted.[76] Although all three are COR their licensing cannot be attributed to binding (place-binding) because (i) /j/ is [–anterior] and thus does not share its place node (or even COR node) with /l, r/ and (ii) there are ill-formed interconstituent clusters whose member consonants share a place node (e.g. */fp, stˢ/, etc.). The licensing of liquid + liquid combinations ceases to be a problem if we assume the fine-tuned sonority scale introduced in section 5.2.3:

(38)    l << r << j

Given (38), the licensing of liquid + liquid interconstituent clusters can be simply attributed to government.

The third group of well-formed interconstituent clusters not licensed by government consists of some nasal + nasal clusters and some obstruent + obstruent clusters. This group differs from the previous two in that only *some* of these clusters are well-formed.

An examination of Table 17 shows that in (non-geminate) stop + stop

---

[75] Compare Kaye, Lowenstamm, and Vergnaud (1990) who assume that the first slot in a geminate is empty and governed by the second.

[76] The clusters /lr/ and /lj/ are subject to assimilations that are unrelated to syllable structure (cf. Chapter 7).

interconstituent clusters the second consonant must not be labial. Clusters whose second consonant is labial are ill-formed/unattested whereas those ending in coronals or velars are well-formed. The working of this 'antilabial constraint' can be seen in the following examples:

(39)   *stop + stop*
    pt   \*tp      kapta '(shoemaker's) last'
    pk   \*kp      lepke 'butterfly'
    bd   \*db      labda 'ball'
    kt = tk      akta 'document', atka 'mite'

The same constraint seems to hold in affricate + stop,[77] nasal + nasal, and stop + fricative clusters. Consider the following examples:

(40)  *a. affricate + stop*
    čk   \*čp      kecske 'goat'
    tˢk   \*tˢp      lecke 'homework'
  *b. nasal + nasal*
    mn   \*nm      himnusz 'hymn'
  *c. stop + fricative*
    pš   \*pf      tapsi 'bunny rabbit', †cupfol[78]
    kš   \*kf      kuksol 'hide, cower', †bukfenc
    ps   \*pf      apszis 'apse', †cupfol 'pluck'
    ks   \*kf      taxi 'id.', †bukfenc 'somersault'

There is a very small number of exceptions to the 'antilabial' constraint in the types of combinations examined. The complete list of the (monomorphemic) exceptional items we have found is as follows: /tʸp/ *pitypang* 'dandelion', *pitypalatty* 'quail's song'; /pf/ *cupfol* 'pluck', *copfos* 'pigtailed'; /tf/ *platform* 'stand'; /tʸf/ *fityfiritty* 'imp'; /kf/ *bakfis* 'young girl', *bikfic* 'kid', *bukfenc* 'somersault', *pakfon* 'German nickel-silver', *ukmukfukk* 'in a jiffy'.

Fricative + stop clusters behave in a more complex way: the 'antilabial' constraint does not work when the first consonant is coronal (41b), but it does if it is non-coronal (41a):

(41)  *fricative + stop*
    *a.*  ft fk \*fp      afta 'thrush', cafka 'whore'
    *b.*  sk = sp      viszket 'itch' = aszpik 'jelly'
        st = sp      posztó 'felt' = aszpik 'jelly'
        št = šp      este 'evening' = püspök 'bishop'

---

[77] Stop + affricate clusters are omitted because there are no labial affricates in Hungarian.
[78] The symbol † marks attested, but phonotactically ill-formed items.

| šk = šp | eskü 'oath' = püspök 'bishop' |
|---------|-------------------------------|
| zd = zb | gazda 'master' = azbeszt 'asbestos' |

We can account for these regularities if we assume that a special kind of licensing (call it Sp-licensing) is needed in order for an interconstituent cluster to be well-formed if it is not licensed by government or binding. A given language may or may not allow Sp-licensing to apply. We assume that in Hungarian, *in general*, Sp-licensing is granted to interconstituent clusters, i.e. it can license coda-onset clusters that are unlicensed by government or binding. Note that this type of licensing is not derivable from government if government is solely based on sonority; nor does it derive from binding.[79] Sp-licensing is thus stipulative and its conditions are language specific.[80]

(42) *Sp-licensing*
Sp-licensed interconstituent clusters are well-formed.

Hungarian, however, imposes certain constraints on Sp-licensing, i.e. it disallows Sp-licensing in some interconstituent configurations. These constraints are discussed and formalized below.

The 'antilabial' effects are due to the following condition on Sp-licensing:

(43) In an interconstituent cluster C1C2 LAB consonants cannot Sp-license the preceding consonant. Condition: C1 $\neq$ [COR, +cont]

As is expected, no antilabial effects can be detected when an interconstituent cluster is licensed by government or root-binding (note that in principle the clusters in question could display antilabial effects because there are labial consonants in the various manner classes that appear in the second position). Consider the following examples:

(44)[81]  a.  *stop + nasal*
etnikum 'ethnic group' = ritmus 'rhythm'
bodnár 'cooper' = ködmön 'sheepskin coat'
b.  *nasal + stop*
cinke 'titmouse' = lámpa 'lamp'
fondorlat 'devious trick' = bomba 'bomb'

---

[79] It could not be interpreted as binding even if we assumed that coronals are placeless because there are well-formed and Sp-licensed interconstituent clusters not containing a coronal: e.g. /pk/ as in *lepke* 'butterfly'. /pk/ is not licensed by government (there is no sonority difference between the segments), and cannot be licensed by binding since /p/ has its own independent place specification.

[80] Hopefully, further research will be able to derive (some of) the effects due to Sp-licensing from general principles.

[81] C + liquid and C + affricate clusters are disregarded because there are no labial liquids or affricates.

   *c.  liquid + stop*
      boldog 'happy' = kolbász 'sausage'
      árkád 'arcade' = Árpád <name>
   *d.  affricate + nasal*
      fecni 'slip of paper' = kecmereg 'crawl'
      kalucsni 'galosh' = pacsmag 'suspicious concoction'
   *e.  fricative + nasal*
      disznó 'pig' = pászma 'ray'
      vézna 'thin' = zuzmó 'lichen'
   *f.  nasal + fricative*
      unszol 'urge' = ténfereg 'loiter'
      emse 'sow' = kámfor 'camphor'
   *g.  liquid + fricative*
      válság 'crisis' = delfin 'dolphin'
      persze 'of course' = férfi 'man'
   *h.  liquid + nasal*
      málna 'raspberry' = elme 'mind'
      barna 'brown' = lárma 'noise'

In addition to the interconstituent clusters that are licensed by government or root-binding there is another group of clusters which could in principle display an 'antilabial' effect, but do not: all clusters consisting of fricatives and/or affricates appear to be ill-formed. There are very few words containing non-analytic fricative/affricate + fricative/affricate clusters. The following is an exhaustive list:[82] *diszciplína* 'subject', *proszcénium* 'fore-stage', *reminiszcencia* 'memory', *ofszájd* 'off-side', *aszfalt* 'asphalt', *atmoszféra* 'atmosphere', *blaszfémia* 'blasphemy', *foszfát* 'phosphate', *foszfor* 'phosphore', *násfa* 'pendant'. Assuming that affricates are contour segments that contain the feature [+continuant], the 'antifricative' constraint can be interpreted as a ban on the occurrence of the features [−son, +cont] under *both* root nodes in an interconstituent cluster. Since, trivially, the constraint only holds if the cluster is not licensed by government or binding, it can be built into Sp-licensing:

(45) A [−son, +cont] segment cannot Sp-license another [−son, +cont] segment in an interconstituent cluster.

Another constraint can be identified if we examine the clusters that contain palatals. It seems that /t$^y$, d$^y$, n$^y$/ make an interconstituent cluster ill-formed irrespective of whether they occur in the first or the second position if the two consonants are not in a governing relation. Consider the following examples:

---

[82] We have disregarded triliteral clusters and clusters containing /v/. Cf. the discussion below.

(46)  dʸm    *dʸd    hagyma 'onion'
      nʸdʸ   *nʸn    ke[nʸ]gyel 'stirrup'
      nʸv    *nʸn    ponyva 'canvas'
      rtʸ    *štʸ    kártya 'card'          (†ostya 'wafer')
      rdʸ    *ždʸ    bárgyú 'stupid'        (†mezsgye 'border')
      rnʸ    *mnʸ    ernyő 'umbrella'       (†nyimnyám 'weakling')

The palatal liquid /j/ is unlike /tʸ, dʸ, nʸ/ in that it forms well-formed interconstituent clusters with any consonant irrespective of whether it occurs in
the first place or the second place. This is to be expected, given that /j/ is at
least minimally sonority-distinct from all the other sonority classes. Thus, all
the examples below are well-formed:

(47)  /j/   se[j]pít 'lisp'              gyapjú 'wool'
            hajcsár 'drover'
            majszol 'munch'             ifjú 'youth'
            hajnal 'dawn'               tömjén 'incense'
            kajla 'scatterbrained'      varjú 'crow'
            ká[j]ha 'stove'

Again, the 'antipalatal' effects can be seen as a result of a constraint on Sp-
licensing:

(48) [COR, –ant] consonants cannot Sp-license another consonant in an
     interconstituent cluster.

There are few exceptions to (48). The following is an exhaustive list of occurring ill-formed items: /tʸp/ *pitypang* 'dandelion', *pitypalatty* 'quail's song'; /tʸk/
*butykos* 'bottle', *fütykös* 'stick', *hetyke* 'proud', *pityke* 'ornamental button',
*pletyka* 'rumour', *potyka* 'carp', *szotyka* 'whore'; /tʸf/ *fityfiritty* 'imp'; /tʸh/
*petyhüdt* 'limp'; /dʸz/, *nagyzol* 'show off'; /stʸ/ *kesztyű* 'glove', *gimnasztyorka*
'Russian military jacket'; /štʸ/ *aggastyán* 'very old man', *bástya* 'bastion',
*borostyán* 'ivy', *hadastyán* 'war veteran of advanced age', *ostya* 'wafer',
*ostyepka* 'a kind of ewe cheese'; /ždʸ/ *mezsgye* 'border', *uzsgyi* 'let's go'; /mnʸ/
*nyámnyám* 'weakling', *nyámnyila* 'weakling', *nyimnyám* 'weakling'.
    In the discussion of interconstituent clusters so far we have disregarded
those containing /x/ or /v/. As can be seen in Table 17, /x/ is free to occur as
the second consonant, but is rare as the first consonant in an interconstituent
cluster (note that it does occur in this position in a few words, e.g. *ihlet* 'inspiration'). This distribution is not due to an interconstituent constraint. We
assume that C+/x/ and /x/+C clusters *are* licensed by government (i.e. that the
sonority difference between /x/ and other sonority classes is sufficient for

government to apply[83]) and the scarcity of preconsonantal /x/ (regardless of whether the coda is part of an interconstituent cluster or not) is accidental. Note that even if the distribution were due to a constraint, it would be relevant to the coda position alone rather than the interconstituent domain.

The behaviour of /v/ is less straightforward. We shall see that it is just as 'two-faced' in its phonotactic behaviour in this position as it is with respect to voicing assimilation (cf. sections 4.1.1 and 7.3). In order to see this let us examine what kind of behaviour we predict with respect to the interconstituent constraints discussed above (i) if /v/ is an obstruent (and has the sonority ranking of a fricative), and (ii) if /v/ has the sonority ranking of a non-nasal sonorant.

First let us suppose that /v/ has the sonority ranking of a fricative. (49a, b) show the predictions about the well-formedness of vC and Cv clusters respectively. Stars mark ill-formed clusters and ✓ marks well-formed ones. */✓ appears if some of the clusters within the class are predicted to be well-formed while others are not.

(49)  a.   vC

| | stop | affricate | fricative | nasal | liquid | |
|---|---|---|---|---|---|---|
| /v/ + | */✓ | * | * | ✓ | ✓ | |

b.   Cv

| | stop | affricate | fricative | nasal | liquid | |
|---|---|---|---|---|---|---|
| | * | * | * | ✓ | ✓ | + /v/ |

If /v/ is a LAB fricative, then both Cv and vC interconstituent clusters are predicted to show antilabial effects. /v/ + LAB stop and stop + /v/ clusters are expected to be ill-formed because they are not licensed by government (the sonority distance between stops and fricatives is too small) and Sp-licensing cannot apply since the second member of the interconstituent cluster is LAB. This prediction is only partly borne out: although /v/ + labial stop clusters do not occur, stop + /v/ clusters are well-formed: e.g. *udvar* 'courtyard', *rögvest* 'at once', *fegyver* 'weapon', *lekvár* 'jam', *borotva* 'razor', *kotyvaszt* 'concoct' (cf. Table 17). In accordance with (45), both vC and Cv clusters should display 'antifricative' effects. This is true of /v/ + fricative/affricate clusters, but (like stop + /v/ clusters) fricative + /v/ clusters are well-formed: e.g. *ösvény* 'path', *özvegy* 'widow(er)', *öszvér* 'mule'. Obstruent + /v/ and /v/ + obstruent clusters are expected to show 'antipalatal' effects since government cannot

---

[83] The ranking of [h] (the realization of /x/ in the onset in Hungarian) in the sonority hierarchy is problematic. It is usually ignored in discussions of the sonority relations between segment classes (cf. for instance, Laver 1994, Ladefoged 1993, Steriade 1982, van der Hulst 1984, Anderson and Ewen 1987) or only mentioned in passing (e.g. Clements 1990 observes that 'the sonority ranking of voiceless approximants is not well-established' (p. 293) and Levin 1985 points out that [h] and [?] may function as obstruents (p. 65)). In the absence of (counter)evidence we simply stipulate that /x/ has the same sonority rank as a fricative.

license the relevant clusters and according to (48) Sp-licensing cannot apply. This again is only partly true since—contrary to the prediction—palatal obstruent + /v/ clusters are well-formed (e.g. *fegyver* 'weapon', *kotyvaszt* 'concoct'). Both /v/ + sonorant and sonorant + /v/ clusters are predicted to be well-formed because these clusters are licensed by government if /v/ is a fricative. This prediction is borne out. To sum up, a fricative interpretation of /v/ makes correct predictions about the well-formedness of interconstituent clusters containing /v/ if (i) the other consonant in the interconstituent cluster is a sonorant and (ii) if /v/ occurs as C1 in an interconstituent cluster C1C2.

Let us now examine what predictions are made and whether they are borne out if /v/ is interpreted as a non-nasal sonorant. Let us assume that the sonority distance between /v/ and the other non-nasal sonorants is great enough for government to apply (i.e. '/v/ << liquids'). (50a, b) show that under this interpretation all interconstituent clusters containing /v/ (vC and Cv alike) are predicted to be well-formed. The reason is that if /v/ has the sonority ranking of a non-nasal sonorant, then government would license all the clusters shown in (50), consequently binding and Sp-licensing would have no effect.

(50)  *a.*    vC

|  | stop | affricate | fricative | nasal | liquid | |
|---|---|---|---|---|---|---|
| /v/ + | ✓ | ✓ | ✓ | ✓ | ✓ | |

     *b.*    Cv

|  | stop | affricate | fricative | nasal | liquid | |
|---|---|---|---|---|---|---|
| | ✓ | ✓ | ✓ | ✓ | ✓ | + /v/ |

This prediction is not correct, however, since (as we have seen above) /v/ + fricative/affricate, /v/ + labial stop[84] and /v/ + palatal obstruent clusters are ill-formed. This suggests that the sonorant interpretation of /v/ makes correct predictions if (i) /v/ occurs as C2 in an interconstituent cluster C1C2; and/or (ii) the other consonant in the cluster is a sonorant.

Thus, we are faced with a 'sonority ranking paradox': /v/ behaves as an obstruent when it occurs as the first member of an interconstituent cluster, but it behaves as a sonorant when it is the second member of an interconstituent cluster.[85] The question is how to express this in terms of licensing. First of all, it is not possible for the same segment to have different sonority rankings depending on the position it occurs in, and we do not want to

---

[84] /v/ + stop clusters do not occur even if the stop is not LAB. In our interpretation this is accidental.

[85] This is completely in agreement with the ambiguous nature of /v/ (cf. sections 4.1.1 and 7.3). It should be noted, however, that /v/ + liquid clusters are rare and /v/ + nasal clusters do not occur. We consider this accidental.

postulate two different underlying /v/'s (a sonorant and an obstruent).[86] Furthermore, it does not help to assume that /v/ is 'asymmetrical' in the sense that—although it is different in terms of sonority from both obstruents and sonorants—it is 'closer' to obstruents than to sonorants (obstruents < /v/ << sonorants), because this would still incorrectly predict antipalatal, antifricative, and antilabial effects in intervocalic obstruent + /v/ clusters. The reverse, i.e. that it is closer to sonorants than to obstruents (obstruents << /v/ < sonorants) does not help either, because it would remove /v/ + obstruent clusters from the purview of the constraints on Sp-licensing and, incorrectly, no antipalatal, antifricative, and antilabial effects would be predicted. Thus, there seem to be two options: we can assume that (i) /v/ has the sonority ranking of a sonorant that is minimally sonority-distant from both the obstruents and the other sonorants (obstruents << /v/ << sonorants) and stipulate that /v/ has to be Sp-licensed when it occurs in a coda which is part of an inter-constituent cluster (even if it is licensed by government); or, alternatively, (ii) /v/ has the sonority ranking of a fricative, but is stipulated to be exempt from the restrictions on Sp-licensing in an onset which is part of an interconstituent cluster. The two solutions are equivalent in that both of them are stipulative. However, since the distribution of /v/ in branching codas suggests that it has the sonority ranking of a fricative (cf. section 5.2.4), we choose the the latter solution and propose the following constraint:

(51) /v/ is Sp-licensed in an onset in an interconstituent cluster.[87]

(51) has the desired effect because while /v/ as C1 in an interconstituent cluster C1C2 remains subject to the 'antilabial' and the 'antipalatal' licensing constraints, it is permitted to occur freely (i.e. unconstrained by these constraints) when it is C2 because it is licensed by (51) in that position. In its present form, (45) cannot prevent /v/ + fricative clusters from being Sp-licensed because /v/ is unspecified for [son] (i.e. it does not have [–son] feature that (45) crucially refers to). Thus, incorrectly, no 'antifricative' effects are predicted. This can be remedied by a minimal modification of (45):

(52) [–son, +cont] segments cannot Sp-license [+cont] segments in an inter-constituent cluster.

Now (52) can revoke Sp-licensing and /v/ + fricative clusters are correctly judged to be unlicensed. Note that the modification has no adverse effect—

---

[86] The two underlying segments would be in complementary distribution, and other phonological processes involving [v] do not require such an analysis; see section 4.1.1.

[87] One might want to build (51) into the constraints on Sp-licensing by exempting /v/ from each of the relevant constraints. That, however, would unnecessarily complicate the constraints while leaving the analysis no less stipulative. It would be more interesting to derive the effect of (51) from the uniqueness of the representation of /v/, the fact that it is unspecified for [son]. We leave this problem for future research.

(52) still prevents interconstituent clusters consisting of fricatives and/or affricates (in any order or combination) from being Sp-licensed. The only difference is that (52) does not allow fricatives to Sp-license [+cont] sonorants. The relevant clusters are well-formed (see Table 17), but they are licensed by government anyway and Sp-licensing is not necessary. Thus the change makes no difference here.

Finally, certain interconstituent clusters are ill-formed in spite of the fact that they appear to be licensed by the constraints discussed above. Specifically, non-homorganic nasal + stop clusters are disallowed although they are licensed by government. This suggests that the licensing of nasals is subject to the following restriction:

(53) Coda nasals must be place-bound when followed by stops.

Note that (53) is not specifically an interconstituent constraint (see section 5.2.4) and that it also holds true of affricates (contour segments whose left 'face' is a stop). There are few exceptions to (53). What follows is a complete list of the exceptional items: *tamtam* 'tomtom', *tömkeleg* 'abundance', *dumdum* 'id', *dínomdánom* 'merry-making', *csámcsog* 'eat noisily', *csemcseg* 'eat noisily'.[88]

### 5.3.2.2. Clusters consisting of more than two members

We have pointed out earlier that intervocalic clusters consisting of more than two consonants are irregular unless an analytic boundary breaks up the cluster.[89] There are such irregular items, but their number is relatively low. In the database there is just one item containing a five-member medial cluster ([ŋkštr] *angström* 'id') and there are only 23 monomorphemic items with a four-member medial cluster (e.g. *szanszkrit* 'Sanskrit', *lajstrom* 'list', *expressz* 'express'). The following monomorphemic four-member clusters occur:

(54) pštr, pstr, jštr, kskl, kskr, kspl, kspr, kstr, nkst, nštr, nskr, rštl

There are 300 monomorphemic items containing a three-member cluster in our database (e.g. *centrum* 'centre', *komplex* 'complex', *export* 'id.', *improvizál* 'improvise'). There are 95 kinds of clusters in these items, but, typically, each type is only 'utilized' in a handful of morphemes. The following monomorphemic three-member clusters occur. The numbers in angled brackets indicate the number of monomorphemic items a given cluster occurs in and in the case of [h] and [ŋ] non-contrastive differences are indicated.

---

[88] These items may not be exceptional at all in that they probably contain internal domain boundaries that fall between the nasal and the stop. They are included here for the sake of completeness.
[89] For the arguments see the discussion in section 5.2.2.

(55)

| | | | | | | | |
|---|---|---|---|---|---|---|---|
| fst | <1> | mfl | <2> | ŋkt | <2> | rkt | <1> |
| xth | <1> | mpl | <15> | ŋkv | <1> | rptˢ | <1> |
| jbn | <1> | mpr | <10> | nšp | <4> | rpr | <1> |
| jdl | <2> | mps | <1> | nšt | <10> | ršl | <4> |
| jgl | <1> | mpt | <1> | nstˢ | <1> | ršp | <1> |
| jšl | <2> | mst | <1> | nsf | <1> | ršr | <1> |
| jst | <1> | mzl | <1> | nsk | <1> | ršt | <1> |
| kstˢ | <2> | ntˢv | <1> | nsp | <3> | rsl | <1> |
| ksh | <3> | nčk | <1> | nst | <4> | rst | <2> |
| ksk | <3> | ndg | <2> | ntl | <1> | rtl | <2> |
| ksl | <1> | ndl | <5> | ntr | <19> | rtn | <1> |
| ksn | <1> | ndr | <8> | pšk | <1> | rtr | <3> |
| ksp | <10> | ndv | <1> | pstˢ | <1> | rtv | <2> |
| kst | <9> | nfl | <4> | psl | <3> | štr | <10> |
| ktr | <6> | nfr | <1> | pst | <2> | štv | <1> |
| lfr | <1> | ŋgl | <8> | ptr | <1> | skr | <1> |
| lft | <1> | ŋgr | <5> | rbl | <2> | skv | <4> |
| lgr | <1> | ŋgv | <3> | rtˢl | <1> | spr | <2> |
| lkl | <1> | ŋktˢ | <4> | rdr | <2> | sth | <2> |
| ľst | <1> | ŋkf | <1> | rdv | <1> | stm | <1> |
| lsk | <1> | ŋkl | <5> | rgl | <2> | str | <27> |
| ltr | <4> | ŋkp | <1> | rxm | <1> | vdb | <1> |
| mbl | <1> | ŋkr | <5> | rkl | <2> | zdr | <1> |
| mbr | <5> | ŋks | <5> | rkm | <1> | | |

The fact that these clusters are irregular does not mean that they do not display certain regularities. Figure (56) informally summarizes some of them:

(56)[90] — C1        C2        C3 —

| | C1 | | C2 | | C3 |
|---|---|---|---|---|---|
| a. | | | C2 ≠ [+son] | | |
| b. | IF C1 = [−son, −cont] THEN C1 ≠ COR | | | | |
| c. | IF C1 = [s, š] | | THEN C2 = [−son, −cont] | | |
| d. | IF C1 = [−son, −cont] | & | C2 = [−son, −cont] | | THEN C3= [+son] |
| e. | C1 = [+son] | | IF C2 = [−son, −cont] | & | C3 = [−son, −cont] |
| f. | IF C1 < C2 | | THEN C2 = [s, š] | | |
| g. | IF C1 ≈ C2 | | THEN C2 = COR | | |

---

[90] Notation: x > y 'x is more sonorous than y'; x < y 'x is less sonorous than y'; x ≈ y 'x and y are equally sonorous'.

In our analysis these regularities (and other possible ones crucially referring to medial –CCC–) are accidental in Hungarian and only reflect a random set of the regularities of the source languages the relevant words were borrowed from.

## 5.4. MORPHEME STRUCTURE: MSCs

In this section we discuss phonotactic constraints that hold within the morpheme. These constraints may or may not be related to syllable structure.

### 5.4.1. Domain-final open syllables and the minimal word/stem

In section 5.2.3 above we pointed out that in general any of the underlying vowels can occur in nuclear position in a syllable. This is not true of open syllables in final position, or more precisely of open syllables at the right edge of a stem. In this position (underlyingly) high ($[-\text{open}_2]$), mid ($[-\text{open}_1$, $+\text{open}_2]$), and low ($[+\text{open}_1]$) vowels behave differently. The restrictions are the following.

Short $[-\text{open}_1, +\text{open}_2]$ vowels (/o, ö/) cannot occur in final position.

(57)             VV               V
        /o/    olló 'scissors', só 'salt'     —
        /ö/    szőlő 'grape', nő 'woman'    —

This constraint is usually assumed to hold in word-final position (cf. Nádasdy 1985, Nádasdy and Siptár 1994, Törkenczy 1994a). However, it is really a constraint on the stem, because /o, ö/ cannot occur at the right edge of an internal analytic domain or immediately before a non-analytic suffix either.[91]

Assuming that unaffixed free morphemes are stems, and that such a stem plus an affix is also a stem, this constraint can be expressed as

(58)

---

[91] The interjection *(no)no* 'well' is a counterexample. Interjections in general do not seem to conform to the phonological constraints of the language (e.g. even syllabic consonants may occur in interjections, which otherwise are unattested in Hungarian: *pszt!* 'hush!'). Note also the behaviour of foreign compounds; see section 5.3.1.

This constraint holds regardless of the number of syllables the stem consists of: monosyllables and polysyllables behave in the same way. (58) is one of the few phonotactic constraints that has an active role in the phonology of Hungarian. The final vowels of loans ending in /o, ö/ are invariably lengthened in Hungarian (e.g. *libretto* [librɛtto:]).[92]

The behaviour of domain-final [–open$_2$] vowels is more complex. (59) shows the distribution of [–open$_2$] vowels at the end of an analytic domain in monosyllabic and polysyllabic words. As can be seen below, three types of items can be distinguished: words in the first column (marked VV) are always pronounced with a long final vowel and those in the third (marked V) invariably have a short final vowel. By contrast, the words in the column marked VV/V may have either long or short final vowels. All native speakers of Standard Hungarian agree in their treatment of the words in columns VV and V, but they may treat those in column VV/V in three different ways. Innovative speakers of ECH have short final vowels in these words (for them there is no difference between the words in VV/V and those in V). Conservative ECH speakers pronounce them with a long final vowel (i.e. for them there *is* a lexical difference between the words in VV/V and those in V). For a third group of speakers (we shall call this group 'intermediate' ECH speakers) the final vowels in the words in column VV/V may be optionally long or short.

(59)[93]

|               |   | VV     | VV/V   | V     |
|---------------|---|--------|--------|-------|
|               | i | sí     | —      | mi    |
| monosyllables | ü | tű     | —      | —     |
|               | u | bú     | —      | —     |
|               | i | futósí | vizaví | buli  |
| polysyllables | ü | kötőtű | keserű | eskü  |
|               | u | mélabú | ágyú   | falu  |

(59) is misleading because it conceals two crucial facts.[94]

(i) Although there are *polysyllabic* words whose final vowel must be pronounced long (in all the three dialects), all of them are compounds or preverb + verb combinations (i.e. they consist of more than one analytic domain) whose final morpheme is a *monosyllabic* free stem: e.g. [[[futó] [sí]], [[kötő] [tű]], [[méla] [bú]]. There are no monomorphemic words, or polymorphemic ones whose final morpheme is a polysyllabic free stem, in this group. Furthermore,

---

[92] Note also that there are no *cseh*-type stems with an /o/ or /ö/ in the stem-final syllable (cf. section 8.2.2).

[93] Glosses: *sí* 'ski', *mi* 'we', *tű* 'pin', *bú* 'sorrow', *futósí* 'cross-country ski', *vizaví* 'opposite, across', *buli* 'party', *kötőtű* 'knitting pin', *keserű* 'sour', *eskü* 'oath', *mélabú* 'spleen', *ágyú* 'cannon', *falu* 'village'.

[94] There is another fact that (59) does not indicate: word-final long /iː/ is extremely rare. There are only 9 such items in our database and with the exception of *rí* 'cry', *sí*, *vizaví*, and *zrí* 'trouble' they are interjections or onomatopoeic words. We have no explanation for this depleted distribution and consider it an accident.

there is not a single polysyllabic item in the other two columns (VV/V or V) whose final morpheme is a monosyllabic free stem. All the polysyllabic words in the latter two columns are either monomorphemic or end in a suffix.

(ii) There is only a very limited number of monosyllabic items that end in a short [–open$_2$] vowel. The complete list is *ki* 'who', *ki* 'out', *mi* 'what', *mi* 'we', *ti* 'you' (pl.), *ni* 'look!'. Of these, *ni* is an interjection and the rest are function words.

Taking (i) and (ii) into consideration and assuming that it applies only to content words, the constraint that governs the distribution of final [–open$_2$] vowels can be (informally) formulated as follows:

(60)   Domain final [–open$_2$] vowels are
   (i)   long in monosyllables and short in polysyllables (Innovative ECH),
   (ii)  long in monosyllables (Conservative/intermediate ECH).

As in (58), the domain in which (60) applies is the stem because [–open$_2$] vowels behave in the same way word finally, at the end of the non-final constituent of a compound, immediately preceding an analytic suffix and immediately preceding a non-analytic suffix. (61) shows this in the innovative ECH dialect (where final [–open$_2$] vowels are always short in polysyllables):

| (61)[95] | *word-final* | *compound* | *analytic suffix* | *non-analytic suffix* |
|---|---|---|---|---|
| monosyllabic | b[u:] | b[u:]-bánat | b[u:]-nak | b[u:]-t |
| | | méla-b[u:] | | |
| polysyllabic | ágy[u] | ágy[u]-talp | ágy[u]-nak | ágy[u]-t |

In accordance with the definition of the stem above, suffix-final [–open$_2$] vowels behave in the same way as [–open$_2$] vowels at the end of lexical stems: compare *tetű* [tɛtü] 'louse' and *jószív-ű* [joːsivü] 'kind-hearted'.

In all the three dialects, monosyllabic words/stems are treated in the same way (i.e. they have long final [–open$_2$] vowels), thus the variation that distinguishes these dialects is confined to polysyllabic words/stems. This fact can be accounted for if we assume that there is a constraint that applies to all dialects and requires that the minimal word/stem should be bimoraic:[96]

(62)   Stem/word$_{min}$ = $\mu\mu$

We assume that all stems (affixed and unaffixed) in the lexicon have to conform to (62).[97] It does not apply to affixes, function words, interjections, and

---

[95] Dashes in (61) only appear to indicate morpheme division. Glosses: *bú* 'sorrow', *búbánat* 'sorrow and remorse', *búnak* 'sorrow+dat', *bút* 'sorrow+acc', *mélabú* 'melancholy', *ágyú* 'cannon', *ágyútalp* 'gun-carriage', *ágyúnak* 'cannon+dat', *ágyút* 'cannon+acc'.
[96] On the mora, cf. section 1.3.
[97] On the minimal word constraint in Hungarian, cf. Csúri (1990) and Törkenczy (1994a).

onomatopoeic words.[98] The minimal stem/word constraint is trivially (vacuously) true of stems ending in [−open$_1$, +open$_2$] (mid) vowels (as stem-final short mid vowels are excluded in general, stems consisting of open monosyllables that end in a mid vowel cannot violate (62)). As pointed out above, stems ending in [−open$_2$] (high) vowels conform to (62). In innovative ECH the distribution of high vowels can be interpreted as the result of a constraint that bans stem-final long high vowels which is blocked if it should violate (62) (we shall return to the formalization of this constraint later).

Let us now examine the behaviour of [+open$_1$] (low) vowels in final position. Stems ending in low vowels also observe (62). There are a (small) number of function words and interjections (*be* 'into', *de* 'but', *he* 'what? <I cannot hear you>', *le* 'down', *ne* 'no(t)', *se* 'either', *te* 'you', *ha* 'if', *ja* 'Now I understand', *na* 'Come on!') and two truly exceptional content words (*fa* 'tree', *ma* 'today') which violate it. Although low vowels are also subject to the minimal word/stem constraint (like high and mid vowels), their distribution in final position is different in several ways. Disregarding the exceptional words listed above, they pattern in the following way:

(63)

| | stem-final ≠ word-final | | word-final | |
|---|---|---|---|---|
| | *monosyllabic* | *polysyllabic* | *monosyllabic* | *polysyllabic* |
| e, a | − | − | − | + |
| eː, aː | + | + | + | + |

(63) shows that (i)—unlike mid vowels—both long and short low vowels occur finally (e.g. *csokoládé* 'chocolate', *teve* 'camel', *burzsoá* 'bourgeois', *apa* 'father'); (ii)—unlike high vowels in innovative ECH—both long and short low vowels occur finally in polysyllabic words: *modulo* the minimal word/stem constraint, the distribution of long and short low vowels is the same in word-final position; and (iii)—unlike high and mid vowels—low vowels behave differently in non-word-final stem-final position and word-final position: only long [eː, aː] can occur before suffixes.[99] This can result in alternations such as *apa* ~ *apá-t* 'father' (nom.~acc.), *teve* ~ *tevé-t* 'camel' (nom.~acc.).

Word-final *á* is rare and final *é* is relatively infrequent. Not counting words containing the suffix -*válvé*, -*nálné*, final *á* occurs in function words, abbreviations, and interjections.[100] Final *é* fares relatively better: disregarding suffixes that end in *é*, it occurs in about 130 stems all of which are loan words.

The question is how to make sense of this distribution. This problem is related to the analysis of stem final *é* ~ *e*, *á* ~ *a* alternations. As pointed out

---

[98] The full list of exceptions to (62) ending in non-low (i.e. [−open$_1$]) vowels is *ki, ki, mi, mi, ni, no, ti*, all of which are function words or interjections.

[99] Some suffixes are exceptional in that they may be preceded by *a* and *e*, e.g. -*ságlség* as in *apa-ság* 'fatherhood'; cf. the discussion of low vowel alternations in sections 3.1.1 and 6.2.

[100] There are about 15 items; *burzsoá* 'bourgeois' and *hajrá* 'spurt' are the only content words with final *á*.

above, of the [+open$_1$] vowels only *é* and *á* can occur before suffixes. Let us assume that the underlying difference between *é ~ e, á ~ a* is only quantitative (cf. section 3.1.1). Then, in an analysis in which all long vowels are underlyingly (pre)associated with two timing slots and short ones with a single timing slot, in principle, these alternations may be interpreted in two ways: as (i) the lengthening of underlying final short [+open$_1$] vowels before suffixes; or (ii) the shortening of underlying final [+open$_1$] vowels word-finally.[101] The choice determines the underlying distribution of final *é* and *á*. If analysis (i) is chosen, the distribution of these vowels in final position is unconstrained by any restriction *specific to [+open$_1$] vowels*: both long and short [+open$_1$] vowels may occur in all final environments underlyingly. Under this analysis the lack/low number of tokens with final low vowels in some environments in (63) is either due to the minimal word/stem constraint, which is independent of [+open$_1$] vowels (and rules out stems that are too short—this would apply to final short [+open$_1$] vowels in monosyllables (*fa* and *ma* are irregular)), or is accidental (thus, the relative infrequency of final *é* and *á* need not be accounted for in the phonology). The surface lack of stem-final *e* and *a* before suffixes is due to a phonological rule that lengthens short low vowels stem-finally if a suffix follows ('Low Vowel Lengthening', cf. Vago 1980*a*, Nádasdy and Siptár 1994, and sections 3.1.1 and 6.2 in this volume). In this analysis the underlying distribution of [+open$_1$] (low) vowels is analogous with that of [–open$_2$] (high) vowels in conservative ECH.

Under the shortening analysis only long low vowels occur stem (and word) finally at the underlying level and the word-final short surface reflexes are derived by rule.[102] According to this analysis there *is* an underlying constraint that is specific to [+open$_1$] vowels: the short ones cannot occur finally. The minimal stem/word constraint would hold without exception. *fa* and *ma* conform to it *underlyingly*. Note, however, that (exceptionally?) the word-final low vowel shortening would not be blocked by the minimal word constraint (*fa* and *ma* surface with a short vowel) while it does seem to apply to high vowels in innovative ECH. In this analysis the underlying distribution of [+open$_1$] vowels is analogous with that of mid vowels. The problem with this analysis is a derivational one: how to prevent non-alternating final long *á* and *é* from undergoing word-final low vowel shortening (e.g. *lé* 'juice', *csokoládé* 'chocolate', *burzsoá* 'bourgeois'). True, the number of these words is low, but, nevertheless, shortening has to be blocked somehow. One possibility is marking these words with arbitrary diacritics in the lexicon.[103] The autosegmental notation allows for a distinction

---

[101] The same is true of a linear analysis where length is a feature [±long].

[102] This analysis has been proposed by e.g. Szépe (1969), Hetzron (1972), Abondolo (1988), Jensen and Stong-Jensen (1989*a*).

[103] Or the diacritic may be 'phonologized': Abondolo (1988) postulates different underlying vowels in these items.

between alternating and non-alternating final long vowels representationally without having to resort to exception features or postulating extra underlying segments.[104]

(64)

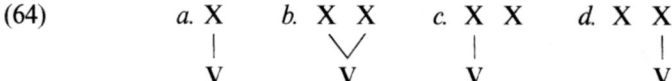

The representation (64a) would be used for non-alternating short low vowels (e.g. *hat* 'six', *nem* 'gender'), (64b) for non-alternating long low vowels (*lé, csokoládé, burzsoá*), and (64c) or (64d) for alternating long low vowels (*apa, teve*). In this case the alternation between long and short low vowels is handled by a rule that spreads the root node of a low vowel to an adjacent floating X slot stem-finally before a suffix, either to the left (if the representation is (64d)) or to the right (if the representation is (64c)). In this treatment, there is a constraint specific to [+open$_1$] vowels: those represented as (64a) cannot occur stem finally in the lexicon. *Fa* and *ma* would be exceptions to the minimal stem/word constraint and the scarcity of final low vowels represented as (64b) would be an accident. The underlying distribution of low vowels would be analogous to that of mid vowels.

It is not our main concern here to decide which analysis handles the low vowel alternations discussed best.[105] The point is that *phonotactically* they are equivalent. The main difference between them is which other class of vowels the final low vowels are grouped together with: if the first analysis is chosen low vowels pattern with high vowels in the conservative dialect; in the latter they pattern with mid vowels. As high vowels in the innovative dialect behave differently from both, the choice between the alternative analyses cannot be made on phonotactic grounds. As we saw in section 3.1.1, we assume in this book that the lengthening analysis is correct. Thus, long and short low vowels can occur freely stem-finally.

To sum up, mid vowels are constrained by (58) and (vacuously) the minimal word/stem requirement (62). (58) holds for all representations, derived or underived. The underlying distribution of low vowels is only constrained by the minimal word/stem requirement. High vowels behave like low ones in the conservative dialect: underlying representations have to conform to (62). In the innovative dialect, however, the minimal word/stem constraint plays an active role. As we have seen above, in this dialect, stem-final long high vowels are banned (cf. 65) *unless* the representation required by (65) should violate the minimal stem/word constraint (62).

---

[104] Similar representations have been proposed by Jensen and Stong-Jensen (1989a), P. Rebrus (personal communication), and Ritter (1995).
[105] For a detailed discussion cf. Nádasdy and Siptár (1998), Siptár (1998a).

(65)

$$* \text{N}$$

$$\begin{array}{c} \text{X} \wedge \text{X} \\ \text{V} \end{array} \quad ]_{\text{stem}}$$

$$[-\text{open}_2]$$

In innovative ECH (65) is a *static* constraint in the sense that there are no alternations between long and short high vowels in this environment. In the intermediate dialect it is possible to argue that there is a rule (66) which shortens stem-final long $[-\text{open}_2]$ vowels optionally because both alternants may surface:

(66)

$$\begin{array}{ccc} \text{N} & & \text{N} \\ \text{X} \wedge \text{X} & \rightarrow & \text{X} \\ \text{V} & & \\ [-\text{open}_2] & & [-\text{open}_2] \end{array} \quad / \underline{\hspace{1em}} ]_{\text{stem}}$$

This rule is blocked if the output should violate the minimal word/stem constraint, which acts as a filter, or a 'derivational constraint' (Kisseberth 1970). There is no alternation evidence in the innovative dialect: stem-final high vowels are simply *always* short except in monosyllables. There is no satisfactory way to express this relationship between (65) and (62) in the present framework. Restricting (65) to stems which are longer than monosyllabic 'does the job', but it should be noted that the minimal stem/word requirement is 'built into' this constraint (and thus is stated twice):[106]

(67)

$$\begin{array}{cc} * \; \sigma & \sigma \\ & | \\ & \text{N} \\ & \wedge \\ \text{X} \; \text{X} & ]_{\text{stem}} \\ \text{V} & \\ [-\text{open}_2] \end{array}$$

Given (67), long high vowels are excluded finally *only* in polysyllabic stems and final short high vowels are banned by the minimal stem/word constraint in monosyllabic ones.

To sum up, all final vowels are subject to the minimal stem/word constraint, mid vowels are also constrained by (58), high vowels are input to the (optional) rule (66) in the intermediate dialect and have to meet (67) in the innovative dialect.

---

[106] In Optimality Theory such a relationship can be expressed in a straightforward manner: the minimal stem/word constraint dominates (65).

## 5.4.2. VVCC: the complexity of the rhyme

In the previous section we discussed the behaviour of open syllables and pointed out that the distribution of vowels is different in medial and final open syllables. Let us now examine the behaviour of closed syllables.

In general, any vowel seems to be possible in a closed syllable (cf. section 5.2.3). However, there are restrictions holding in this environment depending on (i) the position of the syllable in the word and (ii) the morphological complexity of the word. (68) shows the distribution of long and short vowels in word-final syllables closed by a single consonant, in word-final syllables closed by more than one consonant, and word-medially when these syllables occur monomorphemically, i.e. undivided by a morpheme boundary:

| (68)[107] | VC## | VCC## | VC.C |
|---|---|---|---|
| i | hit 'belief' | ring 'sway' | inger 'stimulus' |
| ü | sün 'hedgehog' | csüng 'hang' | kürtő 'funnel' |
| ö | sör 'beer' | gyöngy 'pearl' | ördög 'devil' |
| e | nem 'gender' | szent 'saint' | persze 'of course' |
| u | fut 'run' | must 'grape juice' | undor 'disgust' |
| o | lop 'steal' | gyors 'fast' | boglya 'stack of hay' |
| a | hat 'six' | tart 'hold' | apró 'tiny' |
| iː | sír 'grave' | — | — |
| üː | bűn 'sin' | — | — |
| öː | bőr 'skin' | — | — |
| eː | kém 'spy' | érc 'ore' | érték 'value' |
| uː | rút 'ugly' | — | — |
| oː | kór 'disease' | — | — |
| aː | láp 'marsh' | márt 'dip' | árpa 'barley' |

As can be seen in (68), (a) any short vowel is possible in a closed syllable, and (b) of the long vowels, only /eː/ and /aː/ can occur in non-word-final closed syllables and word-final syllables closed by more than one consonant. This poses two questions: (i) why is there a difference between word-final syllables closed by a single consonant and the other kinds of closed syllables?

---

[107] (68) abstracts away from a few exceptional items. **VCC#:** *tószt* [oː] 'toast', *avitt* [i(ː)] 'obsolete', *blazírt* [i(ː)] 'blasé', *bornírt* [i(ː)] 'narrow-minded', *fasírt* [i(ː)] 'meatball', *múlt* [u(ː)] 'past'. Note that in the words *őrs* [ö] 'squad', *gyűjt* [ü] 'gather', *gyújt* [u] 'light', *nyújt* [u] 'stretch', *sújt* [u] 'hit' the vowels spelt long are pronounced short in ECH. **VC.C:** *ízlés* [i(ː)] 'taste', *sínyli* [i(ː)] 'suffer' (3sg pres. def.), *nőstény* [öː] 'female', *tőzsde* [öː] 'stock exchange', *csúzli* [uː] 'slingshot', *kóstol* [o(ː)] 'taste', *bógnizik* [oː] 'make curves in skating', *bóklászik* [oː] 'loiter', *kókler* [oː] 'impostor', *kóstál* [o(ː)] 'cost', *lófrál* [oː] 'hang around', *sóska* [oː] 'sorrel', *ósdi* [oː] 'old', *ótvar* [oː] 'eczema', *pózna* [oː] 'pole', *ródli* [oː] 'sledge', *ócska* [oː] 'worthless', *-ódzik/-ődzik* <reflexive>. 'Epenthetic' stem forms like *pótlás* [poːtlaːš] 'replacement', *ólmoz* [oːlmoz] 'lead' (verb), etc. are only apparent counterexamples since they do not contain a cluster underlyingly: /poːtV_dlaːš, oːlV_dmoz/ (cf. section 8.1.4.2).

and (ii) why do /eː/ and /aː/ behave differently from the other long vowels?

One might want to answer question (i) by utilizing the notion of extrasyllabicity/extrametricality. If we say that a single word-final consonant is extrametrical in Hungarian at the point where the constraint against long vowels (except /eː/ and /aː/) in closed syllables applies, then it is understandable why there is an asymmetry between word-final VVC sequences vs. word-final and word-medial VVCC sequences. In the first case the word-final consonant is extrasyllabic and therefore the word-final syllable is not closed: *rút* /ruː<t>/. If there is more than one word-final consonant, then rendering the final one extrametrical still leaves a closed syllable behind and thus the constraint on long vowels applies: *VVC<C>. A word-medial vowel followed by two consonants is necessarily subject to the constraint because word-medial consonant clusters are necessarily heterosyllabic[108] and extrametricality cannot apply here because of the Peripherality Condition (Hayes 1980, 1982).[109]

However, the discrepancy between the behaviour of final VVC vs. medial and final VVCC clusters disappears if polymorphemic clusters are considered as well. Any long vowel is possible before a cluster if there is a morpheme boundary after the vowel or between the consonants making up the cluster:

(69)[110]

| | VCC## | VC.C |
|---|---|---|
| iː | sír-t | szív-tam |
| üː | bűn-t | bűn-ben |
| öː | fő-bb | bőr-ben |
| eː | kér-t | kér-ték |
| uː | túr-t | túr-tak |
| oː | kór-t | kór-nak |
| aː | vár-t | vár-tam |

---

[108] Recall that there are no complex onsets in Hungarian (cf. section 5.2.2.) As expected, the constraint applies even in cases when the second member of the interconstituent cluster is more sonorous than the first one: hypothetical (monomorphemic) */muːrta/ and */muːtra/ are equally impossible.

[109] The state of affairs described is reminiscent of that in English where word-final single consonants can be preceded by long vowels, but—disregarding clusters that contain Level 2 suffixes—word-final clusters and some word-medial ones may not. Of course there are important differences (in English *all* long vowels behave in the same way; there are complex onsets (hence *some* medial clusters may follow long vowels: *April*); coronal clusters behave differently from non-coronal ones: *pint*; etc.). The analysis sketched is thus analogous to those presented in Myers (1987), Borowsky (1989), Jensen (1993), Rubach (1996).

[110] Glosses: *bőrben* 'skin' (iness.), *bűnben* 'sin' (iness.), *bűnt* 'sin' (acc.), *főbb* 'main' (comp.), *kért* 'ask' (3sg past indef.), *kérték* 'ask' (3pl past def.)', *kórnak* 'disease' (dat.), *kórt* 'disease' (acc.), *sírt* 'grave' (acc.), *szívtam* 'suck' (1sg past), *túrt* 'dig' (3sg past indef.), *túrtak* 'dig' (3pl past indef.), *várt* 'wait' (3sg past indef.), *vártam* 'wait' (1sg past).

It makes no difference if the intervening morpheme boundary is the edge of an analytical (e.g. *bűn-ben*) or a non-analytical domain (*bűn-t*): the constraint only holds within morphemes. This suggests that the phenomenon discussed is not a constraint on the complexity of the rhyme, but rather on morpheme shape. In other words, it is an MSC and not an SSC. Thus, instead of an extrametricality/extrasyllabicity analysis of the type sketched above,[111] we propose that the distribution of preconsonantal long vowel is simply governed by the following MSC:

(70)    *VVCC
         domain: morpheme
         condition: VV ≠ /eː, aː/

Let us now examine the second question, i.e. why it is just /eː/ and /aː/ that are unconstrained by (70). We have analysed the short–long pairs [ɔ–aː], [ɛ–eː] on a par with the other short–long pairs in the system, i.e. we have assumed that underlyingly, just like the other vowel pairs, they only differ in quantity, not in quality (cf. sections 3.1, 5.4.1). Note, however, that they are special (i.e. differ from all the other pairs in the system) in that they are the only pairs whose members are considerably different[112] in quality at the surface. One may try to explain the special behaviour of /eː, aː/ with respect to (70) by connecting it with the special character of the pairs [ɔ–aː], [ɛ–eː].[113] We argue in Chapter 3 that [ɔ–aː] and [ɛ–eː] are underlyingly [DOR, +open$_1$, +open$_2$] and [COR, +open$_1$, +open$_2$], respectively. The surface quality differences (rounding (and height) in the case of [ɔ–aː] and height in the case of [ɛ–eː]) are the result of phonetic implementation conditioned by the underlying quantity difference: long /eː/ is interpreted phonetically as mid ([eː]) and short /a/ as rounded ([ɔ]). There is no *theoretical* reason not to do this the other way round (cf. Törkenczy 1994a, Polgárdi 1997). One could assume that the underlying difference between the members of the pairs discussed is qualitative (/ɔ–a/, /ɛ–e/) and the surface difference in length is a matter of phonetic implementation which is conditioned by the underlying quality difference (/a, e/ will appear as long at the surface). This move has advantages and disadvantages. On the positive side, the exceptional behaviour of [eː] and [aː] with respect to (70) is no longer a mystery: these vowels are not constrained by (70) because (70) is a constraint on long vowels and [eː] and [aː] are not long underlyingly (the

---

[111] Which is further weakened by the fact that—unlike in English—there is no phenomenon other than preconsonantal vowel length motivating word-final extrametricality.

[112] There are small differences of height between the surface reflexes of the members of some other pairs, notably /oː–o/ and /öː–ö/, the long segments being slightly more closed than the short ones, cf. Chapter 3.

[113] These pairs of vowels are unlike the rest in other ways as well, e.g. they are the only ones that alternate stem finally (cf. the discussion of Low Vowel Lengthening in sections 3.1.1, 5.4.1). It is an interesting idea to suppose that all these phenomena are related and may have a common explanation, but we will not pursue it in this book.

constraint could be restated without the condition). However, in our view, the negative effects are more serious: (i) the vowel inventory would become asymmetrical (only high and *some* mid vowels would have long counterparts); (ii) it would be no longer possible to express Low Vowel Lengthening as a uniform process: it would have to be lowering or raising (depending on which vowel we take as underlying) for /ɛ/ ~ /e/, but rounding or unrounding for /ɔ/ ~ /a/; (iii) as /ɛ/ ~ /e/ and /ɔ/ ~ /a/ could not be analysed as length alternations, the alternations *nyár* 'summer' ~ *nyar-at* 'summer' (acc.), *tél* 'winter' ~ *tel-et* 'winter' (acc.) could not be treated as the same process (Stem Vowel Shortening (cf. section 3.1.2)) as that involving other vowels, e.g. *víz* 'water' ~ *víz-et* 'water' (acc.). Therefore, we shall not reanalyse the representation of the vowels [ɔ–a:], [ɛ–e:] in the way described above (for an additional piece of evidence, cf. section 5.3.1). This means that the condition on (70) remains a stipulation.

PART III

---

# PROCESSES

# PROCESSES INVOLVING VOWELS

In this chapter we will describe a number of phonological processes that involve vowels but are not conditioned by syllable structure.[1] The vowel system of Hungarian, as well as the major alternation types that Hungarian vowels participate in, have been discussed in detail in Chapter 3. In particular, a number of facts concerning vowel harmony (some of which are generally ignored in the literature) were discussed there (see 3.2). In the present chapter (section 6.1), a new type of analysis of those facts will be suggested.

Vowel length alternations (see section 3.1 for data and preliminary discussion) will be treated as lengthening/shortening processes in section 6.2. The processes analysed there will include low vowel lengthening (6.2.1) and stem vowel shortening (6.2.2). A number of surface lengthening and shortening processes characteristic of fast and/or casual speech will be discussed in sections 9.1 and 9.2 below.

## 6.1. VOWEL HARMONY

In what follows, an analysis of Hungarian vowel harmony will be presented, couched in terms of the Clements/Hume feature system (see Clements and Hume 1995). The underlying representation of stem vowels (in general) will be as in (1) (see the introductory paragraphs of Chapter 3 for discussion):

| (1) | | i | ü | u | ö | o | e | a |
|---|---|---|---|---|---|---|---|---|
| | COR | ● | | | | | ● | |
| | LAB | | ● | | ● | | | |
| | DOR | | | ● | | ● | | ● |
| | $open_1$ | − | − | − | − | − | + | + |
| | $open_2$ | − | − | − | + | + | + | + |

Depending on stem class, the place features indicated in (1) are either prelinked to one or several vowels, or are underlyingly represented as floating (unassociated) features. In general, we will assume that a place feature is

---

[1] With respect to processes that *are* conditioned by syllable structure, see Chapter 8 below.

floating (assigned to the whole morpheme rather than anchored to a specific vowel) unless there is positive evidence to the contrary: that is, place features will be prelinked only in cases where this is necessary in order to account for their behaviour (specifically, to encode the exceptional harmonic character of the stems concerned).

Suffix vowels will be represented as in (2):[2]

(2)

| | non-alternating | | | alternating | | | | |
|---|---|---|---|---|---|---|---|---|
| | i | e: | o | ü/u | ö/o | e/a | ö/o/e | ö/o/e/a |
| COR | ● | ● | | | | | | |
| LAB | | | | ● | ● | | | |
| DOR | | | ● | | | | | |
| open₁ | − | − | − | − | − | + | − | |
| open₂ | − | + | + | − | + | + | + | + |

We will assume the following rules in our account of the harmonic alternations:[3]

(3)   a. *Link Place*    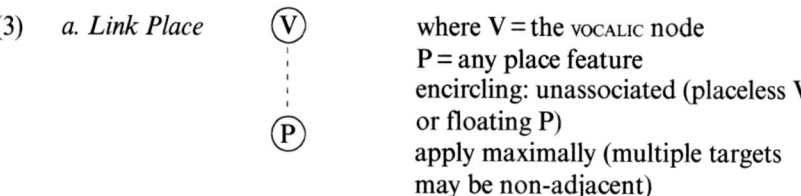    where V = the VOCALIC node
                                                  P = any place feature
                                                  encircling: unassociated (placeless V
                                                  or floating P)
                                                  apply maximally (multiple targets
                                                  may be non-adjacent)

---

[2] The columns in (2) stand for the following classes of suffixes: non-alternating suffixes involving /I/ or /i:/, e.g. infinitival -*ni*, derivational -*it*, cf. (29*a*) in section 3.2.1; non-alternating suffixes involving /e:/, e.g. causal -*ért*, cf. (29*b*) in the same section; non-alternating suffixes involving /o/, e.g. temporal -*kor*, derivational -*ol*, cf. (29*c*) in the same section; alternating suffixes involving /ü ~ u/ or /ü: ~ u:/, e.g. 1pl -*ünk/unk*, adjective forming -*ű/ú*; alternating suffixes involving /ö ~ o/ or /ö: ~ o:/, e.g. derivational -*nök/nok*, ablative -*től/tól*; alternating suffixes involving /e ~ a/ or /e: ~ a:/, e.g. inessive -*ben/ban*, derivational -*ség/ság*; three-way alternating suffixes involving /ö ~ o ~ e/, e.g. allative -*höz/hoz/hez*; four-way alternating suffixes involving /ö ~ o ~ e ~ a/, e.g. plural -*ök/ok/ek/ak*, accusative -*öt/ot/et/at* (on the underlying difference between these two types of suffixes cf. section 8.1.4.3). Note further that the front member of *é/á* alternations (= [e:]) will emerge from the phonology as [COR, +open₁] but will be phonetically implemented as 'front *mid* unrounded' (just like non-alternating *é* that is [COR, +open₂]); the front unrounded member of three-way alternations (= [ɛ]) will emerge as [COR, −open₁] but will be implemented as 'front *low* unrounded'; furthermore, short *a* will be phonetically slightly rounded without ever acquiring the feature LAB, whereas long *á* will be interpreted as 'lowest central unrounded' even though it will be an exact counterpart of *a* as far as phonological features are concerned.

[3] All vowel harmony rules apply in Block 2 (cf. section 1.4). This means that (i) analytic suffixes undergo it just like synthetic suffixes do and (ii) all empty material (including defective vowels and bare X slots but—crucially—*not* including floating features or incompletely specified segments) has disappeared from the representations by this point. The domain of application of vowel harmony is the phonological word. In other words, it applies morpheme internally, across a synthetic suffix boundary, and across an analytic suffix boundary (... ] ...) but not across compound boundaries or word boundaries, i.e. across boundaries of *independent* analytic domains (... ][ ...).

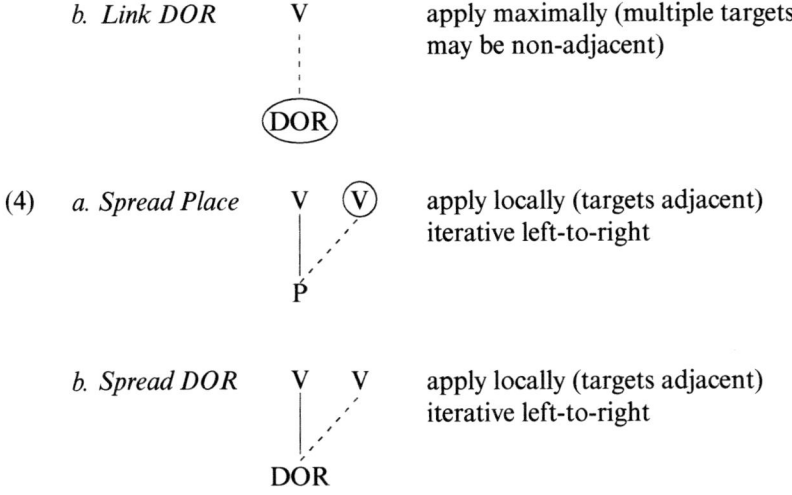

*b. Link DOR*      V          apply maximally (multiple targets
may be non-adjacent)

(4)    *a. Spread Place*    apply locally (targets adjacent)
iterative left-to-right

*b. Spread DOR*     apply locally (targets adjacent)
iterative left-to-right

With respect to the order of application of these rules, let us assume that link-ing rules (3) precede spreading rules (4) but, within each pair, the more spe-cific DOR rule (*b*) takes applicational precedence over the more general 'Place' rule (*a*).

If, after the application of rules (3)–(4), some V nodes are still empty (i.e. dominate no place feature), they will be assigned a COR place feature by the following default operation since placeless vowels are not well-formed at the surface:

(5) *Default COR*

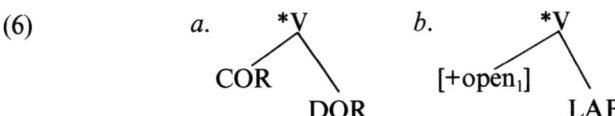

The rules apply as stated, subject to the following general constraints: no vowel may be specified as both COR and DOR (the front and the back of the tongue cannot be simultaneously involved in a Hungarian vowel); and no vowel may be specified as both [+open$_1$] and LAB (no low front rounded vowels in Hungarian):

(6)       *a.*      *V        b.*      *V
         COR          [+open$_1$]
              DOR            LAB

Although the above set of rules and constraints may appear to be overly complicated,[4] it is, in fact, just the outcome of the complex interplay of relatively straightforward principles, some of which may turn out to be universal, whereas others represent an irreducible minimum of information required to account for the asymmetries attested in the system. The underlying principles can be roughly summarized as in (7):

(7) *a.* Linking (i.e. association of a floating feature) is unbounded within a domain.
   *b.* Spreading (i.e. additional association of an anchored feature) is strictly local.
   *c.* Any place feature will link/spread to empty (unassociated) V nodes.
   *d.* DOR will link/spread to any V node.
   *e.* No V node surfaces without a place feature.
   *f.* No V node can accommodate both COR and DOR.
   *g.* No V node can accommodate both [+open$_1$] and LAB.

Let us see how the system works. First, the preliminary classification of stem types given in Table 11 (section 3.2.3.2) will be slightly revised along the following lines:

Simple harmonic stems (IA) will be subdivided into 'pure DOR' and 'COR + DOR' (corresponding to IA–b) on the one hand and 'pure LAB' and 'COR + LAB' (corresponding to IA–f) on the other. Complex harmonic stems (IB) will be labelled 'LAB + DOR' (IB–b) or 'DOR + LAB' (IB–f). Stems of the *kódex* type will be referred to as 'DOR + COR (opaque)' (in Table 11, these were subsumed under IB–f). The class of simple neutral stems (IIA) includes 'pure COR' (IIA–f) and 'DOR + COR (antiharmonic)' (IIA–b). Finally, complex neutral stems are either 'LAB + COR' (IIB–f) or 'DOR + COR (transparent)' (IIB–b).

Table 18 summarizes these correspondences and indicates the mechanism that we will assume to be at work in each case.

Consider the simplest cases first. Pure DOR stems like *ház* 'house', *kalap* 'hat', *koszorú* 'wreath' will be assumed to have a single floating DOR feature that is linked to all their vowels by Link DOR (3*b*). Similarly, pure LAB stems like *tűz* 'fire', *öröm* 'joy', *köszörű* 'grinder' have a single floating LAB; pure COR stems like *víz* 'water', *szegény* 'poor', *rekettye* 'gorse' have a single floating COR. In both cases, Link Place (3*a*) applies to all vowels in the stem:

(8)       *a.* koszorú          *b.* köszörű          *c.* rekettye

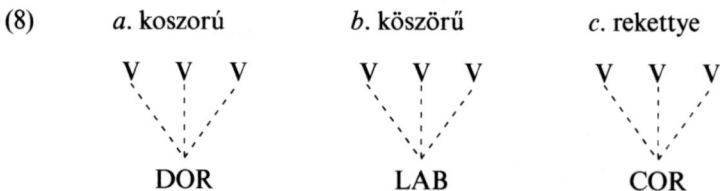

          DOR                   LAB                   COR

---

[4] For instance, the sets of potential inputs to both DOR rules partially overlap with those of potential inputs to the respective 'Place' rules. An example of why this is, in fact, desirable will be discussed in footnote 8.

TABLE 18. *Revised classification of stems and mechanisms of vowel harmony*

| Classification | Representation | Mechanism |
|---|---|---|
| *Simple harmonic* | | |
| IA ⎯ **IA–b** ⟨ | pure DOR: *koszorú* (8*a*) | Link DOR (3*b*), see *ház* (13) |
| | COR + DOR: *telefon* (9*b*) | Link DOR (3*b*), see *piros* (16) |
| **IA–f** ⟨ | pure LAB: *köszörű* (8*b*) | Link Place (3*a*) & Default COR, see *tűz* (14) |
| | COR + LAB: *szemölcs* (11*a*) | Link Place (3*a*), see (19) |
| *Complex harmonic* | | |
| IB ⎯ **IB–b** ⎯ | LAB + DOR: *nüansz* (10*a*) | Spread DOR (4*b*), see (17) |
| **IB–f** ⟨ | DOR + LAB: *sofőr* (11*b*) | Spread Place (4*a*), see (20) |
| | DOR + COR (opaque): *kódex* (12*c*) | Spread Place (4*a*), see (23) |
| *Simple neutral* | | |
| IIA ⎯ **IIA–b** ⎯ | DOR + COR (antiharmonic): *híd* (12*b*) | Link DOR (3*b*), see (22) |
| **IIA–f** ⎯ | pure COR: *rekettye* (8*c*) | Link Place (3*a*), see *víz* (15) |
| *Complex neutral* | | |
| IIB ⎯ **IIB–b** ⎯ | DOR + COR (transparent): *papír* (12*a*) | Link DOR (3*b*), see (21) |
| **IIB–f** ⎯ | LAB + COR: *öreg* (10*b*) | Spread Place (4*a*), see (18) |

COR + DOR stems like *piros* 'red', *beton* 'concrete', *telefon* 'telephone' will be analysed as having a linked COR followed by a floating DOR. Wherever a sequence of coronal vowels is involved before the first dorsal vowel within the stem (as in *telefon*), all coronal vowels are prelinked to the same COR.[5]

---

[5] Alternatively, we could assume that in this class both place features are underlyingly linked to the respective vowels as in (i); we could further assume that the middle vowel in *telefon* receives its COR specification by Spread Place (4*a*) as in (ii). However, this alternative turns out to be wrong if we consider forms like *pirosító* 'rouge' [red-causative-nominalizing suffix] where the DOR has to skip the *-i-* and link onto the *-ó* as in (iii). This is only possible if DOR is lexically unlinked (note that all three place features define separate planes, hence the no-crossing constraint does not prevent DOR from linking to the last V in (iii)). Once this is the case, however, the middle vowel in *telefon* has to be prelinked to COR, otherwise the floating

(9)            *a.* piros             *b.* telefon

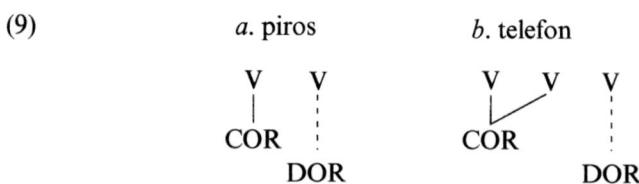

In the representation of LAB + DOR stems like *nüansz* 'nuance' and of LAB + COR stems like *öreg* 'old', both place features are prelinked (in order to prevent various additional stem-internal associations, some of which would be just superfluous, but others would yield wrong outputs like *\*nuansz*, as well as to prevent LAB from linking onto suffix vowels, again just superfluous in some cases but definitely wrong in e.g. *\*öregök* 'old ones').

(10)           *a.* nüansz         *b.* öreg

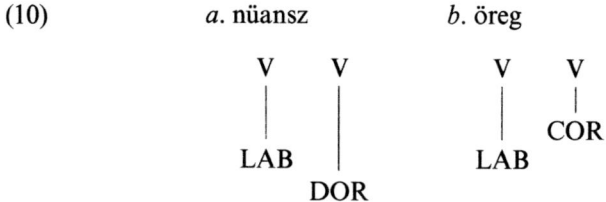

However, COR + LAB stems like *szemölcs* 'wart' differ from those in (10) in that their COR is unlinked (so that it can skip the labial vowel and link up to subsequent suffix vowels, if any).[6] DOR + LAB stems like *sofőr* 'driver' are

---

DOR would wrongly link to it (assuming that Link Place takes precedence over Spread Place), as in (iv).

(i)   piros               (ii)   telefon             (iii) pirosító

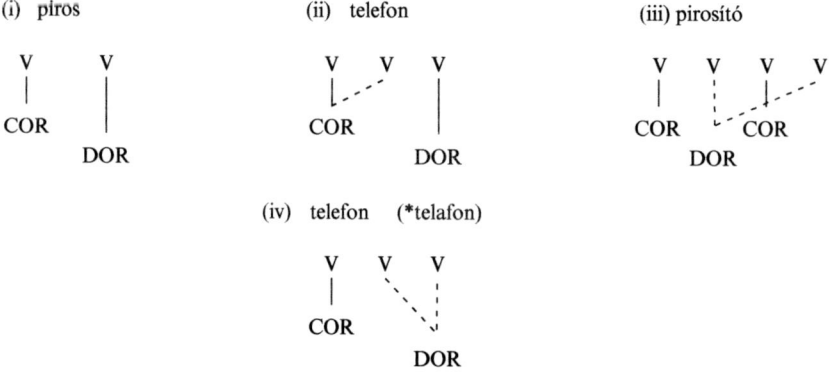

                   (iv)   telefon    (\*telafon)

[6] Alternatively, we could assume that COR is linked in such stems and the COR of subsequent suffix vowels, where needed, is inserted by Default COR (5). However, given our general assumption that place features are floating in the unmarked case dictates the solution suggested in the text. Also, this solution minimizes the use of default operations, which is yet another desirable consequence.

quite exceptional in that their labial vowel is also specified as COR. This has to be the case since otherwise the DOR feature would spread (by rule (4b)) onto the labial vowel; once the vowel in question is also specified as COR, constraint (6a) precludes this possibility.

(11)  *a.* szemölcs    *b.* sofőr

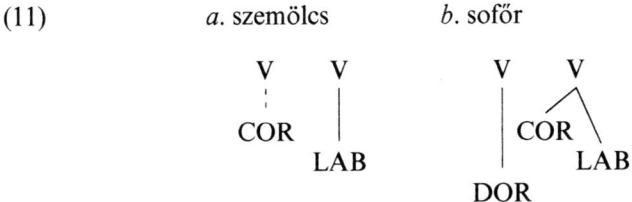

Finally, DOR + COR stems come in three subtypes. The regular case is where the coronal vowel is transparent as in *papír* 'paper', *tányér* 'plate'. Such stems are analysed with an unlinked DOR and a linked COR, somewhat like COR + LAB stems as in (11a). The antiharmonic class (*híd* 'bridge') is similar, except that there is no placeless vowel within the stem for DOR to link up to; it will solely link to the suffix vowel, if there is one (see further below). On the other hand, the opaque type (*kódex* 'codex') differs from the others in that both DOR and COR are linked in it; given that Spread Place can only target a placeless vowel, it cannot apply in a form like this; Spread DOR could in principle apply, were it not for constraint (7a).

(12)  *a.* papír         *b.* híd         *c.* kódex

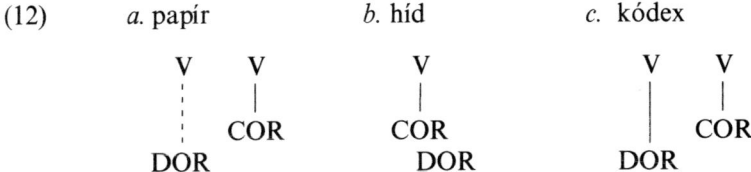

Let us now turn to the crux of the matter: How do all the above assumptions interact and account for the facts of suffix harmony?[7]

In the pure DOR class, whatever type of suffix is added, DOR gets associated to its vowel by Link DOR (3b).[8] Consider the following examples:[9]

---

[7] The issue of lowering stems (cf. 3.1.2.1 and 8.1.3) will be briefly mentioned later in this section and discussed extensively in section 8.1.4.3; four-way suffixes will be ignored in the analysis to follow.

[8] This is where the overlap between Link Place and Link DOR mentioned in footnote 4 comes in handy. If Link DOR were to be formulated so that it only applied to cases unaffected by Link Place (e.g. by restricting it to targets containing a V-place node), a form like *kalapunknak* 'for our hat' would require two separate rule applications: all three *a*'s would be affected by Link Place but the backness of the *u* of possessive -*unk*- would have to be based on Spread DOR. By contrast, with the rules as they are, the floating DOR will be linked to all vowels in *kalapunknak* by Link DOR, irrespective of whether they do or do not have V-place nodes. Whereas in *kalapunknak* an analysis with Link Place cum Spread DOR would only be cumbersome but not impossible, in a form like *házaimtól* 'from my houses' this version would break down altogether. Link Place could only associate DOR to the first two vowels (as the other two already have place

(13)        *a*. ház-unk 'our house'        *b*. ház-tól 'from (the) house'

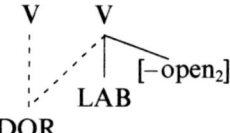    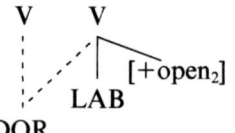

*c*. ház-nak 'for (the) house'    *d*. ház-hoz 'to (the) house'

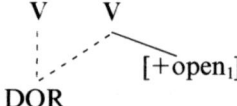    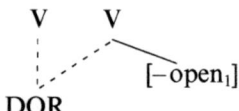

In the pure LAB class, Link Place (3a) applies to all vowels (not underlyingly linked to LAB) except that of *ale* suffixes where constraint (6b) prevents this (see (14c)). Since the suffix vowel in (14c) is still placeless, Default COR (5) subsequently applies to it, deriving *-nek*.

(14)        *a*. tüz-ünk 'our fire'        *b*. tűz-től 'from (the) fire'

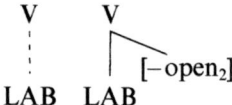    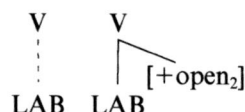

*c*. tűz-nek 'for (the) fire'    *d*. tűz-höz 'to (the) fire'

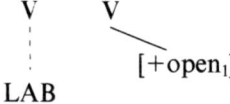    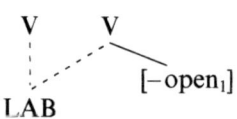

In the pure COR class, stem vowels and placeless suffix vowels get linked to COR by Link Place (3a); LAB suffix vowels are not affected (they surface as front rounded without ever being assigned COR).

features: a COR and a LAB, respectively) whereas Spread DOR could not apply at all. The *i* being linked to COR, it cannot additionally accommodate DOR (cf. (6a)); and the vowel of the case ending is not adjacent to the last dorsal vowel to its left, hence spreading (assumed to be a local operation) is impossible. The result, incorrectly, would be *házaimtól*. On the other hand, Link DOR as in (3b) correctly associates DOR with the appropriate vowels (skipping the *-i-*). The examples to follow will be simpler than these; but the system obviously has to work for plurisyllables and multiply suffixed forms as well.

⁹ For clarity, aperture features of stem vowels will continue to be omitted in the displays. For suffix vowels, relevant (i.e. non-predictable) aperture features will, however, be indicated. Note that V stands for the vocalic node; vowel length is represented on the skeletal tier, not displayed in the examples. The V-place and Aperture nodes are also suppressed for simplicity.

(15)     *a.* viz-ünk 'our water'        *b.* víz-től 'from (the) water'

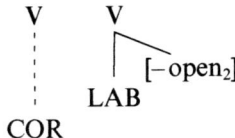    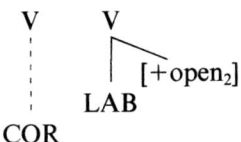

*c.* víz-nek 'for (the) water'    *d.* víz- hez 'to (the) water'

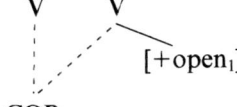    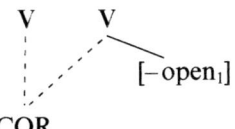

Turning now to the more complex cases, consider COR + DOR stems like *piros* 'red' (16). The rule that applies is Link DOR (3*b*). In the third and fourth cases, we could assume that Link Place (3*a*) applies, but the more specific Link DOR has applicational precedence and gives exactly the same result as Link Place would.

(16)    *a.* piros-unk 'our red one'        *b.* piros-tól 'from (the) red one'

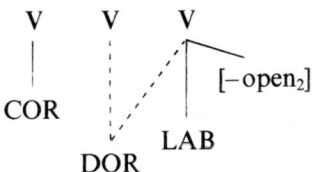    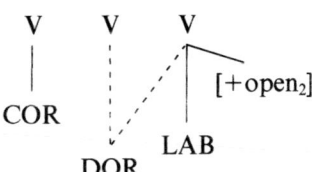

*c.* piros-nak 'for (the) red one'    *d.* piros-hoz 'to (the) red one'

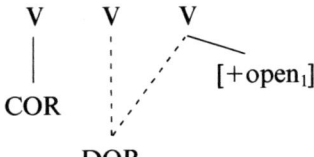    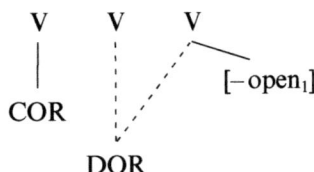

When suffixes are added to LAB + DOR stems like *nüansz* 'nuance', Spread DOR (4*b*) applies to all four types of suffixes:

(17)  *a.* nüansz-unk 'our nuance'          *b.* nüansz-tól 'from (the) nuance'

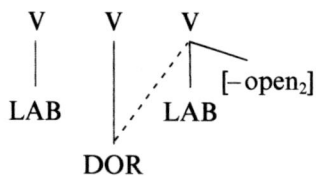
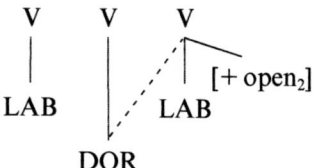

*c.* nüansz-nak 'for (the) nuance'    *d.* nüansz-hoz 'to (the) nuance'

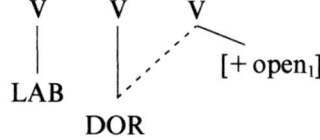
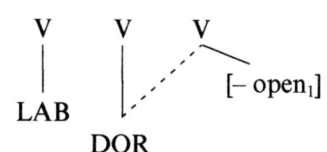

The case of LAB + COR stems like *öreg* 'old' is slightly different. Since COR cannot (and, given our assumptions about phonetic interpretation, need not) spread to a V that is already LAB, no rule applies in the first two cases. In the third and fourth cases COR is free to spread by Spread Place (4*a*).

(18)  *a.* öreg-ünk 'our old one'          *b.* öreg-től 'from (the) old one'

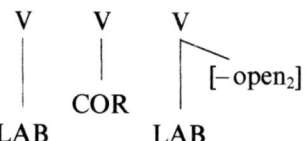

*c.* öreg-nek 'for (the) old one'    *d.* öreg-hez 'to (the) old one'

In cases involving COR + LAB stems like *szemölcs* 'wart', the COR of the stem links to its own V, as well as to placeless suffix vowels by Link Place. The LAB of the stem only spreads in (19*d*); in (19*a, b*) the suffixes are already LAB, whereas in (19*c*) LAB cannot spread due to constraint (6*b*).

(19)  *a.* szemölcs-ünk 'our wart'          *b.* szemölcs-től 'from (the) wart'

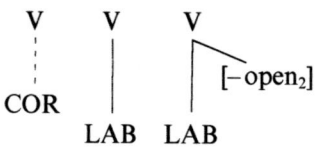
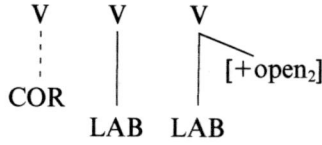

*c.* szemölcs-nek 'for (the) wart'    *d.* szemölcs-höz 'to (the) wart'

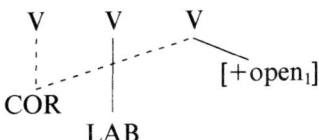    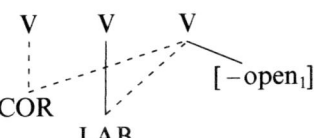

As was pointed out above, the front rounded vowel of DOR + LAB stems like *sofőr* 'driver' is exceptionally specified as [COR, LAB]. This prevents the DOR feature from spreading (by rule (4b)) onto this vowel. The details of suffixation are similar to the previous case: in (20a, b) nothing happens; in (20c) COR spreads; in (20d) both COR and LAB spread. True, we would get the same result if we assumed that only LAB spreads here; but we would need a separate *ad hoc* stipulation to prevent COR from spreading—paradoxically, two spreadings are simpler than one in this case. Similarly, in (19d), COR links vacuously, as it were, to the suffix vowel; but there is no pressing need to prevent this.

(20)   *a.* sofőr-ünk 'our driver'    *b.* sofőr-től 'from (the) driver'

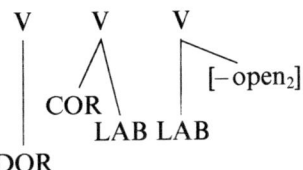    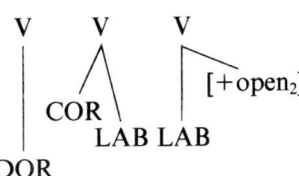

*c.* sofőr-nek 'for (the) driver'    *d.* sofőr-höz 'to (the) driver'

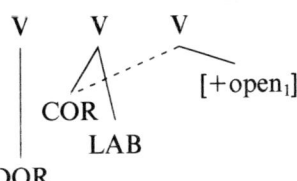    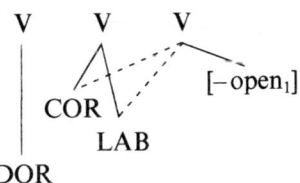

In DOR + COR (transparent) cases like *papír* 'paper' Link DOR (3b) is involved with LAB suffixes; either Link Place (3a) or Link DOR (3b) would give the right result with placeless suffixes (as before, we assume that the more specific Link DOR is at work, but this has no empirical consequence):

(21)   *a.* papír-unk 'our paper'        *b.* papír-tól 'from (the) paper'

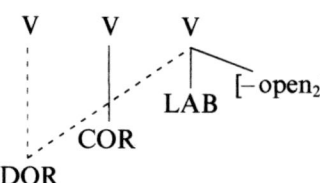 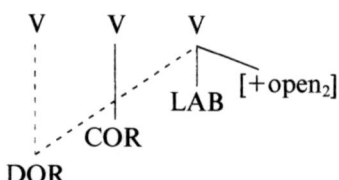

     *c.* papír-nak 'for (the) paper'     *d.* papír-hoz 'to (the) paper'

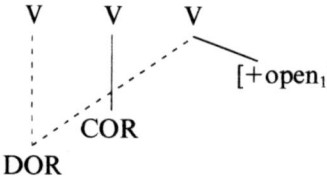

With antiharmonic DOR + COR items, the analysis is the same except that DOR has nowhere to link up within the stem:

(22)     *a.*    hid-unk 'our bridge'        *b.*    híd-tól 'from (the) bridge'

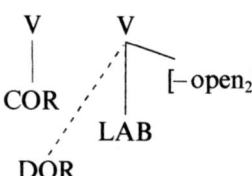 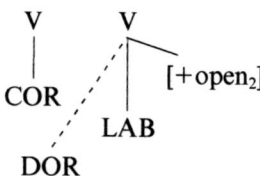

     *c.*    híd-nak 'for (the) bridge'     *d.*    híd-hoz 'to (the) bridge'

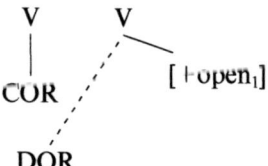 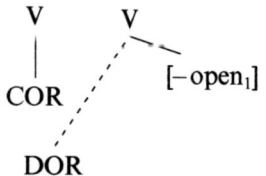

Finally, DOR + COR (opaque) stems like *kódex* 'codex' behave exactly like LAB + COR stems: stem vowels and placeless suffix vowels get linked to COR by Spread Place (4*a*); LAB suffix vowels are not affected.

(23)     *a.*    kódex-ünk 'our codex'       *b.* kódex-től 'from (the) codex'

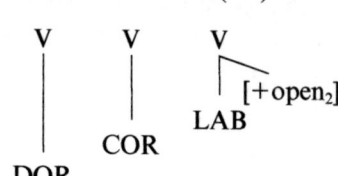

*c.* kódex-nek 'for (the) codex'     *d.* kódex-hez 'to (the) codex'

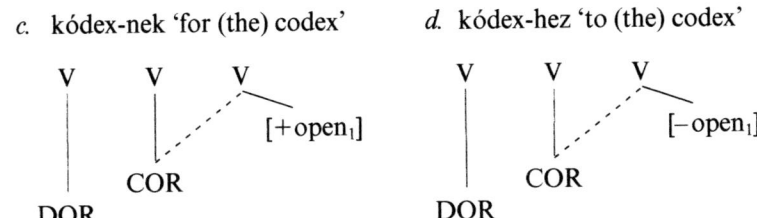

Vacillating stems like *dzsungel* 'jungle' have two underlying representations: one like *papír* and one like *kódex*. Their suffixation goes as in (21) and as in (23), respectively.

Four-way alternating suffixes work exactly like three-way suffixes for non-lowering stems. Lowering stems (cf. section 8.1.3) are special in that they have a floating [+open₁] that links up to the unstable vowel at the stem/suffix boundary if it is unspecified for this feature (this is the case with four-way alternating suffixes; the rule that describes this will be discussed and formulated in 8.1.4.3). The following derivations show the plural accusatives of lowering (24a, b, c) and non-lowering (24d, e, f) stems, respectively (V-place and Aperture nodes are, again, suppressed; OP abbreviates [+open₁]). As we will see in section 8.1.3, the plural morpheme has an OP of its own and contributes this to the subsequent vowel irrespective of whether the stem itself is lowering or not. Note that Lowering must precede Link Place to block the application of the latter to the suffix vowel(s) as (24c) shows.

(24) *a.* há: ház-ak-at     *b.* kez-ek-et     *c.* fül-ek-et
      'houses' (acc.)       'hands' (acc.)      'ears' (acc.)

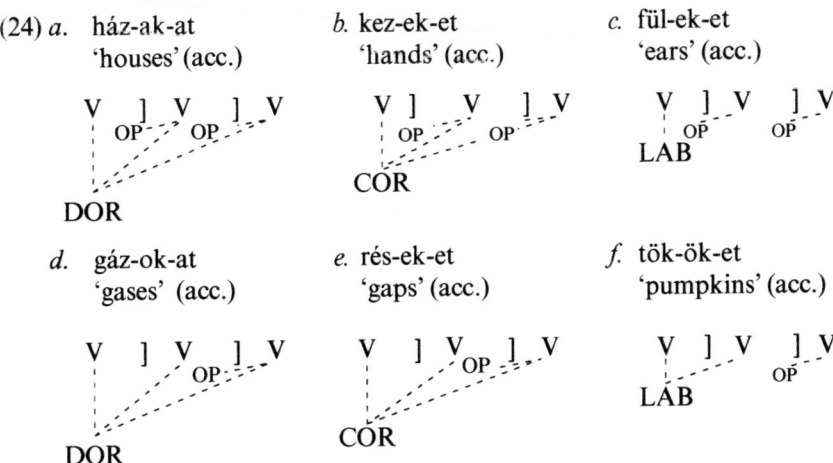

*d.* gáz-ok-at     *e.* rés-ek-et     *f.* tök-ök-et
    'gases' (acc.)       'gaps' (acc.)      'pumpkins' (acc.)

Since in the non-lowering cases the suffix vowels are still not fully specified for aperture, another default rule is required that assigns [–open₁] to them, yielding mid *o/ö*. Recall that [COR,–open₁] is phonetically implemented as 'front *low* unrounded' (= [ɛ]) as usual. Suffix vowels that still lack place features at this point additionally undergo Default COR (5) that specifies -*eket* in (24c) and -*et* in (24f) as coronal.

In this section, we have outlined a novel analysis of Hungarian vowel harmony. As opposed to earlier accounts, in this framework all three place features spread. That is, spreading is not restricted to [+back] (or, the property of backness, roughly corresponding to our DOR) or to the element **I** (that roughly corresponds to COR and the frontness implied by LAB in the present framework). Although such abundance of objects to spread may appear at first glance to result in an overly complicated solution, we have pointed out that this is not the case. If we consider constraints on phonological events to be the primary factors of the analysis and linking/spreading rules are conceived of as merely convenient mnemonic devices that make it easier for us to present the complex interplay of those constraints and certain universal principles, the 'permissive' use of spreading operations turns out to be the simplest account (in the sense that it requires the minimum number of language specific constraints or stipulations).


## 6.2. LENGTHENING AND SHORTENING PROCESSES

As was pointed out in section 3.1.2, vowel length alternations in Hungarian are governed by two types of regularities: Low Vowel Lengthening (LVL) and Stem Vowel Shortening (SVS). The data pertaining to these types of alternations were presented in a static manner (simply as alternation patterns) in 3.1.1 and 3.1.2, respectively. In the following two subsections we will give a dynamic (process-oriented) account of the facts.


### 6.2.1. Low vowel lengthening

Unlike the occurrence of final high and mid vowels, that of word final low vowels is phonologically unrestricted in terms of vowel length (even if long *á* occurs with certain limitations, see section 5.4.1 and footnote 9 in section 3.1.1). However, morpheme final low vowels are invariably long before a suffix. In other words, underlyingly short final low vowels get lengthened if a suffix is added. A sizeable set of examples was provided in (12) of section 3.1.1; for convenience, a few examples will be repeated here:

| (25) | /a/ → /aː/ | alma | 'apple' | almát | 'apple' (acc.) |
|------|-----------|-------|---------|--------|----------------|
|      |           | tartja | 'he holds it' | tartják | 'they hold it' |
|      |           | kutya | 'dog' | kutyául | 'like a dog' |
|      | /e/ → /eː/ | epe | 'bile' | epés | 'bilious' |
|      |           | vitte | 'he carried it' | vitték | 'they carried it' |
|      |           | este | 'evening' | estére | 'by evening' |

This lengthening applies irrespective of whether the base form is morphologically simple or complex, i.e. whether the final low vowel is part of the stem (*alma*) or of some suffix (*tart-ja*). It applies indiscriminately before synthetic suffixes (*almá-t, epé-s*) and before analytic ones (*kutyá-ul, esté-re*): in other words, it applies both in Block 1 and in Block 2; cf. section 1.3. It is also insensitive to what type of segment the following suffix begins with. Furthermore, each word form will be input to low vowel lengthening (LVL) as many times as it contains the appropriate input configuration:

(26)     óra 'watch' → órája 'his watch' → óráját 'his watch' (acc.)
         mese 'tale' → meséje 'his tale' → meséjét 'his tale' (acc.)

Other vowel heights are not affected by a parallel lengthening rule. Mid vowels are always long morpheme-finally (cf. section 5.4.1 and footnote 3 in Chapter 3), whereas the length of final high vowels depends on the length (number of syllables) of the word and/or its word class (content word vs. function word, cf. footnote 2 in the same chapter), but if they are short, they remain short before a suffix, too (e.g. *házi-as* 'house-proud', *eskü-vel* 'with an oath', *kapu-ig* 'as far as the gate').
    Our first approximation to the rule is (27).

(27)     X              X X
         |              \ /
       [–cons]    →    [–cons]    / __ ] Y
         |              |
       [+open₁]        [+open₁]

where ] = morpheme boundary, Y = the first segment of a suffix

Final low vowels that are underlyingly long obviously do not undergo any change since they do not satisfy the structural description of this rule (and they conform to the required output configuration as they stand): *kordé* 'cart'– *kordét* (acc.), *burzsoá* 'bourgeois'– *burzsoát* (acc.).
    The rule can be technically further improved by claiming that it actually involves the insertion of an empty X slot before/after a morpheme final low vowel; the vowel melody would then spread onto this empty slot by convention.[10] (28a) shows one of the possible formulations of this process;

[10] Following and elaborating an idea originally proposed by Hetzron (1972), Jensen and Stong-Jensen (1989a) provide an analysis of the same facts that involves shortening in unsuffixed forms, rather than lengthening in suffixed ones. Their analysis is based on the claim that surface *a/e*-final lexical items include an underying empty X-slot as opposed to invariable *á/é*-final ones that, quite exceptionally in their terms, contain vowel melodies underlyingly associated to two Xs. This further implies that simple low vowels (i.e. ones in which just one timing unit is involved) are banned from lexical representations morpheme finally. For a detailed criticism of that analysis, cf. Siptár (1998a); see section 5.4.1 (especially the paragraphs surrounding (64)) for some discussion of the issues involved.

(28*b*) illustrates the convention whereby the vowel melody spreads onto the empty skeletal slot inserted by (28*a*):

(28)    *a.* $\emptyset \rightarrow$ X / X ____ ] Y          *b.*      X      X
                    |                                       |‚‚‚‚‚‚‚´
                [–cons]                                  [–cons]
                    |
                [+open$_1$]

As was pointed out in section 3.1.1, there are apparent counterexamples to the generalization stated in (28) where something is added to a low-vowel-final lexical item and the vowel remains short. The examples given there are repeated here for convenience (a vertical line is inserted into the examples to help the reader identify the site where LVL fails to occur).

(29)  *a.*  balta | nyél 'hatchet handle', kefe | kötő 'brush-maker';
             haza | megy 'go home', bele | lép 'step into it'
       *b.*  kutya | szerű 'dog-like', mese | szerű 'like a fairy tale';
             macska | féle 'felid', medve | féle 'resembling a bear'
       *c.*  távozta | kor 'on his departure', megérkezte | kor 'on his arrival';
             példa | képp(en) 'for instance', mérce | képp(en) 'as a measure';
             torta | ként 'as a cake', sörte | ként 'as bristles'
       *d.*  haza | i 'domestic', megye | i 'county' (adj.)
       *e.*  katona | ság 'army', fekete | ség 'blackness'

In (29*a*) we find compounds (incuding preverb + verb combinations); in (29*b*) the elements -*szerű* and -*féle* are enclitics (intermediate items between a full compound member and a derivational suffix). We can safely assume that these forms consist of two phonological words each (i.e. they contain two independent analytic domains: ⟦⟦ ... ⟧ ⟦ ... ⟧⟧) and that LVL, just like vowel harmony, applies within (but not across) phonological words. The suffixes in (29*c–d*) are unary suffixes with respect to vowel harmony. One way to account for this would be to claim that they are outside the harmonic domain, i.e. the phonological word. In other words, we could say that all these suffixes start a new independent analytic domain like the items in (29*a–b*).

However, vowel harmony and LVL do not always go hand in hand. On the one hand, the multiplicative suffix -*szor/szer/ször* 'times' does harmonize but does not trigger LVL (although it cannot co-occur with many low-vowel-final stems, perhaps the only instances are names of Greek letters in mathematics, e.g. *lambdaszor* 'lambda times, multiplied by lambda', not \**lambdászor*). Similarly, the noun forming derivational suffix -*ság/ség* as in (29*e*) fails to trigger LVL while it regularly undergoes vowel harmony (and is an extremely productive suffix, attachable to any noun or adjective).

On the other hand, suffixes like terminative -*ig* 'up to, as far as' or causal -*ért* 'for' do not harmonize but lengthen the stem-final low vowel:

*hazáig* 'all the way home', *kedvéért* 'for the sake of'. Note that the lack of vowel harmony alternation need not be based on domain externality: *-ig* and *-ért* would be unary suffixes anyway as they contain a neutral vowel (see section 3.2.1). But then so do the suffixes listed in (29c–d), except for *-kor*. As we saw in section 6.1, the non-alternating harmonic behaviour of *-kor* can also be accounted for without assuming that it is outside the domain of harmony.

Harmonically non-alternating suffixes come in two types. Some of them are opaque to vowel harmony: they start a new harmonic domain. This is true of *-kor* (*távozta-kor-i-ak* 'those coinciding with his departure', *megérkezte-kor-i-ak* 'those coinciding with his arrival', *\*megérkezte-kor-i-ek*) and *-képp* (*mérce-képp-en* 'as a measure', *példa-képp-en* 'for instance', *\*példaképp-an*). Others are transparent: they let harmony pass through them as if they were not there at all. The adjective forming suffix in (29d), as well as *-ig* and *-ért*, are like this, cf. *haza-i-ak* 'those from home' (*\*hazaiek*), *holt-om-ig-lan* 'till I die' (*\*holtomiglen*). With respect to these suffixes (among them, crucially for LVL, with respect to *-i*), non-harmonizing behaviour definitely cannot be explained with reference to domain externality.

All this boils down to the following conclusion: the behaviour of the suffixes in (29c) can best be explained if we take them to be structurally similar to the cases in (29a–b). This is necessary with respect to their behaviour in terms of LVL and does not conflict with their harmonic behaviour (although it would not be strictly necessary on that count alone). However, the suffixes in (29d–e) (as well as multiplicative *-szor/szer/ször*) cannot be analysed similarly since they either undergo vowel harmony or are transparent to it. These are simply exceptions to LVL and have to be marked accordingly in the lexicon. To the best of our knowledge, the literature does not contain any feasible suggestion as to a more principled solution to this problem.

### 6.2.2. Stem vowel shortening

As we saw in section 3.1.2, in a number of stems the vowel (or one of the vowels) is shortened before certain suffixes: *kéz* 'hand'–*kezek* 'hands' (Final Stem Vowel Shortening, FSVS; see Table 9 for examples), *szintézis* 'synthesis'–*szintetikus* 'synthetic' (Internal Stem Vowel Shortening, ISVS; see Table 10).

There are various ways to account for FSVS. The simplest would be to refer the whole thing outside phonology (into morphology) and assume that arbitrary lexical diacritics are attached to both FSVS stems and FSVS suffixes (this is what traditional grammar does in effect). Then the rule would refer to these diacritics:[11]

---

[11] Note that nothing actually forces us to take the long vowel as basic; some descriptions in fact take the short form to be underlying (e.g. *nyár* 'summer' /n'ar-/), call the stems concerned 'lengthening stems', and posit a rule that does the opposite of (30). Whichever way it is done, the phenomenon is strange in that it involves shortening in open syllables or lengthening in closed syllables: just the opposite of what would be expected on cross-linguistic grounds. This might be construed as an argument for a non-phonological treatment.

(30)     $\underset{\text{[\textminus cons]}}{\overset{\text{X X}}{\vee}}$  $\longrightarrow$  $\underset{\text{[\textminus cons]}}{\overset{\text{X}}{|}}$  $/ \underline{\quad}$  $\underset{\text{[+cons]}}{\overset{\text{X}}{|}}$  $]_{\text{FSVS stem}} \ldots ]_{\text{FSVS suffix}}$

However, it is also possible (and perhaps desirable, though opinions differ) to treat the phenomenon within the phonology; i.e. to posit distinct underlying shapes for *nyár* 'summer' (shortening) and *gyár* 'factory' (non-shortening) and let the shortening effect fall out automatically.[12] The orthodox generative solution would be to posit distinct underlying segments for the two [a:]'s (and similarly for all other long vowels). This would involve excessive use of abstractness and absolute neutralization. The current alternative is a non-linear solution where the two stem types have identical underlying vowels (vowel melodies) but length is represented in two different manners. On the skeletal tier both stem types will be represented as XXXX (i.e. four timing slots), and on the melodic tiers our two examples will be /nʸar/ and /dʸar/, respectively. The difference lies in the associations. Several solutions are possible (depending on the principles of association—whether association lines in general are assumed to be underlyingly there, introduced by universal principles or by language specific rules, or any combination of these). Let us consider a relatively straightforward account here (cf. Jensen and Stong-Jensen 1989 for a slightly different proposal). In *gyár*, all skeletal slots are underlyingly associated to some melodic material (31*a*), whereas *nyár* contains an empty slot (31*b*). The melodic content of the preceding vowel may spread onto this empty X (31*c*) unless a deletion rule like (32) removes the latter before spreading had a chance to apply. The resulting representation, (31*d*), will then surface as [nʸɔr].

(31)   *a.* $\underset{\text{dʸ \ a \ r}}{\overset{\text{X X X X}}{|\ \vee\ |}}$     *b.* $\underset{\text{nʸ a \ r}}{\overset{\text{X X X X}}{|\ \ |\ \ \ |}}$     *c.* $\underset{\text{nʸ a \ \ r}}{\overset{\text{X X X X}}{|\ \ |\diagdown\ |}}$     *d.* $\underset{\text{nʸ a r}}{\overset{\text{X X X}}{|\ \ |\ \ |}}$

　　　　'factory'　　　　'summer'　　　　*nyár*　　　　*nyar-*

(32)   $\overset{\text{X}}{\textcircled{X}} \longrightarrow \emptyset\ /\ \underline{\quad}\ \underset{\text{[+cons]}}{\overset{\text{X}}{|}}\ ]\ \text{Y}\ ]$   (where Y is an FSVS suffix)

Is there a way to characterize FSVS suffixes phonologically, too? It would seem that all such suffixes have one thing in common: they are vowel-initial. If this turns out to be true, we could replace the environment of our rule, be it formulated as (30) or as (32), by (33):

---

[12] Ritter (1995) offers a solution along these lines in terms of Government Phonology.

(33)
$$/ \underline{\quad} \quad X \quad ] \quad X$$
$$\quad\quad\quad | \quad\quad\quad\quad |$$
$$\quad\quad [+\text{cons}] \quad [-\text{cons}]$$

However, a number of vowel-initial suffixes do not trigger FSVS: *nyár-on* 'in summer', *kéz-i* 'manual', *gyökér-ig* 'to the root', *víz-ért* 'for water', *kosár-ul* 'as a basket', *szamár-é* 'belonging to a donkey'. Note that most of these have 'stable' initial vowels that appear after vowel-final stems as well: *nő-i* 'feminine', *nő-ig* 'as far as the woman', *nő-ért* 'for a woman', *nő-ül* 'as a woman', *nő-é* 'belonging to a woman', cf. *nő-k* 'women', *nő-s* 'married', *nő-m* 'my wife', etc. Hence, the suffix-initial vowel mentioned in (33) should be restricted to unstable vowels. This could be done in terms of synthetic vs. analytic suffixation (cf. section 1.3). Unfortunately, two problems remain. The first concerns the suffix *-on* 'on': this suffix fails to trigger FSVS (*nyár-on* 'in summer', not *\*nyaron*) but attaches to vowel-final stems in a vowelless form: *nő-n* 'on a woman', and is a synthetic suffix in all other respects, too. The other problem is more serious: there is no single unified way in which the notion of 'unstable vowels' could be formally captured (cf. section 8.1.2.2). Hence, it is not possible to amend (33) so that it refers to all and only unstable vowels. Accordingly, the process of FSVS remains morphologically conditioned at least as far as its context is concerned, as in (32); eventually the best solution might turn out to be to return to the wholly morphological formulation given in (30).

The other type of stem vowel shortening, ISVS, may affect any syllable of the stem, and vowels of any tongue height may be equally involved (see Table 10). This type of shortening is only triggered by derivational suffixes, never by inflections. It can be treated as an extraphonological phenomenon (just like FSVS, cf. (30)), i.e. a process triggered by diacritical marking on both stems and suffixes. ISVS stems then would have to be marked by a different diacritic (say, 'ISVS'), otherwise we would get *\*aktivak*, *\*szlavos* for *aktívak* 'active-pl.', *szlávos* 'Slav-like' (an interesting case is *úr* 'gentleman' that undergoes both SVS processes: *urak* 'gentlemen', *urizál* 'play the gentleman'). However, a phonological account is also possible: *akadémia* 'academy' should be underlyingly represented as *gyár* 'factory' is in (31) whereas *szintézis* 'synthesis' would be represented as *nyár* 'summer' is. The ISVS counterpart of (32) could then refer to bisyllabic, vowel-initial, non-harmonizing suffixes. However, it might also be the case that ISVS is not a process at all, not even a morphological process, but rather just an apparent regularity: the existence in the lexicon of certain sets of lexical items that are semantically (and partly formally) related but not derived from common roots in any productive manner.

This concludes our discussion of lengthening and shortening processes, some real and some perhaps spurious, within the lexical phonology of Hungarian. A number of surface processes mainly but not exclusively characteristic of fast and/or casual speech, including compensatory lengthening and various other optional vowel shortening and lengthening processes resulting in surface vacillation will be discussed separately in Chapter 9.

# 7

# PROCESSES INVOLVING CONSONANTS

The discussion of Hungarian consonants in Chapter 4 was concluded by a summary of the underlying representations we proposed for members of the consonant system. That matrix is repeated here for convenience:

TABLE 19. *The underlying consonant system of Hungarian*

| | Labial | Dental | Palatal | Velar |
|---|---|---|---|---|
| | p b f v m | t d s z tˢ n l r | tʸ dʸ š ž č ǰ nʸ j | k g x |
| R | ● ● ● ● ● | ● ● ● ● ● ● ● ● | ● ● ● ● ● ● ● ● | ● ● ● |
| [cons] | + + + + + | + + + + + + + + | + + + + + + + + | + + |
| [son] | – – – + | – – – – – + + + | – – – – – – + + | – – – |
| [nas] | + | + | + | |
| [lat] | | + | | |
| [cont] | – – + + | – – + + ± + + | – – + + ± ± + | – – + |
| L | ● ● | ● ● | ● ● ● | ● |
| LAB | ● ● ● ● ● | | | |
| COR | | ● ● ● ● ● ● | ● ● ● ● ● ● ● | |
| [ant] | | + + + + + + + | – – – – – – – – | |
| DOR | | | | ● ● ● |

This chapter will be organized as follows. In section 7.1, palatalization rules will be discussed, both lexical and postlexical. In 7.2, various sibilant rules will be presented, including those involved in the imperative of *t*-final verbs, as well as postlexical rules targeting and/or triggered by strident fricatives and affricates. In 7.3, postlexical processes connected with the laryngeal properties (in particular, voicing) of obstruents will be analysed in terms of a single-valued [voice] feature. Finally, in 7.4, we turn to sonorants and briefly discuss processes involving nasals and liquids.

## 7.1. PALATALIZATION

In this book, the term 'palatalization' will refer to a phonological process whereby a palatal consonant, /j/, /tʲ/, /dʲ/, or /nʲ/, affects a preceding (dental) consonant, making it palatal. In particular, we will exclude the fully automatic, low-level, non-neutralizing—and probably non-language-specific—type of 'phonetic palatalization' that is triggered by non-low front vowels and /j/ and that produces more or less palatalized velars/dentals/labials as e.g. in *kín* 'torture' vs. *kár* 'damage' vs. (labialized) *kút* 'well'. This co-articulatory process will be ignored here.

Hungarian phonology has palatalization processes of two different types: lexical palatalization as in *látja* [laːtʲːɔ] 'see' (3sg def.), *hordja* [hordʲɔ] 'carry' (3sg def.), *fonja* [fonʲːɔ] 'braid' (3sg def.), *falja* [fɔjːɔ] 'devour' (3sg def.), and postlexical palatalization as in *átjáró* [aːtʲjaːroː] 'passage', *védjegy* [veːdʲjɛdʲ] 'trade mark', *van joga* [vɔnʲjogɔ] 'he's got the right (to)', *feljön* [fɛjːön] 'comes up'.[1] The most important differences between the two types are that postlexical palatalization is optional, its output is not fused with the trigger (cf. *átjáró* [aːtjaːroː] ~ [aːtʲjaːroː], *[aːtʲːaːroː]) and that the trigger may be a palatal stop or nasal as well as /j/: *két nyúl* [tʲnʲ] 'two rabbits', *van gyufa* [nʲdʲ] 'we've got matches'.

### 7.1.1. Lexical palatalization

The classical generative analysis (Vago 1980a) of this process (as well as its autosegmental reanalysis, Siptár 1994a) involved a wider range of input segments than will turn out to be strictly necessary: in those accounts, the input to lexical palatalization was defined as dental /t d n l/ and—partly vacuously—palatal /tʲ dʲ nʲ/:

(1) | látja | [tʲː] | 'see' (3sg indicative def.) |
|---|---|---|
| | adja | [dʲː] | 'give' (3sg indicative/imperative def.) |
| | kenje | [nʲː] | 'smear' (3sg imperative def.) |
| | falja | [jː] | 'devour' (3sg indicative/imperative def.) |
| | bátyja | [tʲː] | 'his brother' |
| | hagyja | [dʲː] | 'leave' (3sg indicative/imperative def.) |
| | hányja | [nʲː] | 'throw' (3sg indicative/imperative def.) |

In order to be able to include *l* → *j* in the same rule as the other segments, all of which are [–cont], Vago (1980a) assumed that /l/ is [–cont], too. But, as was observed by Olsson (1992), that solution was problematic with respect to the

---

[1] Siptár (1994a) discussed a third type, too: the morphophonological palatalization of /t/ in the imperative of *t*-final verbs (*ütlüss* 'hit', *öntlönts* 'pour', cf. Vago 1980a: 69–73, and see also Vago 1987, 1989, 1991, and Kontra 1992); however, that process will receive a different, more appropriate treatment in the present book (see section 7.2.1).

postlexical behaviour of *lj* sequences. We will discuss this point in section 7.1.2; for the moment let us simply go on assuming that /l/ is [+cont] and that its full assimilation to a subsequent /j/ is to be accounted for by a separate rule of *l*-Palatalization.[2] Recall that the price we had to pay for characterizing /l/ as [+cont] is that we had to introduce an extra manner feature [lat] to distinguish /l/ from /r/.[3] The rule of *l*-Palatalization can now be formulated as in (2):

(2)                      *l-Palatalization*

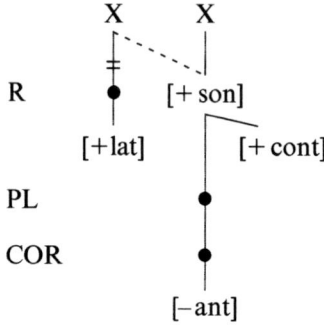

The rest of the process was described in Siptár (1994*a*) as coalescence (mutual assimilation) such that the palatality of /j/ was claimed to spread leftwards whereas all other features of the left-hand segment were to spread onto the /j/, whereby /tj/ and /tʲj/ both became [tʲ:], /dj/ and /dʲj/ both became [dʲ:], and /nj/ and /nʲj/ both became [nʲ:], in a single step. However, that solution was problematic for a number of formal reasons. (To mention just the most serious thing, it involved three spreadings and four delinkings, all in the same rule schema.) Therefore, the process will be broken down into two elementary steps here (somewhat like in the classical generative treatment of Vago 1980*a*): a place assimilation ('palatalization' in the strict sense) and a full assimilation (corresponding to Vago's Palatal Coalescence). Following a

---

[2] Vago's solution implied that the palatalization of /l/ produced a long palatal lateral *[ʎ:] which had to be 'phonetically interpreted' by a ʎ → [j] adjustment rule, e.g. /tolja/ → tolʎʎa → [tojːo] 'push' (3sg ind./imp. def.). Notice that we might wish to ascribe this adjustment to Structure Preservation: the output of the lexical palatalization of /l/ could be assumed to come out as [jː] due to that principle (given that the underlying inventory does not include /ʎ:/). However, in that case the output of the postlexical palatalization of /l/ as in *hol jártál* 'where have you been' should be *[ʎː] (actually, it is also [jː]). Thus, we would need a redundancy rule that remained operative throughout the phonology (including the postlexical component), stating that Hungarian does not have a palatal lateral: if a liquid is [–ant], it must be [–lat] (and, in Vago's system, also [–cont]): [+son, –nas, –ant] → [–lat, –cont].

[3] Alternatively, we could have introduced [trill] for /r/. A third solution is suggested by Olsson (1992). In his system, /l/ (along with /t d n/) is characterized as [+dental], whereas /r/ (along with /s z tˢ dʳ/) as [–dental] (both sets being [+ant, +cor], as well as—redundantly—[–lab, –high, –back]).

suggestion first made by Olsson (1992), we will further assume that the palatalization of /n/ into [nʲ] before /j/ is due to Nasal Place Assimilation (see section 7.4.1). The assimilation of /j/ to [nʲ]—whether underlying or derived by NPA—will, however, be effected by our version of Palatal Coalescence. Our first approximation to the two rules, then, is given in (3) and (4); we assume that after (3) has applied the COR and PL nodes—now dominating identical material—automatically merge by convention; the result of this merger can be seen in the input configuration of (4):

(3)  *Lexical Palatalization*

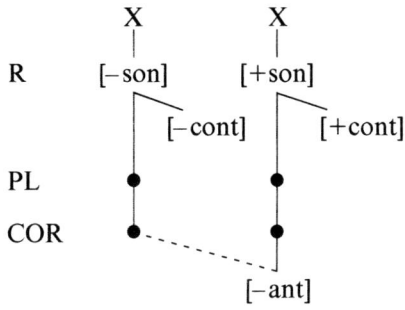

(4)  *Palatal j-Assimilation*

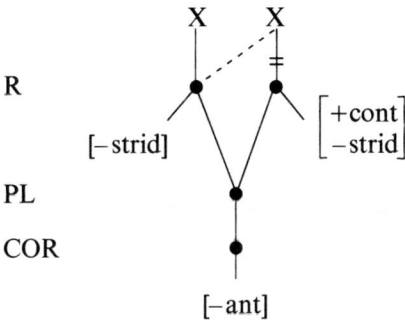

Notice that we use the feature [strid] here although it is not assumed to be an underlying feature in the Hungarian system (see section 4.3). This is simply for presentational reasons and ultimately a more appropriate solution will be proposed in section 7.2.3 below.

## 7.1.2. Postlexical palatalization

The most important difference between the above processes and surface palatalization (of /t d n/ before /tʸ dʸ nʸ jʸ/) is that the latter does not involve coalescence: thus, e.g. *mit jelent* [mitʸjɛlɛnt], *[mitʸ:ɛlɛnt] 'what does it mean', *védjegy* [ve:dʸjɛdʸ], *[ve:dʸ:ɛdʸ] 'trade mark', *van joga* [vɔnʸjogɔ], *[vɔnʸ:ogɔ] 'he's got the right (to)'.

The simplest case of surface palatalization is where both target and trigger are stops or nasals. This is where /t d n/ obligatorily turn into [tʸ dʸ nʸ] before /tʸ dʸ nʸ/. Examples: *két nyom* [tʸnʸ] 'two traces', *vadnyúl* [dʸnʸ] '(wild) hare'; the cases involving /n/ can be relegated to the realm of NPA, as before (*van gyufám* [nʸdʸ] 'I've got matches'). Since the rule is postlexical, its application is not restricted to derived environments. Hence, cases like *satnya* 'stunted' can be considered as derived by morpheme-internal applications of this rule.

Examples like *hat tyúk* [hɔtʸ:u:k] 'six hens', *mit gyártanak* [midʸ:a:rtɔnɔk] 'what do they produce' appear to contradict the claim we made above that surface palatalization does not involve coalescence. Actually, however, these are cases where the linked structures that palatalization yields happen to form geminates either immediately (*hat tyúk*) or via voice assimilation (*mit gyártanak*).

Thus, one branch of surface palatalization can be formulated as follows:

(5)          *Postlexical Palatalization – branch 1 (obligatory)*

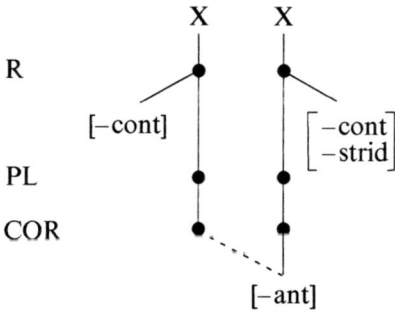

Before /j/, the surface palatalization of /t d/ is optional: e.g. *két játék* [ke:tʸja:te:k] ~ [ke:tja:te:k] 'two games', *szabad jönni* [sɔbɔdʸjön:i] ~ [sɔbɔdjön:i] 'you can come'.[4] This optional rule can be written as follows:

---

[4] Again, the cases involving /n/ are a different matter: *talán jobb* [tɔlã:job:] ~ [tɔla:nʸjob:] ~ [tɔla:njob:] 'perhaps better'. In colloquial speech, a rule of N-vocalization (e.g. *színlap* [si:lɔp] 'playbill', *tonhal* [tõ:fiɔl] 'tuna', *kénsav* [kẽ:šɔv] 'sulphuric acid'; cf. section 7.4) usually bleeds surface palatalization of *nj* sequences (technically: the place assimilation of underspecified N to *j*); in guarded speech, on the other hand, neither rule is normally applied (and N becomes [n] by default). There is, however, a narrow range of cases between [Ṽ:j] and [Vnj] realizations where forms of the [Vnʸj] type surface (via Nasal Place Assimilation).

(6)     *Postlexical Palatalization — branch 2 (optional)*

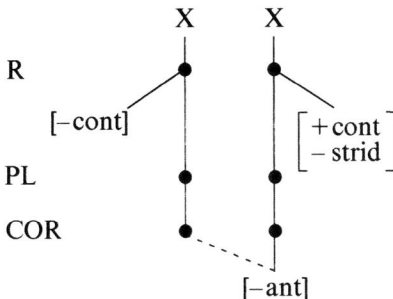

These two rules, (5) and (6), can be collapsed as follows (the parenthesized [(–cont)] is meant to suggest that, with increasing speed and/or casualness, that restriction is removed and the rule applies to /t d/ before /j/ as well as before /tʸ dʸ nʸ/):

(7)     *Postlexical Palatalization*

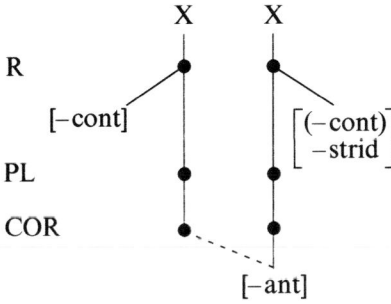

Notice that (7) can also be used in lieu of (3) for word internal (lexical) palatalization (provided we ignore the asymmetry in optionality just noted and say, for simplicity's sake, that all postlexical applications are optional): in fact, there is some evidence that the rule applies lexically also in the context of palatals other than /j/ (e.g. *pillanat* 'moment'–*pillanatnyi* [tʸnʸ] 'momentary'). Thus, a unified rule of Palatalization is proposed for both lexical and postlexical applications; it is, as expected, obligatory lexically and (more or less) optional (i.e. rate/style-dependent) postlexically (see (8)). This rule feeds Palatal *j*-Assimilation (4) lexically but not in the postlexical component (where (4) is no longer applicable).

(8)                    *Palatalization (lexical and postlexical)*

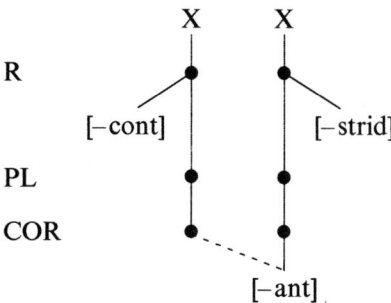

[−ant]

Consider now the case of /l/. Across word boundary (including compound boundary) /l/ remains unaltered before /j/ in guarded speech; whereas in colloquial speech full coalescence takes place just like in lexical palatalization:

(9)    hol jelent meg    [holjε-] ~ [hoj:ε-]    'where did it appear'
       hiteljuttatás     [-εlju-] ~ [-εj:u-]    'granting of credit'
       följut            [följut] ~ [föj:ut]    'reach the top'

In casual speech, /l/ can be simply dropped (as in any preconsonantal context, with or without compensatory lengthening of the preceding vowel: [fö:jut], [föjut]) before palatalization had the chance to apply. Before palatal non-continuants, /l/ has the first and third options, but not the second (whether we interpret it as fusion or as palatalization into [j]):

(10)   fél tyúk    [fe:lt$^y$u:k] ~ [fe:t$^y$uk];    *[fe:t$^y$:uk],    *[fe:jt$^y$uk]  'half a hen'
       ökölnyi     [ököln$^y$i] ~ [ökö:n$^y$i];      *[ökön$^y$:i],     *[ököjn$^y$i]   'fist-sized'
       kopoltyú    [kopolt$^y$u:] ~ [kopo:t$^y$u];   *[kopot$^y$:u],    *[kopojt$^y$u]  'gill'

All this suggests that the palatalization of *l* is a quite separate process and should not be collapsed with Palatalization (8). Vago (1980*a*: 40) assigns the feature value [−cont] to /l/ in order to be able to account for its behaviour in the context of /j/ with the general rule of Palatalization (roughly, our (8)). However, as we have just seen, /l/ does not palatalize before palatals other than /j/, either lexically or postlexically (this observation is due to Olsson 1992). Therefore, /l/ is taken to be [+cont], and has its own 'palatalization' rule (actually, a rule of full assimilation to /j/).[5] In fact, we have given that rule under (2) above; all we have to add at this point is that *l*-Palatalization (2)— just like Palatalization (8)—is obligatory lexically and optional/rate-dependent in its postlexical applications.[6]

---

[5] Vogel and Kenesei (1987, see Vogel and Kenesei 1990: 360 for a brief summary) claim that *l*-palatalization is a diagnostic for prosodic domains in that it applies within, but not across IP's (intonation phrases). Native speaker intuition is controversial on this point: it does not unambiguously support the grammaticality judgements that Vogel and Kenesei's argumentation is based on.

[6] The scope of (2) could be slightly extended to cover the optional full assimilation of /l/ to the other two liquids: to /r/ as in *balról* [bɔr:ol] 'from the left' and (vacuously) to /l/. This generalized rule could then be referred to as Liquid Assimilation and formulated by omitting [−ant] in

## 7.2. SIBILANT RULES

### 7.2.1. The Imperative of *t*-final verbs

Classical generative tradition (going back to Szépe 1969 and fully developed in Vago 1980*a*) maintains that in the imperative of *t*-final verbs a kind of palatalization process takes place: the stem final *t* palatalizes into [š] or [č] to which, subsequently, the imperative *j* fully assimilates, just like in verbs that are underlyingly sibilant-final. This idea was given an autosegmental formulation in Siptár (1994*a*). However, in what follows, we will develop an alternative account in which it is the /j/ that turns into [š] first and what subsequently applies in the [č] cases is an independently motivated affrication rule (eventually, it will prove to be a side-effect of the rule of Palatalization, suitably generalized; see section 7.2.3). This new account is superior in many respects but involves an extra rule of *t*-assimilation in the /tj/ → [š:] cases.

According to the traditional view, then, final *t* in verb stems undergoes three kinds of changes before imperative *j*:

–it surfaces as [š] after short vowels as in *üt* 'hit' / *üss* [üš:] (imp.);

–it becomes [č] after sonorant consonants as in *hajt* 'drive' / *hajts* [hɔjč] (imp.), *önt* 'pour' / *önts* [önč] (imp.), *olt* 'extinguish' / *olts* [olč] (imp.); and

–it deletes after obstruents. That obstruent may be of two sorts: /š/ in *fest* 'paint' and /s/ in verbs of the *oszt* 'divide' type. The corresponding imperatives are *fess* [feš:] and *ossz* [os:].

Of *V:t*-final verbs, *lát* 'see', *bocsát* 'let go', and *lót* (as in *lót-fut* 'rush about', cf. *fut* 'run') pattern with short vowel stems, whereas *fűt* 'heat', *hűt* 'chill', *műt* 'operate on', *szít* 'incite', *tát* 'open wide', and *vét* 'err', as well as hundreds of verb stems involving causative -*ít*, behave like sonorant + *t*-finals. That distribution could be described in three different manners (still within the general idea that this is a palatalization process affecting the stem-final *t*). First, we could rest content with straightforward listing, as we did here. This is the simplest but the least satisfactory solution. Second, we could say that the *lát* set is exceptional, and that long vowel stems regularly change their final /t/ into [č]: this is the traditional account. Finally, we could also claim that the *lát* set plus short vowel stems constitute the regular (vowel + *t*) class and the *fűt* group should somehow be included in the [č] class. One way to do that is by positing an underlying /j/ in the verb stems concerned, one that merges with the preceding vowel (or rather deletes with compensatory lengthening) after the now regular palatalization of /t/ into [č] but before degemination as in *hajts* etc.: /füjt+j/ → füjč: → [fü:č:] 'heat' (imp.) vs

(2). However, in its generalized form, Liquid Assimilation is restricted to casual speech and even there it is rather infrequently applied to *lr* sequences as another casual-speech process, *l*-dropping competes with it for potential inputs (cf. [bɔːrol]). Therefore, we leave (2) in its more restricted formulation as given in the text.

/hajt+j/ → hajč: → [hɔjč] 'drive' (imp.).[7] This solution is not as arbitrary as it might seem: the verbs listed behave *as if* they were /Vjt/-final in a number of other respects as well. For instance, infinitival *-ni* attaches to vowel + *t*-final verbs without a linking vowel (cf. section 8.1.2): *ütni* 'to hit', *futni* 'to run', *látni* 'to see', but requires a linking vowel after consonant + *t*-final ones: *osztani* 'to divide', *hajtani* 'to drive', *váltani* 'to change'. The *fűt* set shares the behaviour of the latter group: *fűteni* 'to heat', *hűteni* 'to chill', *szítani* 'to incite', etc.[8]

The only respect in which this account has a touch of arbitrariness is the quality of the assumed underlying segment: why /j/? The reason is that this is the only eligible sonorant that does not occur on the surface in exactly this environment (see Vago 1991 for discussion). Alternatively, an empty consonant slot can be posited which will do the job without an arbitrarily chosen underlying segment (cf. Vago 1987, 1989, 1991).

The foregoing can be summarized as in (11):

(11)   *a.*   $t \rightarrow š / V \underline{\quad\quad} j$

   *b.*   $t \rightarrow č / \begin{bmatrix} C \\ +son \end{bmatrix} \underline{\quad\quad} j$

   *c.*   $t \rightarrow \emptyset\ [-son] \underline{\quad\quad} j$

In classical generative terms, these three rules can be collapsed in a complex schema (Vago 1980*a*: 71); cf. Kontra (1992) for a criticism of Vago's collapsed statement of the rule.[9]

---

[7] The traditional account (Deme 1961) is followed by Abondolo (1988: 146); cf. also Olsson (1992) for a quite different proposal. The third solution sketched in the text is based on an assumption first made by Vago (1980*a*: 72).

[8] Note, however, that *bocsát* 'let go' is assigned to the *fűt* type by the *-ni* test, and even has a variant *bocsájt* to make things worse, yet its imperative has [š] rather than [č].

[9] In standard Hungarian, (11) only applies in imperative forms. In a stigmatized version of substandard Hungarian ('*suk-sükölés*'), it also applies before *j*-initial personal suffixes, i.e. in indicative forms as well: *lássa* 'he sees it', *hajtsa* 'he drives it' (cf. Kontra 1992 for discussion); the /j/ of possessive suffixes, however, does not trigger it in any variety of Hungarian: *botja* *[boš:ɔ] 'his stick', *lantja* *[lɔnčɔ] 'his lute'. This restriction can be indicated in the rule itself; but it is clearly better to provide some principled explanation for it. The classical generative solution (Vago 1980*a*: 68, 105) relies on the claim that both personal and possessive suffixes have non-underlying (epenthetic) *j*'s (they are epenthesized by two different rules). If the rules of epenthesis are applied after (11), whether by extrinsic ordering or for some other reason, the fact that (11) is restricted to imperative forms is automatically accounted for. (Possessive *j* is actually quite likely to be epenthetic, cf. Kiefer 1985; and the *j* of verbal personal suffixes may well turn out to be so too. The present issue is not whether these segments are underlying or otherwise.) Siptár (1994*a*) suggests that the rules affecting the imperative of *t*-final verbs should be treated as Level 1 lexical rules whereas those involving lexical palatalization (see section 7.1) should be Level 2. This suggestion obviates the need for extrinsic ordering and results in a neat solution in Lexical Phonology terms. The alternative to be proposed below makes even this level-ordering solution superfluous.

The next step in this account involves the fate of imperative /j/ that is now preceded by [š], [č], or [s], respectively. It is natural to assume that it undergoes the same rule of *j*-assimilation that affects it in the case of sibilant-final verbs: /moš/ 'wash' + /j/ → [moš:], /fut/ 'run' + /j/ → fuš + j → [fuš:]. In other words, the three rules in (11) feed that in (12):

(12)    [+strid] *j*
            1      2    →    1    1

Siptár (1994*a*) proposes an autosegmental reanalysis of the foregoing, keeping the general idea that this is a kind of palatalization process but using coalescence rules for /tj/ → [š:] as in *üss* 'hit' (imp.) and for /tj/ → [č:] as in *önts* 'pour' (imp.). The first involves three spreadings and four delinkings, and the second comprises two spreadings and three delinkings, within the same rule. This kind of unrestricted use of autosegmental operations is impossible within the model of phonology we are assuming here. In what follows, an alternative account will be proposed.

For reasons that will become clear in Chapter 8, we take imperative -*j* to be a Block 2 suffix. Therefore, all rules to be proposed in the rest of this section are Block 2 lexical rules. The analytic morpheme boundary that precedes imperative -*j* will not be indicated in the rules because (i) the segmental configurations involved here do not arise either morpheme internally or via synthetic suffixation and (ii) Block 2 rules apply across (i.e. ignore) such boundaries. However, the morpheme boundary *following* imperative -*j* will be included in the rules to restrict their application to imperative forms (cf. the discussion in footnote 9 above; this solution was first proposed in Zsigri 1997). For instance, our version of the deletion rule in (11*c*) will be informally written as (13):[10]

(13)    *t-Deletion*
            t → Ø / [ son]____ j]

In *ossz* /ost + j/ → [os:] 'divide' (imp.), *fess* /fešt + j/ → [fɛš:] 'paint' (imp.), etc., the output of *t*-Deletion (13) will undergo Strident *j*-Assimilation (14), the rule that also accounts for cases like *hozz* /hoz + j/ → [hoz:] 'bring' (imp.), *moss* /moš + j/ → [moš:] 'wash' (imp.), *mássz* /ma:s + j/ → [ma:s:] 'crawl' (imp.) (and roughly corresponds to Vago's rule stated in (12)):

[10] Since this is a straightforward deletion rule (i.e. it deletes the timing slot together with the melody of the stem-final /t/), there is not much point in reformulating it in autosegmental/feature-geometric terms.

(14)        *Strident j-Assimilation*

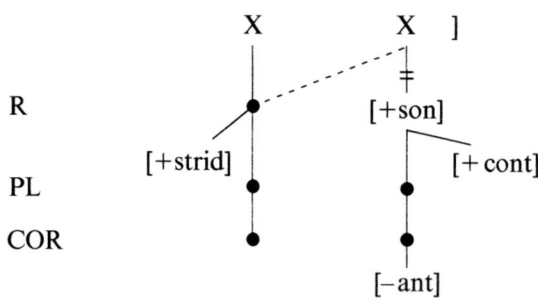

With these preliminaries out of the way, let us turn to the crucial cases.

For items like *üss* 'hit' (imp.), we are *not* adopting a rule that does the job of both (11*a*) and (12) in one go. Rather, we will assume two steps: one that is shared between (11*a*)-type and (11*b*)-type instances, and another one which is specific to the case at hand. The first rule, one that will be used in all cases of what is traditionally called *t*-palatalization, runs as follows:[11]

(15)    *j-Obstruentization*

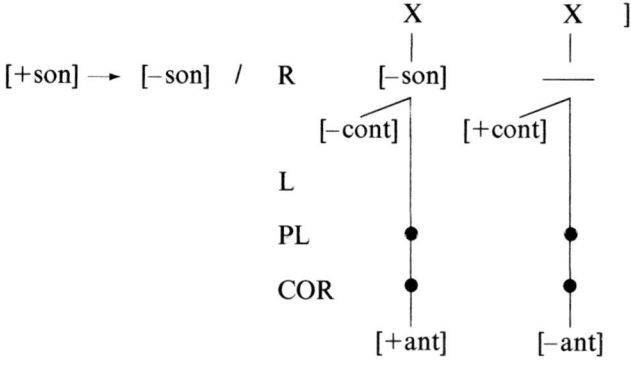

Note that the process under discussion is treated here as a case of progressive voicing assimilation: the rule says, in effect, that the rightmost segment turns into a voiceless palatal fricative.[12] This segment is identical with the underlying representation of /š/; to get its surface shape, it will undergo the

---

[11] The label L with no corresponding node in the diagram is meant to abbreviate a condition to the effect that the obstruent mentioned in the rule is voiceless, i.e. it lacks a laryngeal node.

[12] Given that in our framework there is no [–voice] to spread, rule (15) is formulated as obstruentization—whose output is phonetically interpreted as voiceless (cf. discussion in section 7.3).

redundancy rule of Stridency Spell-out, first given in 4.3 and repeated here for convenience as (16):[13]

(16)                                     *Stridency Spell-out*

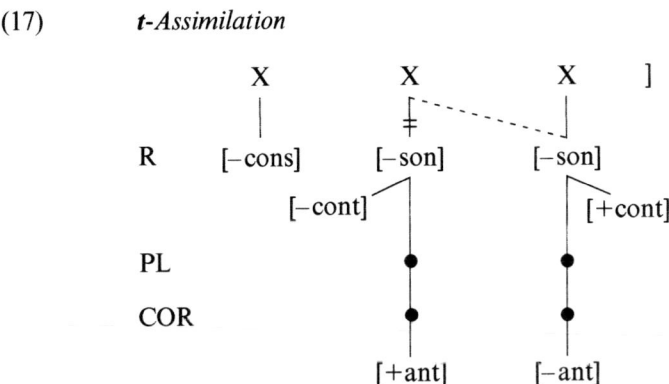

[−son]

| [+cont] [+strid]

COR

Thus, the output of *j*-Obstruentization (15)—as specified further by Stridency Spell-out (16)—will be [tš]. In cases like *önts* 'pour' (imp.), *hajts* 'drive' (imp.), *fűts* 'heat' (imp.), this sequence subsequently undergoes affrication (see section 7.2.3 for further discussion), yielding [č:]; this will be input to degemination (cf. section 9.4) in *önts*, *hajts*, but surface as long in *fűts*. On the other hand, in *üss* 'hit' (imp.) etc., another rule applies that fully assimilates the [t] to the [š] if the former is preceded by a short vowel (we assume that *lát* 'see', *bocsát* 'let go' and *lót(-fut)* 'rush about' are lexically marked to undergo the rule despite their long vowel):

(17)      *t-Assimilation*

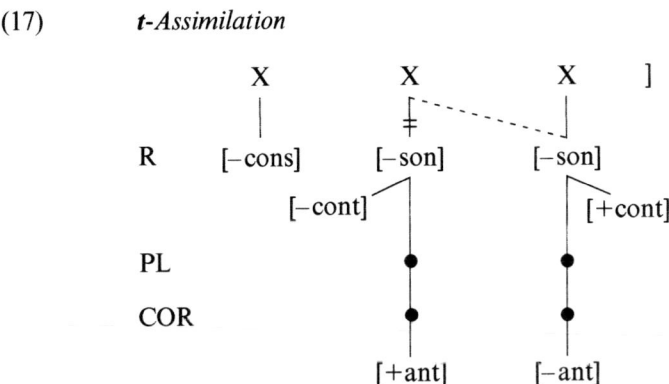

In this section, we have proposed four rules, *t*-Deletion (13), Strident *j*-Assimilation (14), *j*-Obstruentization (15), and *t*-Assimilation (17), all of them lexical rules applying in Block 2, to account for the imperative of *t*-final verbs. The main claim that these rules embody is that it is primarily the imperative /j/, rather than stem-final /t/, that is affected. In other words, the analysis presented here (unlike earlier treatments of the same phenomena) views these data as obstruentization of the /j/ rather than as palatalization of the /t/. Therefore, it obviates the need for level ordering with respect to these rules

[13] This redundancy rule is no longer in operation in the postlexical component. Whenever a /j/ turns into a voiceless obstruent postlexically, it will surface as [ç], rather than [š]. See section 7.3 for an instance of this.

and lexical palatalization (see section 7.1.1), a crucial ingredient of the account in Siptár (1994a).

### 7.2.2. Postlexical sibilant rules

The rules discussed in this section are postlexical and their domain of application increases with speech rate.[14] They include a place assimilation rule and various affrication rules.

The rule of Strident Place Assimilation accounts for cases like *kész-ség* [ʃː] 'readiness', *kis szoba* [sː] 'small room', *egész család* [ʃtʃ] '(the) whole family', *más cipő* [stˢ] 'different shoe' and will be formulated as in (18):

(18)          *Strident Place Assimilation*

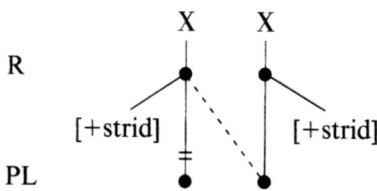

The application of this rule is optional.[15] Word internally (e.g. *nehéz-ség* [sʃ ~ ʃː] 'difficulty', *más-szor* [ʃs ~ sː] 'another time') it is usually avoided in guarded speech but applied in colloquial speech. Across a word boundary, as in *nehéz sors* 'hard lot', *más szempontból* 'from a different point of view', assimilation is normally applied in fast/casual speech only. On the other hand, in semantically opaque cases, the rule applies obligatorily even in the most formal contexts. A case in point is *egészség* [ʃː], *[sʃ] 'health', as opposed to *egész-ség* 'wholeness, totality'; in the latter form, the rule is only applied in casual speech and even there only where no misunderstanding can arise.

The immediate output of Strident Place Assimilation (18) normally undergoes further processes. In cases where both input segments are fricatives, OCP-based fusion of identical nodes (including, in this case, all nodes of the two trees) takes place and the resulting configuration will be identical to that of an underlying geminate, e.g. *kis szoba* [sː] 'small room', *más zene* [zː] 'different music', *hozz sót* [ʃː] 'bring some salt', *húsz zsák* [ʒː] 'twenty sacks', *Balázs szerint* [sː] 'according to Blaise', *egész sereg* [ʃː] 'a whole army', *benéz Zsófi* [ʒː] 'Sophie drops in'. In affricate + fricative cases, the output of (18)

---

[14] Thus, they are what are called 'fast speech rules', cf. Siptár (1991a: 49–53). See also Kaisse (1985), Nespor (1985) for distinctions between rules of external sandhi vs. fast speech rules and between two types of fast speech rules, respectively.

[15] Note that voice assimilation, where necessary, always applies obligatorily on the clusters in the examples to follow; the relative ordering of voice assimilation and (18) is immaterial.

may surface directly but it may also coalesce into a long affricate (see (23) below), e.g. *rácsszerkezet* 'grid structure' ([čs] in guarded speech, [tˢs] in colloquial speech, [tˢ:] in casual and even [tˢ] in fast casual speech), similarly *Kovács Zoltán* <person's name> [ǰz] ~ [dᶻz] ~ [dᶻ(ː)], *polcsor* 'row of shelves' [tˢš] ~ [čš] ~ [č(ː)], *Rácz Zsuzsa* <person's name> [dᶻž] ~ [ǰž] ~ [ǰː]. In fricative + affricate cases, the output of (18) does not undergo further changes: *kis cipő* 'small shoe' [štˢ] ~ [stˢ], *ravasz csel* 'clever trick' [sč] ~ [šč], *gázcsap* 'gas tap' [sč] ~ [šč], *kész dzsungel* 'a real jungle' [zǰ] ~ [žǰ]. Finally, if both input segments are affricates, (18) is rather unlikely to apply: it takes a high degree of speed or casualness for *makacs cápa* 'obstinate shark' to have [tˢː] or for *palóc család* 'Palóc family' to have [čː]. In underlyingly homorganic cases like *kulcscsomó* 'bunch of keys' or *nyolc cövek* 'eight spikes', application of the rule is normally avoided; this is shown by the fact that the cluster remains a fake geminate and is realized as [čč] or [tˢtˢ] rather than [čː] or [tˢː].[16] A colloquial alternative is that the first affricate lenites to a fricative: *kulcscsomó* [šč], *nyolc cövek* [stˢ]. On the other hand, the application of Strident Place Assimilation (18)—if it happens—produces linked structure and the resulting homorganic sequence automatically merges into a single long affricate. Hence, in the non-vacuous cases, we get *makacs cápa* [tˢː] (*[tˢtˢ], *[stˢ]); *palóc család* [čː] (*[čč], *[šč]).

The rest of the sibilant rules all involve what is called 'affrication' and come in three types: stop + fricative, stop + affricate, and affricate + fricative. Let us consider these in turn.

If a dental or palatal stop is followed by a strident fricative, a long affricate may be derived (with voice assimilation, where appropriate):

(19)  ötször      'five times'      /t/  + /s/ → [tˢː]
      ötödször    'the fifth time'  /d/  + /s/ → [tˢː]
      egyszer     'once'            /dʸː/ + /s/ → [tˢː]
      madzag      'string'          /d/  + /z/ → [dᶻː]
      négyzet     'square'          /dʸ/ + /z/ → [dᶻː]
      barátság    'friendship'      /t/  + /š/ → [čː]
      szabadság   'freedom'         /d/  + /š/ → [čː]
      hegység     'mountain'        /dʸ/ + /š/ → [čː]

This coalescence is obligatory within a syllable-final cluster (as in *tudsz* 'you know'); and while in guarded speech it is usually avoided across a syllable

---

[16] With respect to the differential behaviour of true vs. fake geminate affricates cf. section 4.1.2. Note that this reluctance of adjacent identical affricates to undergo merger argues for their interpretation as contour segments, as opposed to [+ strident] stops (cf. Szigetvári 1997 and references therein for the latter position). This is because the [continuant] tier contains a sequence of [−+−+] in the affricate portion of examples like *kulcscsomó* under the representation of affricates assumed here and automatic OCP effects are not predicted. On the other hand, if affricates are strident stops, a fake geminate affricate is in fact a fake geminate stop and its reluctance to be realized as a true geminate remains a mystery.

boundary (except for *t*/*d* + *š* cases that are normally realized as [č:] and tend to be discarded as 'uneducated' if pronounced separately as [tš]), in colloquial speech it is normally applied within words. As speech rate increases, the rule is generalized to apply across word boundaries as in *látszerész* 'optician' [tˢ:], *nem volt soha* 'there never was' [č], *vadzab* 'wild oats' [dᶻ:], *nagy szoba* 'large room' [tˢ:], *főtt sonka* 'cooked ham' [č:], *szakadt zseb* 'torn pocket' [ǰ:], etc.

In the classical generative framework (Vago 1980*a*: 37), two assimilation rules describe this process. One of them, Stop + Fricative Affrication (SFA), turns /s z š ž/ into [tˢ dᶻ č ǰ], respectively, after /t d tʸ dʸ/. The other, Stop + Affricate Affrication (SAA), makes /t d tʸ dʸ/ fully assimilate to a subsequent [tˢ dᶻ č ǰ]. In autosegmental terms, it would be possible to collapse the two statements into one such that it makes /t d tʸ dʸ/ coalesce with /s/ into [tˢ:], with /ž/ into [dᶻ:], with /š/ into [č:], and with /ž/ into [ǰ:]. This latter formulation would be definitely more elegant; the question is whether it would capture the right generalization.

The division of the process into two rules (in addition to the formal reason that classical generative rules are supposed to effect one change at a time) is supported by two further considerations. First: SFA can be applied on its own—we then get outputs that are less formal than [t-s] etc. but less casual than [tˢ:] etc.: [t-tˢ] in *ötször* 'five times', [tʸ-č] in *hegység* 'mountain', [dʸ-dᶻ] in *négyzet* 'square', etc. Secondly: some counterpart of SAA is necessary in the autosegmental framework as well, to account for cases with underlying affricates like *két cica* 'two kittens' [tˢ:], *szomszéd család* 'the family living next door' [č:], *nagy dzsungel* 'large jungle' [ǰ:]. Thus, it is not the case that the classical framework requires two rules for what the autosegmental framework can express with just one; rather, the former theory employs two independently motivated rules to account for the stop + fricative cases.

The specific account that we wish to propose here will be based on a complex interaction of independently motivated rules and principles. Notice, first of all, that no separate rule is needed for some SAA cases (those involving underlying affricates preceded by homorganic stops as in *sötét cella* 'dark cell'): the coalescence of *tc* (/t-tˢ/) into [tˢ:] can be attributed to an automatic OCP effect. This was first proposed for *sötét cella*-type cases by Zsigri (1994). His analysis furthermore claimed that the same automatic effect accounted for cases like *öt csomag* 'five packets':

(20) *a.*  X  X  ⟶  X  X    *b.*  X  X  ⟶  X  X

       |  ∧        ⌞⌐⌐⌝       |  ∧        ⌞⌐⌐⌝

       t  t  s      t  s       t  t  š      t  š

For this idea to work, we must be very careful about what 'hormorganicity' means. In very narrow phonetic terms, no pair of stop + affricate is strictly homorganic in Hungarian. [t] is dental, [tˢ] is alveolar, [č] is palato-alveolar, and [tʲ] is palatal. Zsigri apparently takes the first three of these as (loosely) homorganic (cf. (20a, b)), whereas in terms of the feature system proposed in this book (cf. (14) in section 4.3), /t/ and /tˢ/ are COR, [+ant], whereas /tʲ/ and /č/ are COR, [–ant]. Hence, along with (20a), cases like *egy csomag* 'one packet' are predicted to coalesce via OCP. However, both of these predictions are but partly borne out by the facts. Actually, all possible combinations occur both with separate stop + affricate [t-tˢ], [t-č], [tʲ-tˢ], [tʲ-č] and with a long affricate [tˢ:], [č:], [tˢ:], [č:] (for *öt cella* 'five cells', *öt csomag* 'five packets', *egy cella* 'one cell', *egy csomag* 'one packet', respectively). Since an OCP-effect, if involved at all, cannot be optional, we will assume that the optionality observed here lies in a place assimilation process: if it applies, coalescence will follow automatically, if it does not, the lack of coalescence follows just as automatically. The place assimilation involved can be formalized as follows:

(21)　　　　*Stop + Affricate Place Assimilation (optional)*

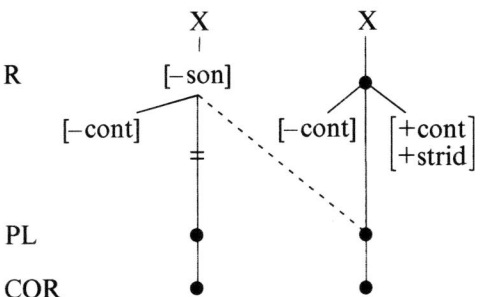

This rule is assumed to be optional; if it applies, it automatically triggers an OCP-based merger of the two [–cont] specifications and of the two root nodes: the net result is a geminate affricate.[17] Returning to the original problem: that of stop + fricative affrication as in *ötször* 'five times' etc., we propose to break it up into elementary steps as in Vago's solution but the individual steps will not be affrication of the fricative and merger of stop + affricate. Rather, we assume that a generalized version of (21), given in (22), assimilates the stop to the place of articulation of the following fricative. This will again trigger the OCP-merger as before, and the result is again a geminate affricate.[18]

---

[17] If the two input segments are not of the same voicing value, voicing assimilation is assumed to apply before the merger (this, in fact, is a prerequisite for the two root nodes to merge): thus, all output geminate affricates will have the voicing value of the underlying affricate, thus: /tǰ/ → [ǰ:], /d č/ → [č:], /dʲ tʲ/ → [tˢ:], etc.

[18] The output of Stop + Strident Place Assimilation (22) may not be technically quite sufficient to trigger OCP-merger if the strident segment is a fricative. Let us assume that Fricative Assimilation (23) below is also involved in such cases; this removes the doubt concerning whether merger should actually be triggered. In what follows, this complication will not be mentioned to simplify the discussion.

(22)                    *Stop + Strident Place Assimilation (optional)*

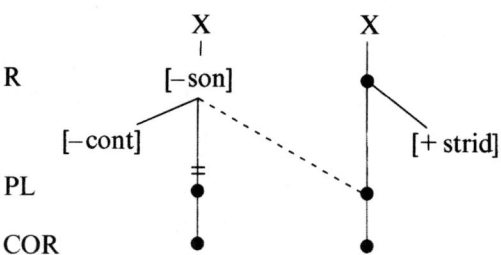

Further possibilities of generalizing this rule in various directions suggest themselves at this point: these will be explored in section 7.2.3 below.

We have not yet accounted for the partial affrication of stop + fricative sequences that Vago's SFA, applied on its own (i.e. without his SAA) would generate as intermediate forms but which, we pointed out, can surface as they are, giving outputs that are less formal than [t-s] etc. but less casual than [tˢː] etc., that is, outputs like [t-tˢ] in *ötször* 'five times', [tʸ-č] in *hegység* 'mountain', [dʸ-dᶻ] in *négyzet* 'square', etc. We propose the following optional postlexical rule for this purpose:

(23)                    *Fricative Affrication (optional)*

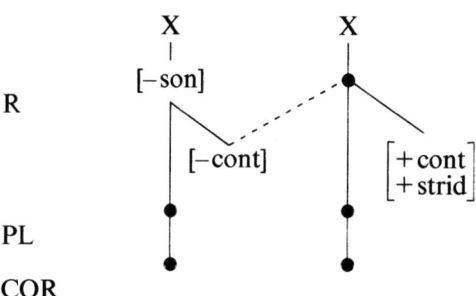

In sum, a form like *ötször* has a number of possibilities. In a very formal style, it can simply surface as [ötsör], with no rule applying to it. Less formally, Fricative Affrication (23) can apply to give [öt-tˢör]. In colloquial speech, Stop + Strident Place Assimilation (22) applies either to the underlying /t-s/ sequence or to the output of (23), both being possible inputs as (22) now stands. The output of Stop + Strident Place Assimilation (22) will then be modified by OCP merger, to give [tˢː]. Similarly, a form like *hegység* 'mountain' may surface as [hɛtʸšeːg] via Voicing Assimilation alone, as [hɛtʸčeːg] via Voicing Assimilation and Fricative Affrication (23), or as [hɛčːeːg] via Voicing Assimilation, Stop + Strident Place Assimilation (22), and OCP. Hence, the order of application of these rules need not be specified (whereas they may be indexed somehow for the level of casualness that goes with each).

The third type of affrication can be observed in cases like *nyolcszor* 'eight times', *makacsság* 'obstinacy', where a fricative turns into the corresponding affricate if a homorganic affricate precedes. Non-homorganic inputs are not affected (e.g. *bohócság* 'clownery' *[t͡sč]), unless Strident Place Assimilation (19) makes them homorganic first (cf. [boho:č:a:g], a possible casual output). Zsigri (1994) suggests that no rule is involved here: the observed process is simply due to OCP again.

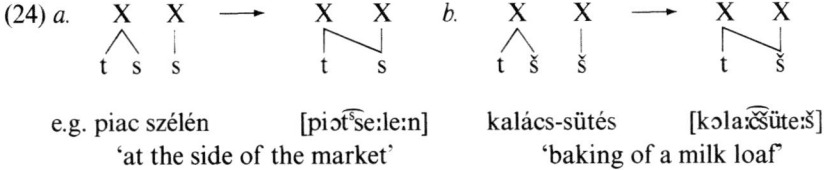

e.g. piac szélén      [pi͡ɔt͡se:le:n]
'at the side of the market'

kalács-sütés      [kɔla͡c͡süte:š]
'baking of a milk loaf'

Note that the condition of homorganicity is automatically met since non-homorganic inputs do not trigger this OCP-based merger. Note furthermore that the output configurations in the stop + affricate cases in (20) are not identical to those in the affricate + fricative cases here. The former suggest a normal geminate affricate of the structure 'long stop phase + short fricative phase', whereas the latter yield a complex segment with the unexpected (but phonetically not unsupported) structure 'short stop phase + long fricative phase' (the transcriptions [t͡ˢs], [c͡š] are meant to reflect this latter structure).

Zsigri (1994) suggests that cases of stop + fricative affrication should be divided into two sets. One set of cases would be based on a lexical rule and would produce normal long affricates (similar to the OCP-structures in (20)), whereas the other would be based on a postlexical rule and would produce a type of long affricate similar to the output structures in (24). His examples include *látszik* [t:] 'seem' vs. *járatszám* [t͡ˢs] 'service number', *virágot szed* [t͡ˢs] 'pick flowers'; *barátság* [c:] 'friendship' vs. *többletsúly* [c͡š] 'excess weight', *szeret sétálni* [c͡š] 'be fond of walking'; *negyedszer* [t:] vs. *ködszitálás* [t͡ˢs] 'misty drizzle', *szabad szemmel* [t͡ˢs] 'with unaided eye'; *fáradság* [c:] 'pains' vs. *padsor* [c͡š] 'row of seats', *szabad sáv* [c͡š] 'unobstructed lane'; and *edzés* [d:] 'training' vs. *rövidzárlat* [d͡z] 'short circuit', *svéd zászló* [d͡z] 'Swedish banner'.[19] Both rules would involve spreading without delinking and would

---

[19] If, as we claim, [d:] is not an underlying segment in Hungarian, it is impossible for a (Block 1) lexical rule to apply in a word like *edzés* for two reasons: the environment is non-derived and, in defiance of the principle of structure preservation, the segment produced is a lexically non-existent one. Therefore, it might seem that either our analysis of [d:] as going back to an underlying cluster, /dz/, should be abandoned, or else both of Zsigri's rules should be seen as postlexical (but then we would lose the way to account for the difference of behaviour between the pairs of examples given in the text, all of which pattern exactly as Zsigri's solution predicts). A third possibility is to assume that his lexical affrication is a 'word level' rule that can apply in

produce the appropriate structures from identical underlying representations; the only difference would be in the direction of spreading:

(25)     *a.*  X     X         *b.*  X     X

                 t     s             t     s

          látszik [tˢ:] 'seem'        látszerész [t͡ˢs] 'optician'

We will not consider the implications of this interesting proposal any further here.

### 7.2.3. Palatalization revisited

In this section, we will explore a number of possibilities for generalizing some of the rules given in previous sections. Consider first (22) and (18), i.e. Stop + Strident Place Assimilation and Strident Place Assimilation, repeated here for convenience as (26*a, b*):

(26)           *a. Stop + Strident Place Assimilation*

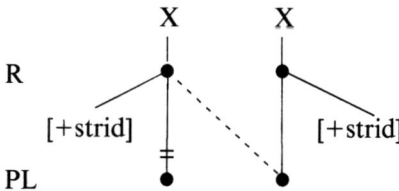

           *b. Strident Place Assimilation*

Given that all strident segments are coronal obstruents in Hungarian, the two rules as they stand partly overlap and partly complement one another. Hence, it is possible to collapse the two rules as in (26*c*); this rule is optional and says

a non-derived environment and need not obey structure preservation. In this case, (25*a*) represents a word-level (Block 2) lexical rule and (25*b*) exemplifies a postlexical one, and both our analysis of [dʲ:] and Zsigri's insight can be maintained.

that all coronal obstruents may assimilate in place to a following strident segment; if the left-hand segment is a stop, this gives us affrication effects (with
OCP), if it is an affricate or a fricative, the rule leads to coalescence in most
cases (see section 7.2.2 for the details):

(26)                    *c.   Coronal Place Assimilation*

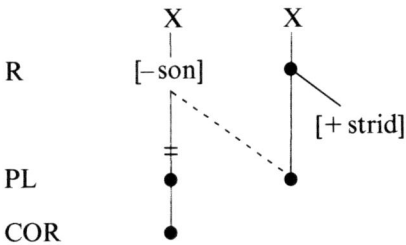

Let us now compare Stop + Strident Place Assimilation (22) with Palatalization (8) and see if a similar move is possible. What happens if we try to collapse these two rules?

(27)                    *a.  [=(22)  Stop + Strident Place Assimilation]*

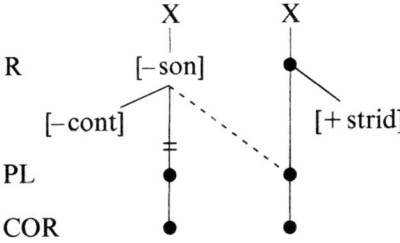

*b.  [=(8) Palatalization]*

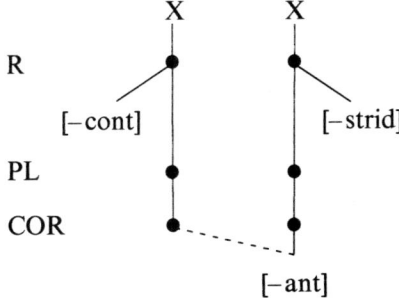

Note that these two rules apply in exactly complementary environments
except that non-strident anterior coronals (*t, d, l, r*) are not involved as a

second segment in either of them. If we tried to collapse the two rules simply by omitting reference to the feature [strid], we would either lose half of the inputs to Stop + Strident Place Assimilation (22)—by restricting the collapsed statement to [–ant] second segments—or else our rule would predict place assimilation in cases like *hagyta* 'left it', *hagyd* 'leave it!', *hagylak* 'I leave you', *hegyről* 'from a hill'. Although [hɔtːɔ] and [hɔdːi] are possible dialectal/non-urban pronunciations for the first two forms, *[hɔdlɔk] or *[hɛdröl] are completely out in any variety of Hungarian. (So is *[hɔdnɔ] for *hagyna* 'would leave', but recall that *n* is assumed to be unspecified for place until the end of the phonology, hence no assimilation is predicted in this type of case.) Therefore, the two rules cannot be literally collapsed; yet it is possible to omit [–strid] from the rule of Palatalization (8). This is a welcome result since [strid], we claimed, is a non-underlying feature for Hungarian.[20] Our final formulation of Palatalization, then, is given in (28):

(28)                  *Palatalization (lexical and postlexical)*

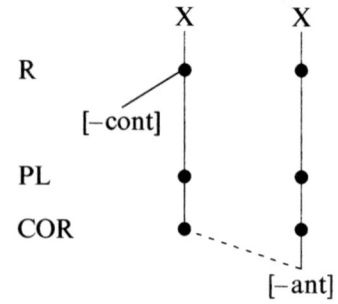

In addition to the effects described in 7.1, this rule now accounts for cases like *önts* 'pour' (imp.), *hajts* 'drive' (imp.), *fűts* 'heat' (imp.), where the output of *j*-Obstruentization (15) undergoes (28) and eventually surfaces as a palatal affricate ([č] or [čː], as the case may be). This solution is far more satisfactory than a possible alternative that would attribute the change [tš] → [č(ː)] in these cases to the postlexical and optional rule of Stop + Strident Place Assimilation (now subsumed under (26c) Coronal Place Assimilation) since the affrication in *önts* etc. is by no means optional. Postlexically, on the other hand, there is some overlap between (26c) and (28): in a case like *két sör* 'two (pints of) beer' the structural description of both rules is satisfied and the same structural change is predicted by both: it is immaterial which of the two applies.

---

[20] Note that reference to [+strid] in (26) is legitimate since the rule is postlexical; but with respect to the rule of Palatalization that is assumed to apply both lexically and postlexically—with somewhat different effects in the two cases—the fact that only underlying features are referred to in it is a precondition for its lexical applications. (Recall that we assume the redundancy rule of Stridency Spell-out to apply at the end of the lexical phonology, not earlier.)

To conclude this section, consider rule (4), the only remaining lexical rule that refers to the feature [strid]. Would a similar generalization be possible here, too?

(29)     [=(4) *Palatal j-Assimilation*]

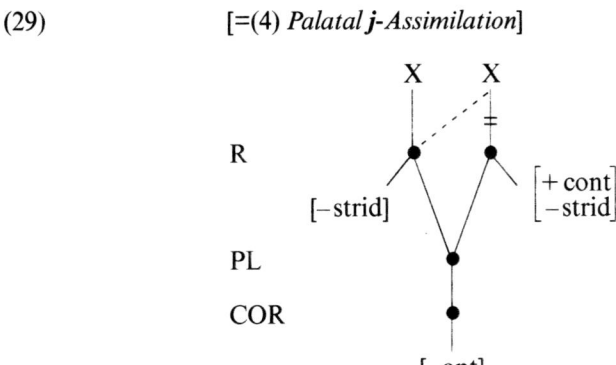

Suppose we simply omit [–strid] on both sides. If we did that, our rule would predict the existence of four types of cases. First, with non-strident segments in both positions, we would get palatal coalescence as in *bátyja* [t<sup>y</sup>ː] 'his brother', *hagyja* [d<sup>y</sup>ː] 'leave' (3sg indicative/imperative def.), *hányja* [n<sup>y</sup>ː] 'throw' (3sg indicative/imperative def.), as well as—vacuously—/jj/ → [jː] as in *hely-jegy* 'seat reservation'. These are the cases the rule was originally designed to account for. Second, for combinations of strident + non-strident, we would get (part of) Strident *j*-Assimilation as in (12): note that this rule is more general than the version we proposed in (14); it applies in non-imperative examples like *mossa* /š+j/ → [šː] 'wash' (3sg def.), too.[21] Third, with strident segments in both positions, we get vacuous instances like /šš/ → [šː] as in *más-ság* 'difference', as well as affricate + fricative coalescence like /čš/ → [čː] as in *makacs-ság* 'obstinacy'. In the latter case we have a slight problem already: such coalescence is optional and would be perfectly accounted for by the mechanism we discussed in (24). But where the idea turns out to be completely unworkable is the fourth type of cases: the combination of non-strident plus strident. For cases like *hegy-ség* 'mountain range', the generalized rule we are considering would give *[hɛt<sup>y</sup>ːeːg] and would bleed the various postlexical rules this form could otherwise undergo: see (22) and (23). Therefore, we have to find a solution that does not employ the feature [strid] but restricts the righthand segment to /j/, to the exclusion of strident non-anterior continuants. The solution is rather straightforward: the right-hand segment must be [+son]. The revised form of this rule (to replace (29)) will then be this:

---

[21] On the other hand, /j/ assimilates to non-anterior strident consonants as well; hence the generalization of (29) we are just considering does not make a rule corresponding to (12) superfluous. Therefore, Strident *j*-Assimilation (14) will be generalized to non-imperatives (by omitting the bracket) and will partly overlap in its domain of application with the rule under discussion.

(30)                          *Palatal j-Assimilation (revised)*

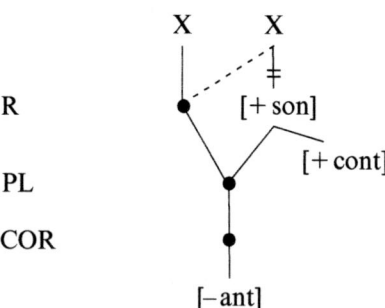

This formulation covers the first and second sets of cases but excludes the other two. There is some overlap between Palatal *j*-Assimilation (30) and Strident *j*-Assimilation (12)/(14), but this is just as harmless as the overlap between Coronal Place Assimilation (26c) and Palatalization (28).

In this section, we have revised our rules of Palatalization and Palatal *j*-Assimilation. The final versions are given in (28) and (30), respectively; in both cases, the rules have become simpler/more general at the cost of partial overlap with other, independently motivated rules. Also, we proposed a collapsed statement of Strident Place Assimilation and the place-assimilation component of some affrication phenomena (see Coronal Place Assimilation (26c)). This concludes our discussion of lexical and postlexical processes affecting sibilants. In the following section, we turn to postlexical processes connected with the laryngeal properties (in particular, voicing) of obstruents.

## 7.3. VOICING ASSIMILATION AND DEVOICING PROCESSES

As we saw in section 4.1.1, Hungarian has a rule of voicing assimilation whereby obstruent clusters come to share the voicing specification of their rightmost member.[22] In a framework with binary [voice], a straightforward rule of Laryngeal Spreading like (31) would neatly account for this state of affairs. The laryngeal node (L) would dominate either [+voice] or [–voice]; the first L of two adjacent obstruents would delink (for any combination of these feature values) and the second would spread onto the vacated root node (R) (see Siptár 1996 for an analysis in these terms):

---

[22] Intramorphemic obstruent clusters are invariably homogeneous with respect to voicing (see section 4.1.1 for numerous examples): the rule discussed here ensures that no heterogeneous clusters arise during derivations, either. The rule is postlexical (it applies across any type of boundary as long as no pause intervenes) but obligatory and non-rate-dependent. See further below for apparent exceptions to the generalization that adjacent obstruents will agree in voicing, whether the cluster is underlying or derived.

(31)                          *Laryngeal Spreading*

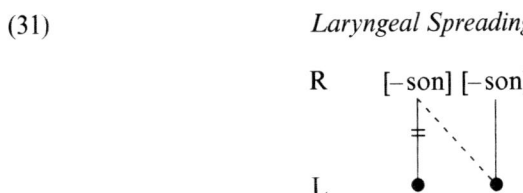

However, in the framework of the present book where [voice] is privative (cf. Lombardi 1995*a* and references cited there), this option is not open to us.[23] Instead, we will follow the course that Mester and Itô (1989) originally proposed for languages where voice assimilation co-occurs with final obstruent devoicing (FOD). Their informal suggestion is shown in (32); since Hungarian has no syllable-final devoicing, we will start from a revised version spelled out in (33).

(32)        *a.* Non-prevocalic obstruents lose their Laryngeal nodes.

            *b.* [voice] spreads to the left.

(33)        *Voicing Assimilation (preliminary formulation)*

The general formulation in (33*a*) could be interpreted in two different manners. The context in which delinking takes place could be taken literally, i.e. it could be restricted to voiced obstruent + voiceless obstruent cases (where the lack of an L node on the second segment is part of the set of conditions under which the rule applies) or else it could be interpreted loosely, i.e. to apply to any pre-obstruent voiced obstruent. Under the second, more general reading, the derivation of e.g. *kád-ban* 'in a bath-tub' would run as follows:

---

[23] Recall that voiceless obstruents and all sonorants (including vowels) are represented without L nodes in this framework and are interpreted as voiceless and voiced respectively in the phonetic implementation module. Voiced obstruents, by contrast, have an L node dominating [voice] and constitute the marked case. Consequently, obstruent voicing can spread but voicelessness (or sonorant voicing) cannot.

(34)

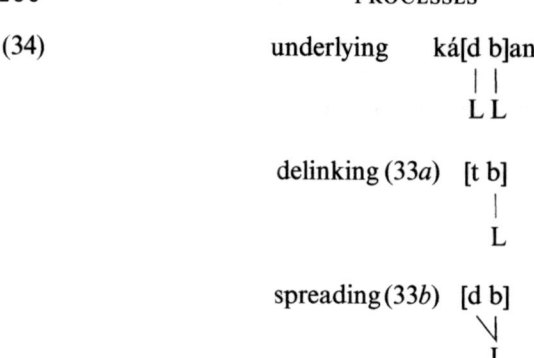

underlying     ká[d b]an

delinking (33a)   [t b]

spreading (33b)   [d b]

This roundabout derivation (also known as the Duke of York gambit, cf. Pullum 1976) can be obviated under the first interpretation (with or without subsequent merger of the two L nodes as a repair to the OCP violation that emerges in such forms). In this case, then, (33a) does not feed (33b) as the two input sets do not intersect. Consider the following derivations:

(35)

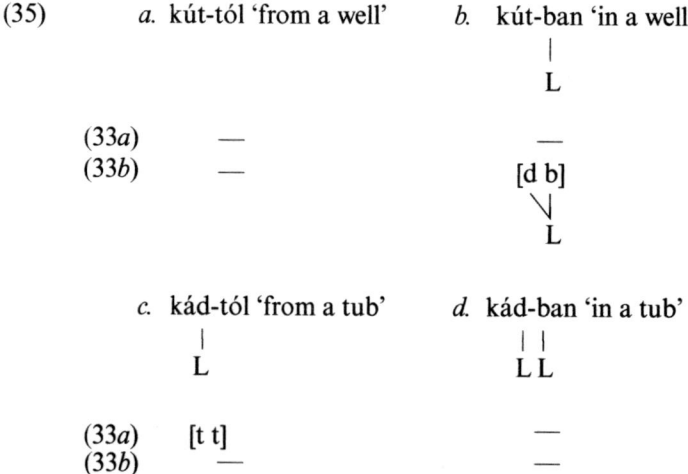

     a. kút-tól 'from a well'     b. kút-ban 'in a well'

(33a)     —        —
(33b)     —        [d b]

     c. kád-tól 'from a tub'     d. kád-ban 'in a tub'

(33a)     [t t]        —
(33b)     —        —

Lombardi (1995b) suggests that the delinking part should not be seen as a rule but rather as a repair operation on representations that violate her Laryngeal Constraint (36):

(36)                    *Lombardi's Laryngeal Constraint*

σ

R ●   [+ son]

L ●

This is a positive constraint saying that a laryngeal node is only licensed in a consonant if it immediately precedes a [+son] segment, i.e. a vowel or a sonorant consonant, *in the same syllable*.[24] This constraint works fine for FOD languages like German but is too strong for Hungarian where voiced obstruents occur before a heterosyllabic sonorant and word finally (before pause) as well. Lombardi points out that 'not all languages show laryngeal neutralization; these are languages that lack the constraint, and thus have no restrictions on where a Laryngeal node can appear' (1995b: 43). But Hungarian is not one of these languages: it *does* have voice neutralization before an obstruent.[25] Lombardi also proposes an additional positive licensing constraint that allows a laryngeal node to be licensed by the word edge that follows it:[26]

(37)         *Lombardi's Word Final Laryngeal Constraint*


         L ]$_w$

However, this still leaves us with the pre-sonorant cases as in *láb-nál* 'at a foot': these do not satisfy either (36) or (37) but the laryngeal node is still licensed (does not delink) in them. It would be possible to amend (36) by omitting the syllable node from it—but it is doubtful if this is an amendment at all since Lombardi's original insight is all but lost if the positive constraint is just this:

(38)         *Laryngeal Constraint (revised)*

         R ●  [+ son]
              |
         L ●

---

[24] Cf. Kenstowicz (1994: 493–8) for discussion. Kenstowicz (wrongly) mentions Hungarian among languages combining (36) and (37), i.e. allowing voiced obstruents syllable-initially as well as in voice-assimilation contexts and word-finally but not before a heterosyllabic sonorant.

[25] Lombardi (1995b: 67) specifically admits in a footnote that Hungarian does not fit into the typology she proposes and raises the possibility that [–voice] may be required, after all, to account for all the facts of Hungarian (that is, all the basic facts, not the complications we turn to further below). She says that voicing assimilation is optional in this language and suggests that this might be the reason why it does not fulfil her predictions. But this is wrong: voicing assimilation is obligatory in Hungarian (especially within words; but even across word boundaries it takes very special circumstances for the speaker to suppress it). We will see that Hungarian voicing assimilation *can* be accounted for with privative [voice]—all we have to do is either to loosen Lombardi's constraint to the point that it loses most of its initial appeal, or to give it up altogether and use delinking and spreading as two straightforward rules.

[26] This constraint would allow a non-devoiced word-final obstruent before a voiceless obstruent in the next word, which is again counterfactual for Hungarian. This problem could be avoided if the constraint referred to utterance edge rather than word edge but this is not likely to be in accordance with Lombardi's original intention.

In sum, rather than use two positive licensing configurations, (38) and a suit-able version of (37), in conjunction with an automatic repair mechanism that delinks all L nodes that are not in either of these configurations, we will sim-ply return to the formulation in (33*a*).

But as we saw in section 4.1.1, two segments behave asymmetrically with respect to this process: /v/ undergoes devoicing (*szívtől* [ft] 'from a heart') but does not trigger voicing (*hatvan* *[dv] 'sixty'), whereas /x/ triggers devoicing (*adhat* [th] 'he may give') but does not undergo voicing before an obstruent. The problem of /x/ was fully discussed in 4.1.1 and we will say nothing more about it here. On the other hand, the problem of /v/ has bear-ing on the rule(s) of voicing assimilation, and it is appropriate to consider it here.

Unlike voiced obstruents, /v/ clusters with voiceless obstruents, including word initial ones (*kvarc* 'quartz', *pitvar* 'porch') and fails to trigger voicing either in the above examples or in heteromorphemic cases (*hatvány* 'power (of a number)', *szép volt* 'it was nice'). But, unlike sonorants, it also occurs last in word final clusters (*érv* 'argument', *könyv* 'book') and undergoes devoicing (*hívsz* [fs] 'you call', *óvtam* [ft] 'I protected it'). In other words, onset /v/'s behave like sonorants while coda /v/'s behave like obstruents. Their phonetic realization (roughly) corresponds to this but there are a few minor hitches that we return to later.

To account for this Janus-faced behaviour of /v/, we suggested in 4.1.1 that it should be underlyingly unspecified for [son] but have a laryngeal node (like voiced obstruents). All we have to do then is to omit [−son] for the first seg-ment in (33*a*); the spreading rule needs no modification but we repeat it here for convenience:

(39)          *Voicing Assimilation (revised)*

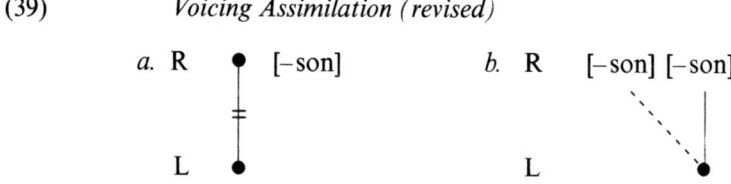

Note that this modification of the delinking rule would be possible even if the special behaviour of /v/ were disregarded: since (apart from /v/) all and only voiced obstruents have an L node, the [−son] of the leftmost segment was redundant anyway.

The rules now work for cases involving /v/ as shown in the following derivations (for other cases, they work as shown in (35) above):

(40)

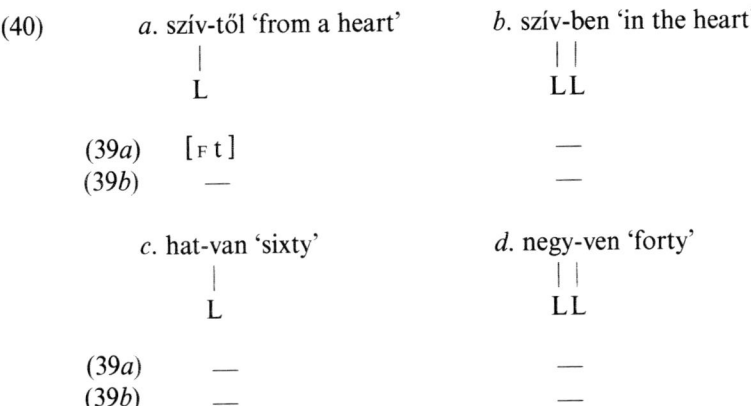

As can be seen, nothing happens in (40*b*, *c*, *d*) whereas the L gets delinked in (40*a*), resulting in a segment lacking an L node but unspecified for [son] (this is informally shown by the symbol F). As we said before, the phonetic implementation module will interpret obstruents lacking an L node as voiceless and make all sonorants (all of which lack an L node) spontaneously voiced. Thus, to get [f] in *szívtől*, we need a fill-in rule located in the phonetic implementation module that specifies it as an obstruent. We will assume that this fill-in rule has a more general format: it turns onset /v/'s into sonorants and coda /v/'s into obstruents by specifying their empty [son] feature as plus and minus, respectively. This will correctly predict that, in general, onset /v/'s will come out as approximant [ʋ] with no (or very little) noise of friction and coda /v/'s as fricative [v] with a normal amount of noise of friction.

In some specific contexts, however, further modification is required. In postconsonantal onsets /v/ is realized as a rather strong (noisy) fricative if the preceding coda consonant is a labial stop: *lopva* 'stealthily', *dobva* 'throwing', although it behaves phonologically as a sonorant (note the lack of voice assimilation in *lopva*). Furthermore, in (exceptional) left edge clusters whose first member is /v/, as in *Wrangler* [vr-], the /v/ is, again, realized as a fricative.[27]

Our rules as developed so far would predict both types of words to come out with approximant [ʋ]'s but what we find phonetically is that they contain proper voiced fricatives with quite strong noise. Notice that it would not do to turn these into obstruents within the phonology since both behave as sonorants with respect to voice assimilation (i.e. the [p] of *lopva* 'stealthily' and the [t] of *két Wrangler* 'two pairs of Wrangler jeans' do not get voiced). But note also that we assumed underlying /v/'s to be unspecified for [son]. This specification might be held responsible for the presence (in [–son] segments) vs.

---

[27] Although this example could be simply dismissed as exceptional, it appears that there is a pattern here. Consider the parallel case of the (equally exceptional) initial cluster in *Hradzsin* [xr-] 'the castle in Prague'. Although /x/ is clearly represented by a glottal glide [h] in onsets and by a velar fricative [x] in codas, in the first position of a word-initial cluster it is the latter that crops up. Note that word initial clusters are not analysed here as branching onsets, cf. section 5.2.2.

absence (in [+son] segments) of fricative noise in [z] or [ž] vs. [n] or [l] (as well
as in [j] as in *kérj* 'ask' (imp.) vs. [j] as in *kéj* 'pleasure'; see further below). We
have assumed that coda /v/'s are specified as fricatives and onset /v/'s are specified as approximants in the phonetic implementation module. Thus, the [v] in
*terv* 'plan' or *révbe* 'to port' will be [−son] whereas the [ʋ] in *pitvar* 'porch' or
*kova* 'flint' will be [+son]. However, an optional (style/rate dependent) phonetic
rule may specify any surface voiced labiodental continuant as [v] rather than
[ʋ], i.e. a fricative rather than an approximant.[28] It is this optional process that
is sharpened into an obligatory switch of (phonetic) status in our two cases:
after a labial stop and in the leftmost position of a word initial edge cluster.

Thus, we propose the following phonetic rules:

(41)                        *a. Coda Obstruentization*

$$[ \ ] \rightarrow [-\text{son}] \ / \ \text{in syllable coda}$$

*b. Postlabial v-Strengthening*

$$[ \ ] \rightarrow [-\text{son}] \ / \ R \qquad [-\text{son}] \quad \underline{\qquad}$$
$$\text{LAB} \qquad \qquad \bullet$$

*c. Left Edge v-Strengthening*

$$[ \ ] \rightarrow [-\text{son}] \ / \quad _w[\overset{|}{X} \qquad \overset{|}{X}$$
$$R \qquad \underline{\quad} \ [+\text{cons}]$$

*d. Optional v-Strengthening*

$$[ \ ] \rightarrow [-\text{son}]$$

*e. Default Sonorancy*

$$[ \ ] \rightarrow [+\text{son}]$$

Rule (41*a*) turns all coda consonants unspecified for [son] into obstruents;
practically this means all coda /v/'s since all other segments are underlyingly
specified as either [+son] or [−son]. Coda /v/'s whose L node had been
delinked are thus interpreted as [f] (*szív-től* 'from a heart'), others as fricative
[v]. Rule (41*b*) strengthens the /v/ of e.g. *lopva* 'stealthily' into a voiced

---

[28] This has to happen in the phonetic component and not earlier, otherwise it would interfere
with voicing assimilation in unwanted ways. Note that voicing assimilation is no longer operative in this module. For instance, occasional word-final phonetic devoicing as in *kezd* [kɛzd̥]
'begin' does not feed voicing assimilation: *[kɛzd]. Similarly, no matter how noisy the [v] in e.g.
*lopva* 'stealthily' is, it will never turn the [p] into a *[b].

fricative [v] but recall that voicing assimilation is no longer in force here. Rule (41c) does the same thing to left edge cluster initial segments (again, only /v/'s are available here—and in extremely few exceptional items like *Wrangler*—since all other consonants have their values for [son]). Rule (41d) optionally strengthens any remaining /v/ into a fricative [v], whereas those that still remain are finally specified as [+son], i.e. approximant [ʋ].

In addition to the general voicing assimilation rule(s) discussed so far, Hungarian has a progressive devoicing process as well, targeting /j/ under special circumstances.

At the surface, Hungarian has no word-final consonant + liquid sequences (except in a few cases with final *l*, provided the first consonant is also a liquid, cf. *görl* 'chorus-girl', *fájl* 'file'). Accordingly, in word-final *Cj* clusters the /j/ turns into a fricative.[29] This fricative will/may subsequently undergo various processes, some of them involving devoicing.

The most often quoted case—*lopj* [pç] 'steal' (imp.), *rakj* [kç] 'put' (imp.), *döfj* [fç] 'steal' (imp.)—appears to be quite simple. All we seem to need is a rule like /j/ → [−son] / C__]ₓ and we get the devoicing effect for free: /j/, being a sonorant, has no L node; turn it into an obstruent without adding an L node and you end up with a voiceless obstruent. However, the issue is rather more complex than that.

There are twelve logical possibilities in terms of context (disregarding cases where a vowel follows and the /j/ is realized as [j]). These are displayed in (42). The columns stand for right context, the rows for left context.[30]

(42)

| | sonorant | voiced obstruent | voiceless obstruent | nothing |
|---|---|---|---|---|
| sonorant | ʝ ~ Ø | ʝ ~ Ø | ç ~ Ø | ʝ |
| voiced obstruent | ʝ ~ Ø | ʝ ~ Ø | ç ~ Ø | ʝ |
| voiceless obstruent | ç ~ Ø | ʝ ~ Ø | ç ~ Ø | ç |

One possibility for each case except the last column (i.e. if anything follows) is to have the reflex of /j/ deleted (see footnote 12 in section 9.5 below). This applies in fast/casual speech and does not bear on the analysis of the rest of the possibilities.

The three cases in which we get [j] before a voiced obstruent could involve

---

[29] These clusters may be underlying as in *szomj* 'thirst', *férj* 'husband', *fürj* 'quail'; or morphologically complex as in *dob-j* 'throw' (imp.).

[30] Examples, going across the table (all verbs are 2sg imp.): *nyomj le* 'push down', *nyomj be* 'push in', *nyomj ki* 'push out', *nyomj* 'push'; *dobj le* 'throw down', *dobj be* 'throw in', *dobj ki* 'throw out', *dobj* 'throw'; *lépj le* 'step down', *lépj be* 'step in', *lépj ki* 'step out', *lépj* 'step'.

L spread (39*b*) assuming, as above, that the /j/ is simply obstruentized first. The five cases in which [ç] is produced before a voiceless obstruent, a sonorant, or utterance finally, could be analysed with no additional process, simply as suggested in the previous paragraph. However, in the remaining cases we have to account for the voiced realization of the palatal fricative. In *dobj* and *dobj le*, we could assume rightward L-spreading, but in *nyomj* and *nyomj le* even this unusual assumption would not help. Therefore, we have to give up the simple idea sketched above and conclude that the /j/ is not just obstruentized: it is turned into a *voiced* obstruent. The rule must delete [+son] and add an L node dominating [voice]. (The missing [–son] can be supplied by Coda Obstruentization (41*a*).)[31]

The voiced fricative thus obtained behaves almost exactly like any voiced fricative: it gets devoiced before a voiceless obstruent (*nyomj ki, dobj ki* [-pçk-], *lépj ki*), and remains unaffected (or is deleted) in most other cases (*nyomj le, nyomj be, nyomj; dobj le, dobj be, dobj; lépj be* [bjb]~[bb]). But there are two cases (*lépj le, lépj*) where we need an extra rule that delinks the L node which the obstruentization rule has just supplied:

(43)                          *Final Fricative Devoicing*

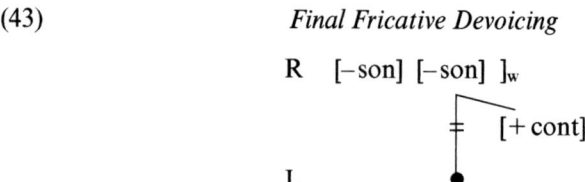

Note that the rule as formulated here delinks the L in *lépj ki* and *lépj be* as well; in *lépj ki* it is immaterial which rule applies, Voicing Assimilation/Delinking (39*a*) or Final Fricative Devoicing (43), the result is the same; in *lépj be*, the more specific (43) has a chance to apply first and delinks the L node of /j/; then the more general Voicing Assimilation/Spreading (39*b*) applies twice, deriving [pçb] → [pjb] → [bjb]. This order of application is vital for cases like *lépj* and *lépj le*: if (39*b*) was allowed to apply to these forms, it would derive *[leːbj].

Finally, the reader may wonder why Final Fricative Devoicing (43) does not mention /j/ specifically but applies indiscriminately to any word final voiceless obstruent + voiced fricative sequence. The reason is simple: such sequences do not occur elsewhere in the language. Morpheme internal obstruent clusters are always homogeneous in terms of voicing (cf. section 4.1.1); and the only single-consonant suffix that is a voiced obstruent is *-d* as in *rakd* [rɔgd] 'put it!'. The specification [+cont] in (43) is enough to exclude *-d* as a possible input segment to this rule; having escaped devoicing, the *-d* then spreads its L node onto the stem final voiceless obstruent as usual.

---

[31] Alternatively, we could assume that /j/ underlyingly contains a laryngeal node which remains redundant in the vast majority of cases and comes to play a role when the /j/ is obstruentized.

## 7.4. PROCESSES INVOLVING NASALS AND LIQUIDS

In this section, we turn to sonorants and review various assimilation and deletion rules, most of which are postlexical, optional, and rate/style-dependent. One notable exception is nasal place assimilation that has both lexical and postlexical applications and is obligatory in some contexts.

### 7.4.1. Nasal place assimilation and related processes

As we argued in section 4.2, the class of underlying nasals includes /m/, /nʲ/, and a general nasal consonant unspecified for place, /N/. This segment undergoes various processes in the appropriate contexts, see below. If none of these apply, the underspecified nasal receives phonetic implementation as dental [n].

Preconsonantal /N/ will, in general, undergo place assimilation of the following form:

(44)                    *Nasal Place Assimilation*

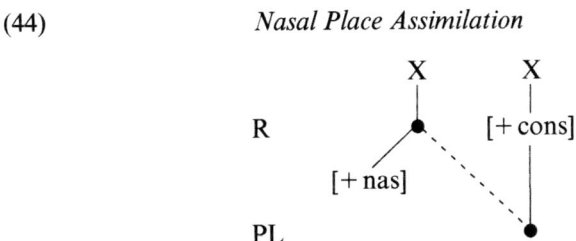

This rule applies both lexically and postlexically. In its lexical applications, when the context is palatal /j/ as in *ken-j* 'smear' (imp.) or *szán-ja* 'his sleigh', the resulting *nʲj* sequence undergoes Palatal *j*-Assimilation (30) (see section 7.2.3) and surfaces as [nʲ:]. Before labials, the nasal becomes labial: e.g. *szín-ben* /N + b/ → [mb] 'in colour', *fon-va* /N + v/ → [ɱv] 'braiding' (participle).[32] The rule applies before dentals and velars (e.g. *men-t* /N + t/ → [nt] 'went', *üzen-get* /N + g/ → [ŋg] 'keep sending messages'), and in non-derived contexts, too:

(45)   /koNp/        [komp]        komp        'ferry'
       /loNb/        [lomb]        lomb        'foliage'
       /troNf/       [troɱf]       tromf       'trump'
       /poNt/        [pont]        pont        'dot'

[32] The fact that the nasal is realized as labiodental in the latter example is a matter of phonetic implementation. In particular, all that matters within the phonology is that the nasal shares the place of articulation of the subsequent consonant. Labial fricatives—and approximants, as in the present example, cf. section 7.3 for discussion—are implemented as labiodental; hence the preceding nasal will be labiodental, too. Conversely, labial stops (and 'independent' labial nasals) are phonetically bilabial: hence the nasal in *szinben* 'in colour' (and the whole nasal cluster in e.g. *fennmarad* 'stay up') will be bilabial.

| /goNd/ | [gond] | gond | 'anxiety' |
|--------|--------|------|-----------|
| /koNtˢ/ | [kontˢ] | konc | 'loot' |
| /roNč/ | [roṇč] | roncs | 'wreck' |
| /poNtʸ/ | [ponʸtʸ] | ponty | 'carp' |
| /roNdʸ/ | [ronʸdʸ] | rongy | 'rag' |
| /čoNk/ | [čoŋk] | csonk | 'stump' |
| /goNg/ | [goŋg] | gong | 'gong' |
| /la:Npa/ | [la:mpɔ] | lámpa | 'lamp' |
| /eNber/ | [ɛmbɛr] | ember | 'man' |
| /niNfa/ | [nimfɔ] | nimfa | 'nymph' |
| /seNved/ | [sɛɱvɛd] | szenved | 'suffer' |
| /tiNta/ | [tintɔ] | tinta | 'ink' |
| /roNda/ | [rondɔ] | ronda | 'ugly' |
| /piNtˢe/ | [pintˢɛ] | pince | 'cellar' |
| /kaNčal/ | [kɔṇčɔl] | kancsal | 'cross-eyed' |
| /fiNja/ | [fiṇjɔ] | findzsa | 'cup' |
| /piNtʸö:ke/ | [pinʸtʸö:kɛ] | pintyőke | 'finch' |
| /aNdʸal/ | [ɔnʸdʸɔl] | angyal | 'angel' |
| /šoNka/ | [šoŋkɔ] | sonka | 'ham' |
| /heNger/ | [hɛŋgɛr] | henger | 'cylinder' |

Across compound boundary (e.g. *szénpor* [mp] ~ [np] 'coal-dust') and in phrasal contexts (e.g. *nagyon káros* [ŋk] ~ [nk] 'very harmful'), the rule of nasal place assimilation is optional. If it does not apply (and no other rule is applicable), [n] will surface by default.

However, the postlexical application of Nasal Place Assimilation (44) may be bled by a more specific (but optional) rule that also involves the placeless nasal. It applies before continuant consonants (/f v s z š ž x l r j/) and turns vowel + N sequences into long nasalized vowels as in *tanszer* [tɔ̃:sɛr] 'school equipment', *ínség* [ĩ:šeːg] 'misery', *tonhal* [tõ:fiɔl] 'tuna', *bűnjel* [bũ:jɛl] 'corpus delicti',[33] etc. (cf. section 9.2 for discussion). That rule can be formulated as follows (note that the first segment mentioned in the rule is in the nucleus which may be branching or non-branching; the third segment is typically in onset position as in the above examples but is not necessarily so, e.g. *elegáns* [-ã:š] 'well-dressed'):

---

[33] It is instructive to compare *bűnjel* with *tűnj el* [nʸ:] 'disappear' (imp.). In the latter form, Nasal Place Assimilation (44) applies lexically and feeds Palatal *j*-Assimilation (30), hence its only possible pronunciation is [tü:nʸːɛl]. With *bűnjel*, on the other hand, the application of (44) is optional and (30) is inapplicable because that rule is no longer in force in the postlexical module. Furthermore, this form satisfies the conditions of Nasal Deletion/Vowel Nasalization (46) and, if Nasal Place Assimilation (44) applies, those of Nasal Assimilation (47) as well. Thus, it has the following pronunciations: [bü:nj:ɛl] if no rule applies (and /N/ is spelt out as [n] by default); [bü:nʸjɛl] if (44) applies; [bü:j:ɛl] if (44) and (47) apply; and, probably most often, [bü:jɛl] if (46) applies to it. The only impossible rendering is the one that rhymes with *tűnj el*: *[bü:nʸːɛl].

(46)     *Nasal Deletion/Vowel Nasalization (optional)*

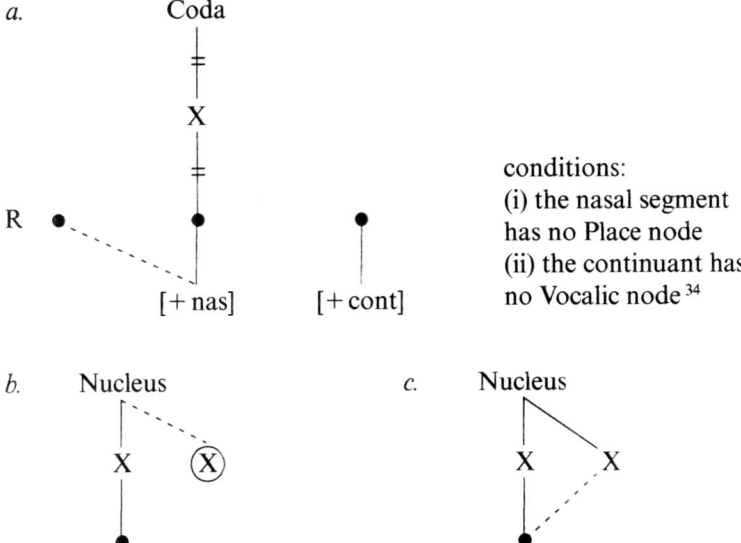

*a.*     Coda

conditions:
(i) the nasal segment
has no Place node
(ii) the continuant has
no Vocalic node [34]

*b.* Nucleus          *c.* Nucleus

If the nucleus is branching (the vowel is long), nothing else happens after the application of (46*a*); the resulting unaffiliated empty X has no effect on the pronunciation of the string. However, in cases where the nucleus is non-branching to begin with, the empty X joins the preceding nucleus (cf. (46*b*)); subsequently, all features (in particular, place and aperture features) of the leftmost segment will be shared between the two nuclear slots; specifically, it will be assumed that the first root node spreads onto the second slot by convention (46*c*).

In cases where Nasal Deletion/Vowel Nasalization (46) does not apply (either because the context consonant is non-continuant or just because the speaker chooses to ignore this option), Nasal Place Assimilation (44) may take place. In a subset of these cases, a further rule becomes applicable. If the nasal is followed by a liquid, it may fully assimilate to that liquid in casual speech as in *olyan lassú* [lː] 'so slow', *olyan rossz* [rː] 'so bad', *olyan jó* [jː] 'so good'. Given that the place assimilation of /N/ to subsequent nasals as in *olyan magas* [mː] 'so high' and *olyan nyakas* [nʲː] 'so obstinate' automatically produces full assimilation, the rule under consideration can be formulated more generally, to apply in the context of any sonorant. On the other hand, /Nv/ does not assimilate into [vː] (cf. *olyan vékony* 'so thin'). One way to avoid this would be to restrict the rule to *coronal* sonorants (in which case /m/

---

[34] The reason why the second condition is expressed in these terms, rather than specifying the context consonant as [+cons] is that [h] has to be included which is assumed to be [–cons] (at the surface). Notice that Nasal Place Assimilation (44) correctly excludes [h] as source of spreading: /N/ does not become glottal before [h].

would also be removed as potential trigger). However, note that the feature [son] is left unspecified for /v/ until the phonetic implementation module (cf. section 7.3). Assuming that the full assimilation discussed here is a phonological rule (although very late and style-dependent), it will apply at a point where no instance of /v/ has been specified as [+son] yet. We will formulate the rule as in (47):

(47)  *Nasal Assimilation (optional)*

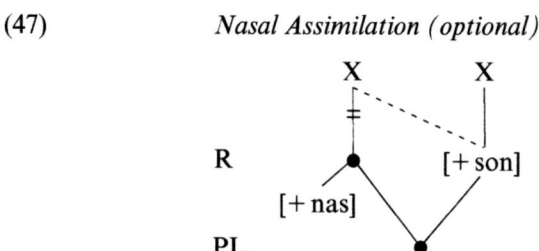

Formulated in this way, Nasal Assimilation (47) is fed by Nasal Place Assimilation (44). Alternatively, we could make it precede (and bleed) (44) by restricting it to placeless nasals (as opposed to shared-place nasals). But in that case a separate rule would be needed for cases like *szegény jó* /nʸj/ → [jː] 'poor good'. That rule would look much like (47) except that it would be restricted to [–ant] coronals (assuming that the two adjacent place nodes dominating identical material have been merged by OCP). Thus, we would have two rules shown in (48):

(48)  a. *Placeless Nasal Assimilation*          b. *Palatal Nasal Assimilation*

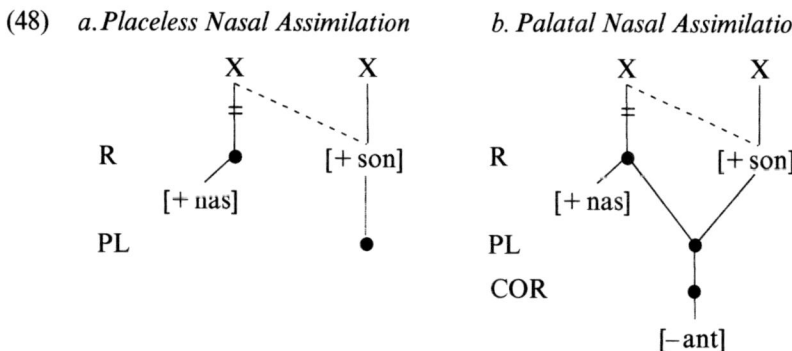

For simplicity, we prefer the analysis involving (47) to the alternative just sketched that involves both (48*a*) and (48*b*).

Turning to the other two nasal consonants, we can state that both may undergo changes that resemble, more or less, the rules discussed so far.

With respect to place assimilation, let us note that /m/ may be realized as labiodental [ɱ] before [f] and [v]/[ʋ]: *somfa* [ɱf] 'dogwood', *nem fog* 'will not'; *hamvak* [ɱʋ] 'ashes' (cf. *hamu* 'ash'), *romváros* 'ruined city', *három veréb* 'three sparrows'. This is reminiscent of the /Nf/, /Nv/ cases briefly referred to in footnote 32 above. We suggested that in cases like *kámfor* /Nf/ → [mf] 'camphor', *honfoglalás* 'conquest' [compound], *vén folyó* 'old river', *Szinva* <geographical name>, *honvágy* 'homesickness', *olyan vékony* 'so thin', the /N/ gets associated to the *flv* and the shared place of the whole cluster is sub-sequently implemented as labiodental. However, in the *m* + *flv* cases two separate (though identical) place specifications are involved. In order for both to be interpreted phonetically as labiodental, as is the case in colloquial speech, we have to assume that they undergo (OCP-triggered) merger first (just like *nʸj* clusters of the *szegény jó* 'poor good' type referred to above). Underlying /m/ does not assimilate to other places of articulation (e.g. *teremt* [mt], *[nt] 'create', *három kör* [mk], *[ŋk] 'three circles'), does not delete with vowel nasalization (e.g. *szomszéd* [ms], *[õːs] 'neighbour', *nem helyes* [mɦ], *[ɛ̃ːɦ] 'not right') and does not fully assimilate to sonorants (e.g. *homlok* [ml], *[lː] 'forehead', *kémnő* [mn], *[nː] 'female spy').

Underlying /nʸ/ is not place-assimilated in standard Hungarian (e.g. *leányka* [nʸk], *[ŋk] 'little girl'), although in some dialects it is; but it may undergo Nasal Assimilation (47) in casual speech (e.g. *vékony jég* [jː] 'thin ice') and has a weakening process of its own before continuants, somewhat similar to Nasal Deletion/Vowel Nasalization (46a): *hányszor* [j̃s] 'how many times', *reménység* [j̃š] 'hope', *viszonylag* [j̃l] 'relatively', *aranyhaj* [j̃ɦ] 'golden hair' (also as an alternative rendering of *vékony jég* [j̃j] 'thin ice'). This rule will be formulated as in (49). Note that nasality is not transferred to the preceding vowel in this case and the nasal segment is not deleted; rather it is weakened into a nasalized palatal approximant:

(49)   *Palatal Nasal Weakening (optional)*

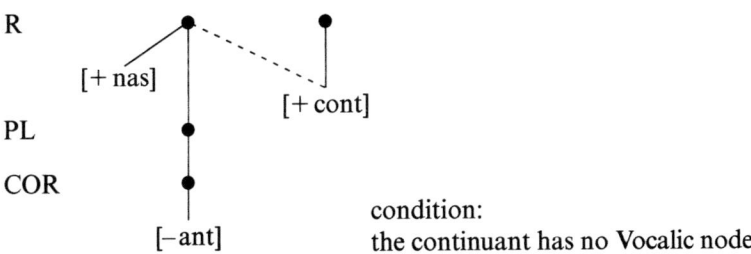

condition:
the continuant has no Vocalic node

Given that all (postlexical) processes discussed in this section are optional (except nasal place assimilation morpheme internally), most of our examples have several possible pronunciations. The following table shows this:

| (50) | no rule | default [n] | (44) | (46) | (47) | (49) |
|---|---|---|---|---|---|---|
| én 'I' | — | [eːn] | — | — | — | — |
| sonka 'ham' | — | — | [šoŋkɔ] | — | — | — |
| szénpor 'coal dust' | — | [seːnpor] | [seːmpor] | — | — | — |
| inség 'misery' | — | — | [iːn̪šeːg] | [ĩːšeːg] | — | — |
| tonhal 'tuna' | — | [tonɦɔl] | — | [tõːɦɔl] | — | — |
| honfoglalás 'conquest' | — | [honfog-] | [hoɱfog-] | [hõːfog-] | — | — |
| olyan lassú 'so slow' | — | [ojɔnlɔšːu] | [ojɔnlɔšːu] | [ojɔ̃ːlɔšːu] | [ojɔlːɔšːu] | — |
| bűnjel 'corpus delicti' | — | [büːnjɛl] | [büːnʲjɛl] | [bũːjɛl] | [büːjːɛl] | [büːj̃jɛl] |
| szomszéd 'neighbour' | [somseːd] | — | — | — | — | — |
| kámfor 'camphor' | [kaːɱfor] | — | — | — | — | — |
| aranyhaj 'golden hair' | [ɔrɔnʲɦɔj] | — | — | — | — | [ɔrɔj̃ɦɔj] |
| viszonylag 'relatively' | [visonʲlɔg] | — | — | — | — | [visoj̃lɔg] |
| vékony jég 'thin ice' | [-konʲjeːg] | — | — | — | [-kojːeːg] | [-koj̃jeːg] |

### 7.4.2. Liquid deletion

It was argued in section 4.2 that the class of liquids includes, in Hungarian, /l/, /r/, and /j/. The arguments listed there (with respect to the status of /j/) can be supplemented by rule (47) in the previous section (Nasal Assimilation) that treats all three segments in exactly the same manner.

Another process for the purposes of which /j/ patterns with the other liquids is liquid deletion, to be discussed in more detail in section 9.2. As we will see there, it is /l/ that is deleted the most easily of the three, e.g. *balra* %[bɔrɔ] 'to the left', *elvisz* %[ɛːvis] 'take away', *el kell menni* %[ɛːkɛːmɛnːi] 'one must leave'. The deletion of /r/ (*egyszer csak* %[ɛtˢːɛːčɔk] 'after a while') is usually observed in casual speech only but in *arra* 'that way', *erre* 'this way', *merre* 'which way' it applies in colloquial (and even in moderately formal) speech, too: [ɔːrɔ], [ɛːrɛ], [mɛːrɛ]. /j/ is primarily dropped after front vowels: *gyűjt* [dʲüːt] 'collect', *szíjra* [siːrɔ] 'on a leash', *mélység* [meːšeːg] 'depth', *felejt-hetetlen* [fɛlɛːt(h)ɛtɛtlɛn] 'unforgettable'. The rule we give in (51) abstracts away from this last observation but refers to the difference between /l/ and the other two liquids by the parentheses surrounding [+lat] in it. This is meant to suggest that, with increased speed and/or degree of casualness, this feature specification is dropped from the rule and it becomes applicable to all three liquids. The parenthesized X shows that the rule applies in branching codas as well, deleting the first coda consonant if it is a liquid. In cases where the preceding vowel is short, the application of the deletion rule automatically triggers that of the redundancy rules stated in (46b) and (46c) above; taken together, (51) and (46b–c) account for the compensatory lengthening effect observed in the examples involving underlying short vowels.

(51)                      *Liquid Deletion (optional)*

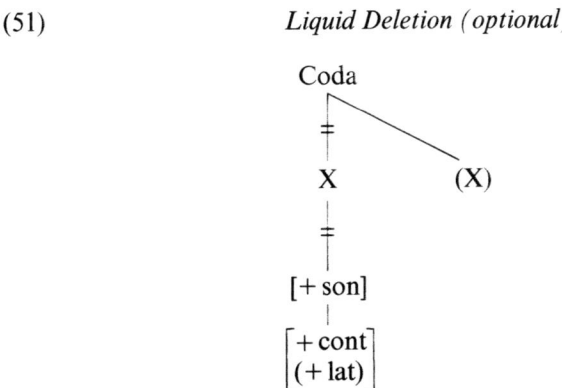

This concludes our discussion of processes involving consonants. Some rules we have presented here already referred to syllable constituents (Nucleus, Coda, etc.). In the following chapter, we turn to a number of further processes triggered by or affecting syllable structure.

# PROCESSES CONDITIONED BY SYLLABLE STRUCTURE

## 8.1. VOWEL ~ ZERO ALTERNATIONS

Hungarian has an intricate system of vowel ~ zero alternations. We shall see as we proceed that not all of them can be analysed in the same way phonologically. Henceforward, we shall informally refer to any vowel that alternates with zero as 'unstable' and denote it with the symbol $V_u$. This term is meant to be neutral with respect to whether a given vowel ~ zero alternation is considered to be the result of epenthesis, vowel deletion, or some other phonological mechanism. In this section first we examine the distribution and the quality of unstable vowels and then we present an analysis of vowel ~ zero alternations.

The unstable vowels may occur stem-internally, i.e. inside a stem, and stem-externally, i.e. between a stem and a suffix.

### 8.1.1. Stem-internal unstable vowels: 'epenthetic' stems

The stem-internal occurrence of unstable vowels is restricted to a non-productive class of stems traditionally called 'epenthetic' (e.g. Vago 1980a),[1] e.g. *bokor* 'bush' (compare *bokr-ok* 'bush' (pl.)). Although we shall refer to them by the traditional name, we make no claim here about (and actually will argue against) their epenthetic character. In general, the unstable vowel of these stems is phonetically expressed if they occur in isolation, or before consonant-initial suffixes, but it does not appear at the surface if a vowel-initial suffix follows the stem.[2] This is shown in (1):

---

[1] Although it is a non-productive class, it is rather populous: it contains about 150 verbs and 250 noun stems (some of which are morphologically complex). There are also a few adjectives and numerals that are 'epenthetic'.

[2] Some-vowel initial suffixes behave differently (e.g. terminative -*ig*, causal-final -*ért*: *bokor-ig*, *bokor-ért* and not *\*bokr-ig*, *\*bokr-ért*). These will be discussed in section 8.1.4.6.

(1)　_#　　　　　　　_C-initial suffix　　　　　_V-initial suffix

　　bokor　　　　　　bokor-ban 'bush' (iness.)　　bokr-ok
　　retek 'radish'　　retek-ben 'radish' (iness.)　retk-ek 'radish' (pl.)
　　kölyök 'kid'　　　kölyök-ben 'kid' (iness.)　　kölyk-ök 'kid' (pl.)

The unstable vowels in 'epenthetic' stems are regularly short and mid ($[-open_1, +open_2]$) whose frontness and rounding is determined by vowel harmony.[3] Thus, they exhibit a ternary alternation: o/ö/e (see the examples in (1)) when they are phonetically realized. The front unrounded alternant is phonetically low, but this is due to phonetic implementation (see section 6.1). There are only seven 'epenthetic' stems in which the height of the unstable vowel is irregular (high or low). These are shown in (2):

(2) a vacak 'something worthless'　　vack-ot 'something worthless' (acc.)
　　　kazal 'haystack'　　　　　　　kazl-at 'haystack' (acc.)
　　　ajak 'lip'　　　　　　　　　　ajk-at 'lip' (acc.)
　　　fogazz 'teethe!'　　　　　　　fogz-ás 'teething'

　i　őriz 'guard' (3sg pres. indef.)　　őrz-i 'guards' (3sg pres. def.)

　ü　becsül 'estimate' (3sg pres. indef.)　becsl-és 'estimate' (noun)

　u　bajusz 'moustache'　　　　　　bajsz-ot 'moustache' (acc.)

Stem-internal unstable vowels always occur in the last syllable of the stem, but the only vowel of a monosyllabic stem may not be unstable. If we examine the final $C_i V_u C_j$ string of 'epenthetic' stems (where $V_u$ denotes the unstable vowel, and $C_i$ and $C_j$ are consonants of any kind flanking the unstable vowel), it becomes clear that here the vowel ~ zero alternation is not phonotactically motivated. The reason is that the stem-final consonant cluster $C_i C_j$ that appears in the stem alternant whose unstable vowel is phonetically unexpressed is not always a phonotactically ill-formed final cluster (cf. Törkenczy 1992). Of course, there are consonants separated by a stem-internal unstable vowel that would make up an ill-formed cluster finally (e.g. the cluster /kr/ that occurs in bokr-ok, for instance), but this is not always the case. (3) shows some examples where the $C_i C_j$ cluster corresponding to the consonants flanking the unstable vowel in an 'epenthetic' stem is a possible word-final cluster:

---

[3] In a few 'epenthetic' stems with a back unstable vowel, the vowel in the syllable preceding the unstable vowel is front i, í, or é: e.g. szirom 'petal', szirm-ok 'petal' (pl.), kínoz 'torture' (3sg pres. indef.), kínz-ás 'torture' (noun), céloz 'aim' (3sg pres. indef.), célzás 'aiming'. These behave exactly like other antiharmonic stems with respect to vowel harmony and have analogous underlying representations (i.e. they have an unlinked DOR that can link up to the unstable vowel if it is realized, cf. section 6.1).

(3)[4]

| $C_iV_uC_j$ | | $C_iC_j\#$ |
| --- | --- | --- |
| viszonoz | viszonz-om | vonz |
| telek | telk-ek | halk |
| füröd-nek | fürd-és | kard |
| szerez | szerz-ünk | borz |
| torony | torny-ok | szörny |
| majom | majm-ok | slejm |
| bagoly | bagly-ok | fogj |

Note that most of the $C_iC_j$ clusters in (3) are not only well-formed as word final clusters, but are perfectly well-formed syllable codas as well (/gj/ as in *fogj* is the only one that is not a well-formed branching coda, although it is a well-formed word-final cluster with /j/ in the appendix).[5] Naturally, not all well-formed codas or word-final consonant clusters appear divided by an unstable vowel in these stems. Nevertheless, intuitively, even the non-occurring types seem possible. For example, there is no 'epenthetic' stem with /n/ as $C_i$ and /č/ as $C_j$ in its $C_iV_uC_j$ string—but such a stem is perfectly possible phonologically: hypothetical *penecs* could be an 'epenthetic' stem with *pencs-* as an alternant before vowel-initial suffixes. Thus, we conclude that vowel ~ zero alternation is not phonotactically motivated stem-internally. Still, there are certain phonotactic restrictions on the shape of 'epenthetic' stems: (i) their unstable vowel is never preceded or followed by a consonant cluster (*$CCV_uC$, *$CV_uCC$);[6] (ii) the consonants flanking the unstable vowel are never identical (*$C_iV_uC_j$ if $C_i=C_j$);[7] and (iii) if they are both obstruents, they are either both voiced or both voiceless *$C_iV_uC_j$, if $C_i$, $C_j$ = [−son] and only one of them has a laryngeal node).

It is a lexical property of a stem if it is 'epenthetic' or not, i.e. whether it has an unstable vowel or not. There are near-identical pairs of stems such

---

[4] Glosses: *viszonoz* 'reciprocate' (3sg pres. indef.), *viszonz-om* (1st pres. indef.), *vonz* 'attract' (3sg pres. indef.), *telek* 'land', *telk-ek* 'land' (pl.), *halk* 'quiet', *füröd-nek* 'bathe' (3pl pres. indef.), *fürd-és* 'bathing', *kard* 'sword', *szerez* 'acquire' (3sg pres. indef.), *szerz-ünk* 'acquire' (1pl pres. indef.), *borz* 'badger', *torony* 'tower', *torny-ok* 'tower' (pl.), *szörny* 'monster', *majom* 'monkey', *majm-ok* 'monkey' (pl.), *slejm* 'phlegm', *bagoly* 'owl', *bagly-ok* 'owl' (pl.), *fogj* 'hold!'

[5] We have not included occurring *irregular* $C_iC_j\#$ clusters (cf. section 5.2.4.2) in (3) that also occur in $C_iV_uC_j$ strings. Several such examples exist: e.g. /tk/ *retek*, *retk-ek*, *Detk* ‹place name›; /čk/ *mocsok* 'filth', *mocsk-os* 'filthy', *Recsk* ‹place name›; /sk/ *piszok* 'dirt', *piszk-os* 'dirty', *maszk* 'mask'.

[6] There is a single exception: the vowel of the denominal verb-forming suffix *-Vz* is unstable after the cluster-final stem *hang* 'sound': *hang-oz-tat* 'proclaim' (3sg pres. indef.), *hang-z-om* 'sound' (1sg pres. indef.). The same suffix-initial vowel is always stable after other cluster-final stems: *folt-oz-om* 'patch' (1sg pres. def.), *rend-ez-em* 'put in order' (1sg pres. def.), *bors-oz-om* 'pepper' (1sg pres. def.) and not *folt-z-om*, *rend-z-em*, *bors-z-om*.

It has been suggested (P. Rebrus, personal communication) that this latter constraint is more general and requires that $C_i$ and $C_j$ should not be homorganic. Given the feature system we use, this claim is not true. In the following examples of epenthetic stems the consonants flanking $V_u$ have the same place: /t$V_u$r/ *bátor* 'brave', /s$V_u$n/ *vászon* 'canvas', /t$V_u$l/ *ismétel* 'repeat', /z$V_u$l/ *közöl* 'inform' etc.

that one of the members of a given pair has an unstable vowel where the other member has a stable one. Compare the 'epenthetic' stems in (4a) with those in (4b) whose last vowel is stable:

(4)          $-C_iV_uC_j$                                    $-C_iVC_j$
  a. terem 'hall'      term-ek (pl.)       b. perem 'edge'       perem-ek (pl.)
     vödör 'bucket'    vödr-ök (pl.)          csődör 'stallion'  csődör-ök (pl.)
     szobor 'statue'   szobr-ok (pl.)         tábor 'camp'       tábor-ok (pl.)
     torony 'tower'    torny-ok (pl.)         szurony 'bayonet'  szurony-ok (pl.)

The data described above suggest that—unless arbitrary lexical marking is involved—stem-internal vowel ~ zero alternation is neither due to epenthesis nor to the deletion of a vowel represented on a par with vowels that do not alternate with zero.[8]

The unstable vowel cannot be epenthetic because if the vowelless alternant of an 'epenthetic' stem is taken to be underlying, the epenthesis site cannot be predicted, i.e. the clusters that are supposedly broken up by epenthesis cannot be distinguished from those that are not (compare 'epenthetic' *torony* and non-alternating cluster-final *szörny*).

Stems like *csukl-ik* 'hiccup' (pres. 3sg indef.), *bűzl-ik* 'stink' (pres. 3sg indef.), *vedl-ik* 'slough' (pres. 3sg indef.), etc. (Hetzron 1975) provide a further argument against the epenthesis analysis. These are bound stems that end in ill-formed coda clusters. They can only occur before (surface) vowel-initial suffixes, and their stem-final clusters are never broken up. Before (surface) vowel-inital suffixes, they look like 'epenthetic' stems (compare *csukl-ás* 'hiccup' (noun) with 'epenthetic' *vezekl-és* 'penitence'). However, their paradigms are defective in that they simply do not have the forms in which a suffix (or the lack of it) would render the stem-final cluster unsyllabifiable. Crucially, their stem-final clusters cannot be repaired by epenthesis (*csukol-j* 'hiccup!' (imp.)). In contrast to these defective stems, 'epenthetic' ones do have the corresponding forms (*vezekel-j* 'repent!' (imp.)). If we want to distinguish defective stems from 'epenthetic' ones representationally, both types cannot be cluster-final, because then, the ill-formed final clusters that are to be broken up by epenthesis cannot be distinguished from those that are not.

On the other hand, the vowelless alternant of an 'epenthetic' stem cannot be derived by deletion from an underlying CVC-final form (where V is represented like other vowels) because (i) it would not be possible to distinguish CVC-final stems that exhibit vowel ~ zero alternation from those that do not (compare 'epenthetic' *torony* and non-alternating *szurony*), and (ii) it would not be possible to explain why the quality/quantity of the unstable vowel is predictable.

---

[8] Both epenthesis and deletion have been proposed in the literature (cf. Vago 1980a, Jensen and Stong-Jensen 1988, 1989b (epenthesis); Kornai 1990a (vowel deletion)).

Any successful analysis will have to be able to make a three-way distinction between triplets of stems like *torony–szurony–szörny*, and must distinguish 'epenthetic' stems from 'defective' CC-final ones like *csukl-*.

### 8.1.2. Stem-external vowel ~ zero alternations: stem-final unstable vowels and 'linking' vowels

Several (types of) suffixes are involved in stem-external vowel ~ zero alternation. Given that we have defined stem-external vowel~zero alternation as occurring at the boundary between a stem and a suffix, there are two logical possibilities: (i) the stem-final vowel may be unstable, or (ii) the suffix-initial vowel may be unstable.[9] Both (i) and (ii) occur in Hungarian.

#### 8.1.2.1. Stem-final vowel ~ zero alternations

Stem-final vowel ~ zero alternations only occur before the deadjectival verb-forming suffixes *-ít* (*barn-ít* 'make brown', cf. *barna* 'brown'), *-ul/-ül* (*laz-ul* 'become loose', cf. *laza* 'loose'), and *-odik/-edik/-ödik* (*szomor-odik* 'become sad', cf. *szomorú* 'sad'; *feket-edik*[10] 'become black', cf. *fekete* 'black'). As can be seen, adjectival stems lose their final vowel before these suffixes. It is the special property of these suffixes (and not of the stems) that they cause the loss of the stem-final vowels since the same stems retain their final vowels before other vowel-initial suffixes. Consider the comparative forms *barná-bb*, *lazá-bb*, *szomorú-bb*, *feketé-bb*, etc. where the stem-final vowels are retained and the initial unstable vowel of the comparative suffix *-V$_u$bb* does not appear at the surface.[11] In the case of stem-final vowel ~ zero alternation there is no restriction on the quality or quantity of the unstable vowel. It can be high (*gömböly-ödik* 'become spherical', cf. *gömbölyű* 'spherical'), mid (*fak-ít* 'make pale', cf. *fakó* 'pale'), or low (*laz-ul*

---

[9] Unstable vowels do not occur in non-initial position in a suffix. Noun-forming (derivational) *-alom/-elem* might be considered a suffix with an internal unstable vowel: *vigalom* 'merrymaking', *vigalm-at* 'merry-making' (acc.), cf. *víg* 'merry'; *félelem* 'fear', *félelm-et* 'fear' (acc.), cf. *fél* 'be frightened'. This suffix, however, is no longer productive and the morphological complexity of the stems containing it has become obscured. Therefore we do not consider this 'ending' a suffix in the stems in which it occurs. (It would also be irregular if analysed as a suffix since it is clearly derivational, not adjective-forming, but nevertheless it is lowering (cf. section 8.1.3 below).)

[10] The fact that the form is *feketedik* and not *\*feketédik* is evidence that it is indeed the stem-final vowel that deletes and not the suffix-initial one. If the suffix-initial vowel had deleted, then the stem-final *e* would have had to lengthen by Low Vowel Lengthening (cf. section 3.1.1) giving *\*feketédik*.

[11] There are four stems that irregularly lose their stem-final vowels before the comparative suffix and some other vowel-initial suffixes: *könnyű* 'easy' (*könny-ebb* 'easier', *könny-en* 'easily'), *ifjú* 'young' (*ifj-abb* 'younger', *ifj-an* 'as a youth'), *hosszú* 'long' (*hossz-abb*, 'longer', *hossz-an* 'at length'), *lassú* 'slow' (*lass-abb* 'slower', *lass-an* 'slowly'). There are also three exceptional nominal stems that lose their final vowels before some vowel-initial suffixes: *borjú* 'calf' (*borj-ak* 'calf' (pl.)), *varjú* 'crow' (*varj-ak* 'crow' (pl.)), *ifjú* 'youth' (*ifj-ak* 'youth' (pl.)).

cf. *laza*), and it may equally be short or long.[12] The unpredictability of the unstable vowel suggests that the mechanism responsible for this alternation is deletion.

### 8.1.2.2. *Suffix-initial vowel ~ zero alternation*

Suffix-initial vowel ~ zero alternation is more common and more complex than the stem-final one. It is not restricted to a handful of suffixes: many suffixes begin with an unstable vowel (called a 'linking' vowel traditionally).[13] Two types of suffixes may be distinguished depending on what motivates the suffix-initial vowel ~ zero alternation.

Type A. The initial unstable vowel of Type A suffixes is only unrealized when they are added to a vowel-final stem; a phonetically realized vowel is always present after a consonant-final stem, regardless of the identity of the stem-final consonant. Suffixes of this type include *-ok/-ek/-ök* 'pl.', *-on/-en/-ön* 'spr.', *-om/-em/-öm* '1sg poss.', etc. Verb roots end in a consonant before vowel-initial suffixes in Hungarian. Therefore, on the basis of the presence/absence of vowel ~ zero alternation alone, in principle, a verbal suffix with an initial stable vowel is indistinguishable from a Type A suffix added to a verb root, since the latter would also not display vowel ~ zero alternation. We shall consider vowel-initial verbal suffixes as Type A if their initial vowel is short *o/e/ö* (i.e. the typical unstable vowel), e.g. *-ok/-ek/-ök* '1sg pres. indef.', *-od/-ed/-öd* '2sg pres. def.', *-om/-em/-öm* '1sg pres. def'. This is confirmed by their behaviour after a relative verb stem that ends in a vowel; compare *lát-om* 'see' (1sg pres. def.) and *lát-ná-m* 'see' (1sg pres. cond def.); *ül-ök* 'sit' (1sg pres. indef.) and *ül-né-k* 'sit' (1sg pres. cond. indef.). The behaviour of Type A suffixes is illustrated in (5):

|     |       | *pl.* | *spr.* | *1sg poss.* |
|-----|-------|-------|--------|-------------|
| (5) | C-final stem | lány 'girl' | lány-ok | lány-on | lány-om |
|     | V-final stem | holló 'raven' | holló-k | holló-n | holló-m |

In these suffixes the presence of a realized unstable vowel is not motivated by the phonotactics of final consonant clusters, i.e. the vowel is not there to 'repair' otherwise ill-formed clusters. This can be seen in (6) where the realized suffix-initial unstable vowel appears to 'break up' well-formed word-final clusters all of which are also well-formed codas.

---

[12] Note that (i) these suffixes may not be added to any adjective and (ii) after some vowel-final adjectives a *-s-* ([š]) is inserted before the suffix, e.g. *olcsó-s-ít* 'make cheap', *karcsú-s-ít* 'make slim', *állandó-s-ul* 'become constant'.

[13] The literature is divided concerning the morphological affiliation of these vowels: there is no agreement as to whether linking vowels are part of the stem, or part of the suffix, or a separate entity between the stem and the suffix (see Antal (1977) and Papp (1975) and references therein). We will take them to be part of the suffix.

(6)     $C_i$-$V_u$$C_j$                               -$C_i$$C_j$
        dal-ok 'song' (pl.)                     halk 'quiet'
        bor-ok 'wine' (pl.)                     park 'id.'
        tan-ok 'tenet' (pl.)                    tank 'id.'
        kér-ed 'ask' (2sg pres. def.)           térd 'knee'
        ostor-oz 'whip' (3sg pres. indef.)      borz 'badger'
        csalán-os 'nettle' (adj.)               gáláns 'gallant'

As there is no phonotactic interaction between the consonants flanking the unstable vowel, it is reasonable to assume that the mechanism responsible for the vowel ~ zero alternation here is not epenthesis.

Type B. The vowel ~ zero alternation in Type B suffixes is phonotactically motivated. The accusative (-$V_u$$t$) is the suffix that unquestionably belongs here. In this suffix the unstable vowel is phonetically unrealized if the suffixal consonant can syllabify as (part of) a well-formed coda, i.e. no linking vowel appears after vowels (7a), and stem-final consonants with which $t$ can form a branching coda (7b). Otherwise (7c), there is a vowel preceding the $t$ at the surface:[14]

(7) a.  holló-t 'raven'     b.  ón-t 'tin'         c.  nyom-ot 'trace'
        kocsi-t 'cart'          lány-t 'girl'          pad-ot 'bench'
        tevé-t 'camel'          dal-t 'song'           kép-et 'picture'
        kapu-t 'gate'           sör-t 'beer'           tök-öt 'pumpkin'
        fő-t 'head'             baj-t 'trouble'        hegy-et 'hill'
        tű-t 'needle'           rés-t 'gap'            ív-et 'arc'
        anyá-t 'mother'         kosz-t 'dirt'          zsiráf-ot 'giraffe'
        sí-t 'ski'              gőz-t 'steam'          doh-ot 'must'
        menü-t 'menu'           varázs-t 'magic'       rab-ot 'prisoner'

Interestingly, a linking vowel is present after an 'epenthetic' stem even if the final consonant *can* form a licit coda with the following *t*—compare (8a) and (8b).

(8) a.  *non-epenthetic stems*      b.  *'epenthetic' stems*
        ón-t                            haszn-ot 'profit'
        lány-t                          torny-ot 'tower'
        dal-t                           öbl-öt 'bay'
        sör-t                           ökr-öt 'ox'
        baj-t                           bagly-ot 'owl'
        rés-t                           —
        kosz-t                          bajsz-ot 'moustache'
        gőz-t                           —
        varázs-t                        —

---

[14]  In some marked cases the -*t* can also syllabify as an appendix, cf. section 5.2.4.3.

Some of the words in (8b) have alternative forms without the linking vowel, in which case the stem internal unstable vowel is phonetically expressed, e.g. *bajusz-t/bajsz-ot, öböl-t/öbl-öt*. The conditions on this variation are idiosyncratic and often unclear (some stems do not show any variation, others only with certain suffixes and not with others; there is variation across speakers, etc).[15]

Disregarding the few exceptional cases discussed in section 5.2.4.3, regularly, the accusative attaches to stems that end in a consonant cluster with a linking vowel (e.g. *rajz-ot* 'drawing' acc.). This follows from the fact that the coda in Hungarian is maximally binary branching. It has to be pointed out, however, that stem-final geminates are peculiar in that some of them behave as if they were single consonants: there is no linking vowel after the stem-final geminates /ss, šš, zz, nn, nʸnʸ, ll, rr, jj/, i.e. those geminates whose short counterparts can form a licit coda with /t/ (e.g. *idill-t* [idilt] 'idyll' (acc.), *finn-t* [fint] 'Finnish' (acc.), *plüss-t* [plüšt] 'plush' (acc.), *dzsessz-t* [ʒɛst] 'jazz' (acc.)).[16]

Note that, although geminates are well-formed branching codas, there is a linking vowel present when the accusative is added to a *t*-final stem: *bot-ot* 'stick' (acc.), *rét-et* 'meadow' (acc.), *öt-öt* 'five' (acc.), and not *bot-t*, *öt-t*, *rét-t*.

The past tense suffix also belongs to Type B in that the realization of its unstable vowel is also phonotactically motivated, but in some respects it behaves differently from the accusative. The differences are as follows:

(i) Its consonantal part is realized as geminate *-tt* when it is not adjacent to another consonant (regardless of whether the preceding vowel is part of the stem)[17] or if it is an unstable vowel that appears at the surface: *lő-tt* 'shoot' (3sg past. indef.), *dob-ott* 'throw' (3sg past. indef.). This is usually explained as a result of the difference between the representation of the accusative and the past suffix (e.g. Vago 1980a).

(ii) The conditions as to when the unstable vowel appears at the surface after consonant-final stems are somewhat different: it can only be unexpressed if the stem-final consonant and *-t* can form a licit branching coda *whose first term is a sonorant*. Thus, as opposed to the accusative (9a), only a subset of the possible complex codas are available for syllabification for the past tense suffix (9b):

---

[15] Type A suffixes typically do not cause this kind of variation in 'epenthetic' stems. The unstable vowel of the stem is not realized before such a suffix: *bagly-ok* 'owls' and not *bagoly-ok*.

[16] The geminates later degeminate by post-lexical degemination (cf. section 9.4).

[17] 'v-adding' stems behave differently before the accusative and the past tense suffix. The final *v* of these stems does not appear before the past tense suffix (e.g. *lő-tt* 'shoot' (3sg past indef.), *fő-tt* 'cook' (3sg past), *ri-tt* 'cry' (3sg past)), but it does before the accusative (e.g. *lov-at* 'horse' (acc.), *köv-et* 'stone' (acc.)). Cf. section 8.2.1 on 'v-adding' stems.

(9) a.   *accusative*                    b.   *past*
     ón-t                           ken-t 'smear'
     lány-t                         hány-t 'vomit'
     dal-t                          él-t 'live'
     sör-t                          vár-t 'wait'
     baj-t                          fúj-t 'blow'
     rés-t                                          kés-ett 'be late'
     kosz-t                                         csempész-ett 'smuggle'
     gőz-t                                          főz-ött 'cook'
     varázs-t                                       —

This additional restriction can be seen as an idiosyncratic property of the
past tense suffix which is not related to the phonological mechanism governing the behaviour of the unstable vowel in Type B suffixes.[18]

(iii) The suffix-initial vowel ~ zero alternation of the past tense morpheme
is not only sensitive to the phonological material that precedes the suffix, but
also to that which follows it: there is no linking vowel at the surface if the suffix is followed by a vowel(-initial suffix).[19]

(10)                      *3sg past indef.*        *3sg past def.*
     nyom 'push'          nyom-ott                 nyom-t-a
     rak 'put'            rak-ott                  rak-t-a
     vés 'chisel'         vés-ett                  vés-t-e

The linking vowel is *sometimes* present after cluster-final stems even if a
vowel-initial suffix follows the past suffix. Three types of behaviour may be
distinguished: (i) after some CC-final stems the linking vowel is optional
(11a); (ii) after others it is compulsory (11b); and (iii) some CC-final stems do
not permit a linking vowel in this environment (11c).

(11)                      *3sg past indef.*        *3pl past indef*
a.   fing 'fart'          fing-ott                 fing-ott-ak / fing-t-ak
     mond 'say'           mond-ott                 mond-ott-ak / mond-t-ak
     told 'lengthen'      told-ott                 told-ott-ak / told-t-ak

---

[18] Another idiosyncratic property of the past tense suffix is that no linking vowel appears
after two irregular (bound) verb stems: *feküd-* 'lie' and *alud-* 'sleep' (*feküd-t* (3sg past), *alud-t* (3sg
past) and after verbs that belong to a class of stems ending in *adled*: e.g. *szalad-t* 'run' (3sg past),
*marad-t* 'remain' (3sg past), *mered-t* 'stand out' (3sg past), *reped-t* 'burst' (3sg past). These are
pronounced with final geminate [tt] as a result of Voicing Assimilation, cf. section 7.3. Note that
some verbs ending in *adled* do not belong to this class: e.g. *ad-ott* 'give' (3sg past), *fed-ett* 'cover'
(3sg past), *fogad-ott* 'receive' (3sg past), *szenved-ett* 'suffer' (3sg past), *tagad-ott* 'deny' (3sg past).
Essentially, the linking vowel does not appear if the final *adled* string can be regarded as a suffix
(cf. Rebrus 2000).
[19] It is not possible to test whether this behaviour of the past tense suffix is really different
from that of the accusative since the latter may not be followed by another suffix.

| b. | fest 'paint' | fest-ett | fest-ett-ek | *fest-t-ek |
|---|---|---|---|---|
| | látsz-ik 'seem' | látsz-ott | látsz-ott-ak | *látsz-t-ak |
| | csukl-ik 'hiccup' | csukl-ott | csukl-ott-ak | *csukl-t-ak |
| | old 'solve' | old-ott | old-ott-ak | *old-t-ak |
| c. | küld 'send' | küld-ött | küld-t-ek | *küld-ött-ek |
| | kezd 'begin' | kezd-ett | kezd-t-ek | *kezd-ett-ek |
| | fedd 'scold' | fedd-ett | fedd-t-ek | *fedd-ett-ek |

The phonological shape of a given verb stem only partly determines its behaviour. The linking vowel must be present after the stem-final cluster if it ends in a *t* (e.g. *fest-ett-ek*) or if it is not a possible branching coda (e.g. *csukl-ott-ak*). Otherwise, it is phonologically unpredictable whether a linking vowel is compulsory, optional, or disallowed after a given CC-final verb stem in this environment. Indeed, after some stem-final clusters (e.g. /ld/), all three kinds of behaviour are attested, i.e. some stems have both forms (*áld-t-ak/áld-ott-ak* 'bless' (3pl past indef.)) while others have only one, either with or without the linking vowel (*küld-t-ek/*küld-ött-ek* 'send' (3pl past indef.) vs. *\*száguld-t-ak/száguld-ott-ak* 'speed' (3pl past indef.)). This suggests that *phonotactically* the strings VC1C2VttV and VC1C2tV are both well-formed provided that C2 ≠ /t/ and C1C2 is a well-formed coda. Individual CC-final stems that end in a well-formed coda[20] whose second term is not /t/ must bear an arbitrary lexical mark as to which string they require or whether they permit both.

The past participle -*Vtt/-t* behaves similarly to the past tense suffix as the conditions on the appearance of the unstable vowel are usually the same: *köt-ött* 'bind', *megtér-t* 'convert' (3sg past indef. or past participle), *köt-ött-ek*, *megtér-t-ek* (3pl past indef. or past participle+pl.). Compare *Ő megtért* '(S)he converted' and *Ő megtért ember* '(S)he is a converted person'. The past participle, however, is sometimes irregular in that the linking vowel shows up unexpectedly even if the conditions for its non-occurrence are met: e.g. *ír-ott* 'written' vs. *ír-t* 'write' (3sg past indef.), *ad-ott-ak* 'give' (past participle+pl.) vs. *ad-t-ak* 'give' (3pl past indef.). This behaviour is phonologically unpredictable and is a lexical property of the stem.

There are suffixes in Hungarian whose initial unstable vowel is followed by a consonant cluster (e.g. associative -*ostul/-estül/-östül*, distributive -*onként/-enként/-önként*, distributive-temporal -*onta/-ente/-önte*, diminutive -*ocska/*

---

[20] Verb stems may end in clusters that are not well-formed codas. With one exception (*metsz* 'etch' (3sg pres. indef.)) they are bound stems. These stems are defective (cf. section 8.1.1) and typically belong to the -*ik* class of verbs that have the suffix -*ik* in present indefinite 3rd person singular instead of zero. Examples include *bűzl-ik* 'stink', *vedl-ik* 'slough', *áraml-ik* 'flow', *vonagl-ik* 'writhe', *habz-ik* 'foam', *játsz-ik* 'play', etc. (It must be pointed out that not all -*ik* verbs end in clusters that are ill-formed as a coda, e.g. *álmod-ik* 'dream', *hull-ik* 'fall', *csikland-ik* 'tickle', etc.) There are two defective stems that are not -*ik* verbs (both are used in the definite conjugation only): *kétl-* 'doubt' and *sínyl-* 'suffer'; e.g. *kétl-em* (1sg pres. def.), *sínyl-i* (3sg pres. def.). Special thanks are due to Attila Novák and Nóra Wenszky for these two elusive items.

*-ecskel-öcske*, and *-osdil-esdil-ösdi* [-oždi/-eždi/-öždi] 'game of/about'). They behave like Type A suffixes with respect to vowel ~ zero alternation, i.e. they attach to all consonant-final stems with a linking vowel (e.g. *dob-ostul* 'together with the drum', *fej-enként* 'per head', *nap-onta* 'every day', *tök-öcske* 'little pumpkin', *indián-osdi* 'game in which children play (American) Indians'), but have no linking vowel after vowel-final stems (e.g. *kutyá-stul* 'together with the dog', *falu-nként* 'village by village', *falu-cska* 'little village', *katoná-sdi* 'game in which children play soldiers'). Note, however, that since internal and final CCC clusters are ill-formed,[21] the alternation may equally be seen as phonotactically motivated, in which case they may be regarded as Type B suffixes.

As in the case of Type B suffixes there is phonotactic interaction between the stem-final consonant and the suffixal one, the mechanism underlying the vowel ~ zero alternation and/or the representation of these suffixes must be different from that of Type A suffixes.

Disregarding the special case of lowering (to be discussed in section 8.1.3 below), in suffix-initial vowel ~ zero alternation, the quality of the unstable vowel is identical to that involved in stem internal vowel ~ zero alternation. There is a single suffix the height of whose unstable vowel is irregularly high instead of mid: possessive 1pl *-unkl-ünk* (e.g. *bot-unk* 'our stick' vs. *fá-nk* 'our tree').

### 8.1.3. Lowering

We have seen above that the quality of the unstable vowel is regularly mid both stem-internally and suffix-initially. In the latter case, however, the linking vowel is sometimes low instead of mid (compare *ház ak* 'house' (pl.) with *gáz-ok* 'gas' (pl.)). We shall refer to this phenomenon as the lowering of unstable vowels. The back alternant of a lowered linking vowel is low *a* [ɔ] instead of *o*. The front alternant is low [ɛ] whose lowness is not the result of phonetic implementation (which interprets mid *e* as low as well, cf. section 6.1), but is due to lowering. Since lowering makes the linking vowel low, rounding harmony cannot apply to it because Hungarian has no low front rounded vowels (compare *szüz-ek* 'virgin' (pl.) with *büz-ök* 'smell' (pl.)). Therefore, the lowered linking vowel only shows the binary alternation *ale* instead of the usual unlowered ternary one *olelö* (cf. section 6.1). In ECH the lowering effect on the quality of the linking vowel can be detected after stems whose last nucleus is back or antiharmonic (e.g. *ház-ak, hid-ak* 'bridge' (pl.)) and after stems whose last nucleus is a front rounded vowel (e.g. *szüz-ek*,

---

[21] Except some final CCC clusters whose final consonant is syllabified into the appendix (cf. section 5.2.4.3) and also CCC clusters in general that are divided by an analytic morphological domain boundary (cf. section 5.2.2). On internal CCC clusters created by the past suffix cf. section 8.1.4.4.

*tőgy-ek* 'udder' (pl.)). After a (non-antiharmonic) stem whose last nucleus is a front unrounded vowel, this lowering effect can only be detected in dialects that retain the distinction between mid [e] and low [ɛ] at the surface; compare *kép-et* [keːpɛt] 'picture' (acc.) and *gyep-et* [dʸepɛt] 'lawn' (acc.). This distinction is lost in ECH. The lowered and the unlowered front unrounded linking vowel are equally realized as [ɛ]: [keːpɛt, dʸepɛt]. Thus, vowel quality is not a clue for lowering after these stems. There is, however, a lowering effect that is (partially) independent of vowel quality. After stems that cause lowering (henceforward 'lowering stems'), Type B suffixes occur with a (lowered) linking vowel even if the stem ends in a consonant that they can normally attach to *without* a linking vowel. Thus, the following generalization seems to hold:[22]

(12)    Lowering requires a phonetically expressed unstable vowel.

This is shown with the accusative in (13):[23]

(13)    | *normal stem* | *lowering stem* |
|---|---|
| ón-t | tehen-et 'cow' |
| lány-t | hány-at 'how many' |
| dal-t | hal-at 'fish' |
| sör-t | ár-at 'price' |
| baj-t | haj-at 'hair' |
| rés-t | has-at 'stomach' |
| kosz-t | mesz-et 'lime' |
| gőz-t | méz-et 'honey' |
| varázs-t | darazs-at 'wasp' |

This lowering effect makes it possible to detect lowering even when the quality of the linking vowel is not a clue, i.e. in ECH after stems whose last nucleus is a front unrounded vowel, e.g. *mesz-et*, *méz-et*.

Normally, both Type A and Type B suffixes attach to vowel-final stems without a linking vowel. Given (12), however, a linking vowel is expected to occur after vowel-final lowering stems as well. (14) shows the behaviour of the plural and the modal (both Type A) after vowel-final lowering stems:[24]

---

[22] There are a few irregular stems that *are* lowering, but nevertheless the accusative attaches to them without a phonetically expressed unstable vowel, e.g. *báj* 'charm' (*báj-ak* (pl.), *báj-t* (acc.)), *szakáll* 'beard' (*szakáll-ak* (pl.), *szakáll-(a)t* (acc.)), *(hegy)oldal* '(hill)side' (*(hegy)oldal-ak* (pl.), *(hegy)oldal-t* (acc.)). For a complete list cf. Papp (1975).

[23] The past morpheme does not show this effect. This, however, is due to the fact that there happen to be no lowering verb stems.

[24] Glosses: *városi* 'urban', *pesti* 'Budapest' (adj.), *sátáni* 'satanic', *emberi* 'human', *szomorú* 'sad', *keserű* 'bitter', *bántó* 'annoying', *sértő* 'insulting'.

(14)    *plural*                    *modal*
        városi-ak                   sátáni-an
        pesti-ek                    emberi-en
        szomorú-ak                  szomorú-an
        keserű-ek                   keserű-en
        bántó-(a)k                  bántó-(a)n
        sértő-(e)k                  sértő-(e)n

As can be seen, the generalization expressed in (12) holds for the items in (14). The only special property of vowel-final lowering stems is that the linking vowel of the plural and the modal suffix is optional after mid vowels. Note that there is no linking vowel if the stem-final vowel is *e, a, é, á*: *fekete* 'black': *feke-té-k, feketé-n; durva* 'rough': *durvá-k, durvá-n; ordenáré* 'vulgar': *ordenáré-k, ordenáré-n; burzsoá* 'bourgeois': *burzsoá-k, burzsoá-n*.[25] Arguably, this is independent of lowering because the lack of a low linking vowel here can be viewed as an OCP effect. Given the representations of the relevant vowels (cf. section 3.1.1), forms like *\*feketé-ek* or *\*durvá-ak* would be OCP violations. Thus, one could suggest that, with some idiosyncratic variation, generally, (12) also holds for vowel-final stems unless another constraint (such as the OCP) is violated.

However, the behaviour of some other suffixes with initial unstable vowels (such as the accusative and the superessive) appears to be different in that they attach to the same stems without a linking vowel:

(15)    *accusative*                *superessive*
        városi-t                    városi-n
        pesti-t                     pesti-n
        sátáni-t                    sátáni-n
        emberi-t                    emberi-n
        szomorú-t                   szomorú-n
        keserű-t                    keserű-n
        bántó-t                     bántó-n
        sértő-t                     sértő-n

One may try to salvage (12) as a general statement in the following way. The stems in (14) are all adjectives. If we assume that (in contrast to the plural and the modal) case endings can only attach to nouns, then the stems in (15) must be nominal stems since the superessive and the accusative are case endings. Most adjectives are lowering, but only some nouns are, and usually a noun stem corresponding to a lowering adjectival stem is non-lowering.[26] Compare,

---

[25] The length of the stem final vowels in forms like *feketé-k, durvá-k* etc. is due to Low Vowel Lengthening (cf. section 3.1.1).

[26] It has to be pointed out that adjectives are not turned into nouns by zero derivation, but some adjectives have lexicalized nominal counterparts. Thus, not every adjective has a corresponding (homophonous) noun.

for instance, *vörös-ek* 'red' (pl.) and *vörös-ök* 'communist' (pl.), *komikus-ak* 'comical' (pl.) and *komikus-ok* 'comedian' (pl.), *szárnyas-ak* 'winged' (pl.) and *szárnyas-ok* 'poultry' (pl.), etc. One could claim then that the stems in (15) are simply non-lowering nouns that have corresponding adjectives that are lowering. Given this assumption, their behaviour in (15) conforms to (12).[27]

Unfortunately, however, the argument above is untenable. (i) There is evidence that case endings *can* attach to adjectives since they can follow (overt) adjective-forming derivational suffixes: e.g. *só-s-at* 'salty', (acc.) (where *-s* is a denominal adjective-forming derivational suffix).[28] (ii) The initial unstable vowel of some suffixes is unexpressed even after vowel-final stems that are unquestionably adjectives (and lowering). Consider the comparative suffix *-V$_u$bb*, which is a Type A suffix just like the plural (e.g. *piros-abb*), and can only attach to adjectives, but whose unstable vowel never appears phonetically after lowering (or non-lowering) vowel-final stems (compare *szomorú-bb*, *keserű-bb* with *szomorú-ak*, *keserű-ek*).

To sum up, (12) must be restricted to consonant-final stems:

(16)    Lowering requires a phonetically expressed unstable vowel.
        Condition: V$_u$ is preceded by a C

After vowel-final lowering stems, suffixes with an unstable initial vowel behave idiosyncratically. Some suffixes attach to these stems with a (lowered) linking vowel, others without a linking vowel. As it is unpredictable which suffix will behave in which way, lexical marking must be involved. It must be pointed out, though, that the behaviour of linking vowels after vowel-final stems is not completely unrelated to lowering since—although the linking vowel does not always appear after a vowel-final lowering stem—if a linking vowel does appear after a vowel-final stem, it is always a lowered one (i.e. there is never a linking vowel after a vowel-final non-lowering stem).

Nominal vowel-final stems are non-lowering in general. The only exceptions are *férfi* 'man' and *-fi* '-man' (as in e.g. *hadfi* 'warrior'): *férfi-ak* 'man' (pl.) in which the linking vowel is present (compare *férfi-t* (acc.)).

According to its source, two types of lowering may be distinguished (cf. Rebrus and Polgárdi 1997):

(i) The source may be the preceding stem, i.e. after stems belonging to an arbitrary (closed) class (that of 'lowering stems', cf. Vago 1980*a*) the immediately following linking vowel is low. Data in (17) show that it is indeed the stem that is the source of lowering in this case, since the linking vowel of one

---

[27] This line of reasoning has been pointed out to us by Péter Rebrus (personal communication).

[28] It is not possible to analyse *sósat* as *só-s-ø-at* (where *-ø* is a deadjectival noun-forming derivational suffix) because derivational suffixes are non-lowering unless they are adjective-forming (see the discussion at the end of this section).

and the same suffix shows up as mid after non-lowering stems and as low after lowering ones:[29]

(17)

|  |  | *plural* | *accusative* | *gloss* |
|---|---|---|---|---|
| *a.* normal stems | bot | bot-ok | bot-ot | 'stick' |
|  | gyep | gyep-ek | gyep-et | 'lawn' |
|  | köd | köd-ök | köd-öt | 'fog' |
| *b.* lowering stems | fog | fog-ak | fog-at | 'tooth' |
|  | kép | kép-ek | kép-et | 'picture'[30] |
|  | szög | szög-ek | szög-et | 'nail' |

Some suffixes cause lowering as well, i.e. there are suffixes that turn any stem into a lowering one: *gáz-ok-at* 'gas-pl-acc.', *bűn-öm-et* 'sin-my-acc.'. Compare *hat-od-ot* 'sixth' (acc.) where the fraction forming suffix *-od/-ed/-öd* does not cause lowering. We shall return to the problem of multiple suffixation and lowering below.

(ii) Some suffixes appear to be 'self-lowering' in that the source of the lowering of the suffix-initial unstable vowel is the suffix itself. The suffixes involved are *-sz/-asz/-esz* (2sg pres. indef.), *-ni/-ani/-eni* (inf.) and *-lak/-lek/ -alak/-elek* (1sg$^s$ 2sg$^o$), *-nak/-nek/-anak/-enek* (3pl pres. indef.), and *-na/-ne/ -ana/-ene* (cond.). The initial linking vowel of these suffixes is low even after stems that are demonstrably non-lowering. This is shown in (18) where the non-self-lowering suffix *-k/-ok/-ek/-ök* (1sg pres. indef.) is included to show the contrast with their behaviour:

(18)

|  | *2sg pres. indef.* | *infinitive* | *1sg$^s$ 2sg$^o$ pres.* | *(1sg pres. indef.)* |
|---|---|---|---|---|
| mond 'say' | mond-(a)sz | mond-ani | mond-(a)lak | (mond-ok) |
| sért 'hurt' | sért-esz | sért-eni | sért-(e)lek | (sért-ek) |
| küld 'send' | küld-esz | küld-eni | küld-(e)lek | (küld-ök) |

In general, the initial unstable vowel of these suffixes appears if the stem ends in more than one consonant. Note, however, that they often behave idiosyncratically. In some forms the linking vowel is (unexpectedly) optional (e.g. *mond-(a)sz* vs. *sért-esz*). After some stems ending in a geminate, the linking vowel does not appear (e.g. *áll-ni* 'to stand' vs. *hall-ani* 'to hear'). On the other hand, it does appear after some stems that end in a long vowel followed by a single consonant (*bocsát-ani* 'to forgive' vs. *lát-ni* 'to see'). We argued in section 5.2.4.3 that—despite the phonological conditioning—the selection of the allomorphs of 2sg pres. indef. *-sz* is morpho-

---

[29] The linking vowel of the superessive behaves differently: it remains mid (and displays a ternary alternation) even after lowering stems: *ház-on* 'house' (spr.), *méz-en* 'honey' (spr.), *szűz-ön* 'virgin' (spr.).

[30] Here, the lowering effect can only be detected in dialects that retain the distinction between mid [e] and low [ɛ].

logical rather than phonological. Similarly, we suggest that it is allomorphy that is involved in what appears to be low vowel ~ zero alternation in 'self-lowering' suffixes. Considering the selection morphological has significant advantages. It makes it possible to maintain the generalization in (16) and this way phonological lowering will always have a local source: the preceding (relative or absolute) stem. Also, if the alternation is morphological, idiosyncrasies of the kind described above are more likely to occur.

Since 'self-lowering' is non-phonological, phonological lowering always spreads from the stem. Lowering can only influence the linking vowel *immediately* following the lowering stem. Consider *nyolc-ad-ot* 'eighth' (acc.) where *nyolc* is a lowering stem, but the accusative surfaces with an unlowered vowel (*-ot*). As we have pointed out above, the source of lowering may be the relative stem, i.e. some suffixes may be lowering, and can turn a non-lowering stem into a lowering one. While lowering is a lexical property of (nominal) stems,[31] it is not (completely) unpredictable which suffix is lowering and which one is not. The claim in Vago (1980*a*) that lowering is positionally predictable, i.e. that the unstable vowel of first position suffixes may or may not lower (depending on whether the stem is lowering or not), but it is always low later (i.e. when not immediately adjacent to the stem) is correct as far as it goes, provided that we define 'stem' (as Vago does) as a complex of 'root (+ derivational suffixes)'; in other words, as long as 'first position' is defined as that after the (last) derivational suffix of a word form.[32] According to Rebrus and Polgárdi (1997) derivational suffixes are non-lowering and inflectional ones are lowering. They point out contrasting pairs of examples like *un-tat-om* 'bore' (1sg pres. def.) and *un-t-am* 'be bored' (1sg past def.). In the former word, the causative derivational suffix (*-tat*) does not make the unstable vowel of the personal suffix low, but in the latter, the inflectional past tense suffix does. While it is true that inflectional suffixes are all lowering,[33] the correlation between lowering and derivational suffixes is more complicated. Superficially, it seems that Rebrus and Polgárdi's claim does not hold for all derivational suffixes: they may be lowering or non-lowering—compare *só-s-ak* 'salty' (pl.) (where *-s* is a denominal adjective forming derivational suffix) and *harc-os-ok* 'warrior' (pl.) (where *-Vs* is a noun-forming derivational suffix). This, however, is due to a factor independent of the inflectional/derivational character of suffixes. It can be explained with reference to the part-of-speech distribution of lowering. Note that the derivational suffix above that appears to be lowering is adjective-forming. In fact, all the derivational suffixes that lower are adjective-forming as well (e.g. *tanul-ékony-ak* 'teachable'

---

[31] See, however, the part-of-speech distribution of lowering to be discussed below.

[32] This amounts to the claim that all inflectional suffixes are (whereas derivational suffixes may or may not be) lowering.

[33] The lowering effect of inflectional suffixes that are never followed by another suffix (e.g. case endings) never actually manifests itself, but we assume that they are lowering as well.

(pl.), *ír-ott-at* 'written' (acc.)).[34] As adjectives are generally lowering,[35] we claim that these suffixes lower *because* they are adjective-forming. The distribution of lowering items in other word classes is the following. Nouns and pronouns include a large (but closed) set of lowering stems. It is unpredictable which nominal stems are lowering so they must be marked as such in the lexicon. New items entering the language are invariably of the unmarked, non-lowering type (e.g. *szponzor-ok* 'sponsors', *kűr-ök* 'free exercises (in figure skating)'). Absolute verb stems are never lowering. The attested distribution of lowering in stems and suffixes follows if we assume that they are non-lowering by default, but some are marked as lowering individually/idiosyncratically, while others are assigned lowering status by two morphological redundancy rules:

(19)   *a.*  Inflectional suffixes are lowering.
       *b.*  Adjectives are lowering.

In addition, stems that belong to certain (more or less irregular) morphological classes are always lowering. For instance nominal 'v-adding' stems and 'shortening' stems all lower: e.g. *ló* 'horse' *lov-at* 'horse' (acc.), *madár* 'bird' *madar-at* 'bird' (acc.).[36] Thus, (19) may be extended with rules referring to specific morphological classes. To sum up, lowering is partly predictable on a morphological basis.

Only an unstable vowel may be the target of lowering. Stable suffix initial vowels never lower. As pointed out above, the superessive suffix is special because it is unstable, but it does not lower. The three types of suffixes are illustrated in (20). The plural is lowerable and unstable, the causal-final is non-lowerable and stable, and the superessive is non-lowerable and unstable:[37]

| (20) | *plural* | *causal-final* | *superessive* | *gloss* |
|---|---|---|---|---|
| normal stem | fá-k | fá-ért | fá-n | 'tree' |
|  | gáz-ok | gáz-ért | gáz-on | 'gas' |
| lowering stem | ház-ak | ház-ért | ház-on | 'house' |

[34] The privative suffix *-(V)tlan/-(V)tlen/-talan/-telen* and *-van/-ven* '-ty' are exceptional. The former is adjective forming, but does not lower: e.g. *tanul-atlan-ok* 'uneducated' (pl.), and the latter is derivational, not adjective forming, but does lower: e.g. *nyolc-van-at* 'eighty' (acc.).

[35] There are a few (irregular) exceptions, e.g. *nagy-ok* 'big' (pl.), *agg-ok* 'old' (pl.).

[36] 'Epenthetic' stems are not involved in such an implicational relationship. They may be lowering or non-lowering: compare *marok* 'fist', *mark-om* 'my fist' with *farok* 'tail', *fark-am* 'my tail'.

[37] The fourth type does not exist: there are no suffixes with an initial lowerable stable vowel.

### 8.1.4. Analysis

*8.1.4.1. Syllabification—full vowels and defective vowels*

Hungarian vowel ~ zero alternation is partly due to an underlying difference between *full vowels (Vf)* and *defective vowels (Vd)*. Defective vowels are empty in the sense that they only consist of a skeletal slot without any segmental melody.[38] By contrast, full vowels *minimally* have a VOCALIC node (and a ROOT node). This is shown in (21) below (where non-essential structure between the ROOT and the VOCALIC node is suppressed and the symbol △ denotes a structure of any complexity, or nil):

(21)                   *a. full vowel*              *b. defective vowel*

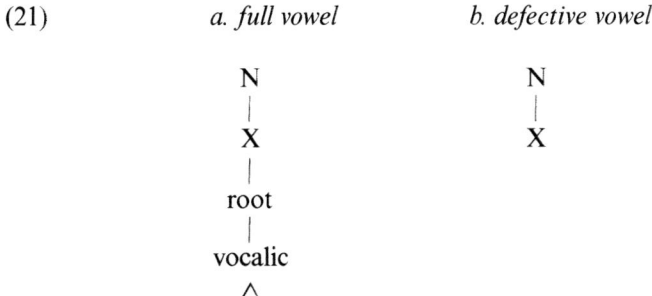

Note that defective vowels (as opposed to full ones) will have to be marked underlyingly in some way to syllabify as nuclei since otherwise the syllabification algorithm will not be able to identify them as such. We simply assume that they are prelinked to nucleus nodes.

Defective vowels are not interpreted phonetically unless they receive a vocalic node (i.e. are turned into full vowels) in the course of the derivation. This default process only targets a *licensed* $V_d$, i.e. one that is incorporated into a syllable, and turns it into a minimally full vowel. Default V achieves this by assigning [+open₂] to licensed defective vowels. Higher nodes (including the vocalic node and the root node) are automatically appended to ensure well-formedness (Sagey 1986, Clements and Hume 1995). This is shown in (22), where irrelevant structure has been suppressed.[39]

---

[38] On empty vowels see Anderson (1982), Spencer (1986), Kenstowicz and Rubach (1987), Kaye, Lowenstamm, and Vergnaud (1990), Kaye (1990), Charette (1991).

[39] The parentheses enclosing the root node and the vocalic node in (22) indicate that these nodes have been automatically appended. In this figure encircling indicates that a node dominates no structure.

(22)                          *Default V*

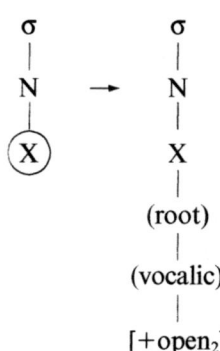

(root)
|
(vocalic)
|
$[+open_2]$

Vowel Harmony and the other default processes apply to the output of Default V to derive the correct surface vowel quality.

In accordance with standard assumptions about prosodic licensing, we assume that prosodically unlicensed material does not receive phonetic interpretation (cf. for instance Selkirk 1982, Itô 1986, 1989). Thus, unsyllabified defective vowels might persist up to the level of surface representation without being phonetically realized (i.e. they need not be stray-erased). Alternatively, they may be assumed to delete at some point in the derivation (possibly at the end of Block 1 or lexical phonology). Note that if unsyllabified $V_d$'s persist in the postlexical phonology, then the locality conditions of postlexical processes (e.g. Degemination, Voicing Assimilation) have to be determined in such a way that they ignore defective vowels since in a string $C1V_dC2$, C1 and C2 should count as adjacent. In section 8.1.4.5 we argue that they are invisible in Block 2. We express this by stipulating that they are erased at the end of Block 1. This will also simplify the statement of postlexical phonological rules.

We also assume that defective vowels are restricted in occurrence compared to full ones. Notably, they can only occur in singly closed syllables. Thus, the following constraints are added to the well-formedness conditions defining the Hungarian syllable template:

(23)              *a.*    $* V_d]_\sigma$              *b.*    $* V_dCC]_\sigma$

Disregarding non-essential structure, full vowels minimally have a vocalic node. Minimally full vowels receive place and aperture values by vowel harmony and the default processes described in section 6.1.

In addition to the lexical difference between full and defective vowels, vowel ~ zero alternation is also due to syllabification. We follow Itô (1986, 1989) and assume that syllabification is a template-matching algorithm. Template matching is directional (left-to-right or right-to-left), maximal (i.e.

the syllable template is filled up with segmental material maximally), and is constrained by the Onset Principle (i.e. onsetless syllables are avoided if possible). Syllabification and epenthesis are not separate processes in that syllabification can build degenerate syllables, i.e. syllables that contain nodes dominating empty X-slots. Thus, syllabification may overparse segmental material by inserting empty positions. However, we only allow overparsing by empty *nuclear* positions. (24) is intended as a language-specific restriction on syllabification in Hungarian:[40]

(24)     Empty onset or coda positions may not be created in the course of syllabification.

Syllabification may be non-exhaustive (cf. Hyman 1990, Kenstowicz 1994). We assume that this can happen under the special condition given below:

(25)     *Non-exhaustiveness*
         Defective vowels may remain unparsed into syllables.

Thus, a representation is well-formed even if it contains unparsed defective vowels.

As a result of (24) and (25) syllabification will skip *lexical* empty nuclear positions (defective vowels) should it be impossible for them to syllabify in a singly closed syllable. This can happen word-finally, prevocalically, or before a single consonant followed by a full vowel:

(26)     *a.*  $\ldots CV_fCV_d\# \rightarrow \ldots \{CV_fC\}V_d\#$
         *b.*  $\ldots V_fCV_dV_f \rightarrow \ldots V_fC\}V_d\{V_f\}$
         *c.*  $\ldots V_fCV_dCV_f \rightarrow \ldots V_fC\}V_d\{CV_f\}$

As is shown in (26) (where the syllable edges are indicated by curly brackets) the empty nuclear positions in question remain unaffiliated syllabically. They are not 'rescued' by syllabification creating a coda position after them because this is excluded by (24).[41] Non-exhaustiveness together with the Onset Principle (cf. Itô 1989) ensures that ... $V_fCV_dCV_f$ ... strings syllabify as in (26c) rather than like this *... $V_f\}\{CV_dC\}\{V_f$..., i.e. since defective vowels may be left unparsed, it is more important to obey the Onset Principle than to parse a defective vowel.

---

[40] Compare Itô's analysis of Axininca Campa (Itô 1989) where syllabification inserts empty onset positions as well.

[41] A constraint disallowing completely empty syllables (i.e. syllables that have a defective vowel and an empty onset and/or coda) would have the same effect as (24).

An empty nuclear position created by syllabification is representationally identical with a lexically empty position. Both are defective vowels in the sense defined above and may be turned into full vowels by (22).

In Hungarian, syllabification proceeds from right to left and is continuous, i.e. it (re)applies after morphological and phonological operations. Resyllabification is permitted, i.e. prosodic structure is erased if the nucleus is deleted along with its X-slot (cf. Hayes 1989) and the coda of a stem-final syllable becomes available for (re)syllabification if a vowel-initial suffix is added (compare Levin 1985).[42]

In addition to prosodic structure, vowel ~ zero alternation is also sensitive to morphological structure. In section 1.3 a distinction was made between analytic and synthetic suffixation and it was pointed out that phonotactic constraints do not apply across the boundary of an analytic domain. The distinction between these two kinds of suffixes is crucial in the interpretation of vowel ~ zero alternation. We assume that both Type A and Type B suffixes are synthetic; the behaviour of analytic suffixes (e.g. *-ig*, *-ért*, *-d*, etc.) will be discussed in section 8.1.4.5.

### 8.1.4.2. *Major stems and 'epenthetic' stems—Type A and Type B suffixes*

In section 8.1 we saw that vowels can alternate with zero stem-internally, stem-finally, and suffix-initially. Of these, stem-final vowel ~ zero alternation is phonologically irregular in that (i) only an arbitrary set of suffixes trigger it in all the stems to which they are attached (cf. section 8.1.2.1), and (ii) an arbitrary set of stems undergo it before an arbitrary set of suffixes (only some of which belong to the set referred to in (i), cf. section 8.1.2.1 footnote 11). Therefore, we assume that stem-final vowel ~ zero alternation is essentially morphological and we shall disregard it in the analysis below.

The stems and suffixes showing regular (phonological) vowel ~ zero alternation have the following underlying representations. We claim that 'epenthetic' stems do not end in consonant clusters ($C_iC_j$) as is usually assumed (e.g. Vago 1980a, Jensen and Stong-Jensen 1988, 1989b, Törkenczy 1994a, 1995), but contain a defective vowel in their final syllable that ends in a single consonant: $-C_iV_dC_j\#$ (compare Törkenczy 1992 and Ritter 1995).[43] Thus, the three-way distinction between the last syllables of the triplets described in section 8.1.1 is made representationally in the following way: $-CV_dC$ (*torony* 'tower'), $-CV_fC$ (*szurony* 'bayonet'), $-CC$ (*szörny* 'monster'). Type A suffixes have an underlying initial full vowel (e.g. $-V_jk$ 'pl.') and Type B suffixes are underlyingly consonant-initial (e.g. *-t* 'acc.').

Let us now examine the relationship between vowel ~ zero alternation and syllabification. Figure (27) shows how consonant-final non-lowering

---

[42] Analytic suffixes behave differently; see the discussion below.

[43] Both these treatments are formulated in a Government Phonology framework.

non-epenthetic stems (i.e. major stems, cf. section 2.4) are syllabified when Type A suffixes (27a) and Type B suffixes are attached to them:[44]

(27) *a.*   keːp-V$_f$k        {keː} {pV$_f$k}        képek 'picture' (pl.)
       bor-V$_f$k        {bo} {rV$_f$k}        borok 'wine' (pl.)
   *b.*   keːp-t         {keː} {pV$_d$t}        képet 'picture' (acc.)
       bor-t          {bort}            bort 'wine' (acc.)

The syllabifications follow from the Hungarian syllable templates and right-to-left template matching and non-exhaustiveness. Type A suffixes are insensitive to the identity of the stem-final consonant because they are vowel-initial and thus they can always form a well-formed syllable with the stem-final consonant. Type B suffixes, on the other hand, are phonotactically sensitive to the stem-final consonant. The reason is that they are consonant-initial and unsyllabifiable in themselves: accusative *-t* syllabifies with the stem-final consonant just in case they can form a licit coda (*bort*). Syllabification creates a degenerate syllable, i.e. a syllable with an empty nuclear position V$_d$ if the cluster is not syllabifiable as the coda of the last syllable of the stem (*képV$_d$t*). Thus, the vowel ~ zero alternation ('epenthesis' in this case) is due to syllabification.

The behaviour of Type A suffixes after vowel-final stems shows that, in addition to syllabification, there is a rule which is responsible for the vowel ~ zero alternations. This rule eliminates hiatus by (i) deleting a defective vowel when it is adjacent to a full one, and (ii) deleting a full vowel (together with its X-slot) when it follows another full vowel. It is necessary to delete the X-slot too, because simply deleting the segmental melody would only turn a full vowel into a defective one. This rule can be formulated as in (28a, b) where V is the vocalic node.

(28)            *Hiatus*

          *a.*    N              N
                  |              |
                 (X) → Ø %      X              (mirror image)
                                 |
                                 V


          *b.*    N              N
                  |              |
                  X → Ø /       X  __

---

[44] In the transcriptions the vowels denoted by phonetic characters and the symbol V$_f$ are all full vowels. V$_f$ is a *minimal* full vowel. The real difference in syllabification is between all full vowels vs. V$_d$.

The rule is formulated in as general a form as possible. (28b) deletes the second of any two adjacent nuclei. It is the elsewhere part of (28) and thus only applies if the more specific (28a) cannot. There are four possible ways in which a full vowel and a defective one may combine in hiatus: $\underline{V}_dV_f$, $V_f\underline{V}_d$, $V_f\underline{V}_f$ and $V_d\underline{V}_d$, where the underlined vowel is the one which is deleted by (28).[45] (28) shows derived environment effects (cf. Vago 1980a). It does not delete a postvocalic vowel in monomorphemic items, where hiatus is tolerated: e.g. [oaːzis] *oázis* 'oasis'.[46] Note that analytic vowel-initial suffixes retain their initial vowel: e.g. *kapu-ig* 'to the gate'. Their behaviour will be discussed later in this chapter. For obvious reasons, (28) does not apply to vowel-final stems suffixed by a Type B suffix. These suffixes simply syllabify with the stem-final vowel: kapu-t → {ka}{put} *kaput* 'gate' (acc.). Type A suffixes, on the other hand, lose their initial full vowel by (28) and then syllabify with the stem-final vowel: kapu-$V_f$k → {ka}{pu}{$V_f$k} → {ka}{puk} *kapuk* 'gate' (pl.). The deletion of defective vowels in hiatus will be discussed in detail below in section 8.1.4.3.

Multiple suffixation may create strings of Type A and Type B suffixes.[47] Type B suffixes do not combine with other Type B suffixes (*T$_B$+T$_B$). All the other possible combinations of a Type A and a Type B suffix are attested: T$_A$+T$_A$ (*nagy-obb-ak* 'big' (comp.+pl.)), T$_A$+T$_B$ (*kép-ek-et* 'picture' (pl.+acc.)) and T$_B$+T$_A$ (*kap-t-am* 'get' (past+1sg)). Figure (29) shows how the first two sequences syllabify when attached to major stems:[48]

(29)    T$_A$+T$_A$     nad$^y$-$V_f$bb-$V_f$k     {na}{d$^y V_f$b}{b$V_f$k}     nagyobbak
        T$_A$+T$_B$     keːp-$V_f$k-t     {keː}{p$V_f$}{k$V_d$t}     képeket

The syllabification of these forms is straightforward. In the second example a degenerate syllable is created because /kt/ is not a possible (regular) coda.

In the examples discussed above empty nuclear positions (defective vowels) are created in the course of syllabification. We have also pointed out, however, that defective vowels are underlyingly present in 'epenthetic' stems. The vowel ~ zero alternation in epenthetic stems is not the result of overparsing by syllabification, but of the special constraints (23a, b) on syllables whose

---

[45] Of the four combinations $V_dV_d$ does not arise.

[46] On postlexical hiatus filling see section 9.3.

[47] In the discussion of syllabification we shall first temporarily disregard the representation and the syllabification of lowering stems and suffixes for expository reasons. We shall deal with lowering in detail later in section 8.1.4.3. Due to the representation of lowering stems/suffixes some of the syllabifications that follow will have to be modified.

[48] We shall discuss the third type of suffix combination (T$_B$+T$_A$) together with the past tense suffix later in section 8.1.4.4. The double consonant in (29) means a true geminate (i.e. two timing slots associated with a single root) and not adjacent identical melodies.

nucleus is a defective vowel. Figure (30) shows how the words *szurony, szörny,* and *torony* syllabify in isolation (30a), when suffixed by Type A suffixes (30b) and Type B suffixes (30c):

(30) *a.*   $suron^y$         $\{su\}\{ron^y\}$              szurony
            $sörn^y$          $\{sörn^y\}$                   szörny
            $torV_dn^y$       $\{to\}\{rV_dn^y\}$            torony
     *b.*   $suron^y\text{-}V_fk$   $\{su\}\{ro\}\{n^yV_fk\}$      szurony-ok
            $sörn^y\text{-}V_fk$    $\{sör\}\{n^yV_fk\}$          szörny-ek
            $torV_dn^y\text{-}V_fk$ $\{tor\}V_d\{n^yV_fk\}$       torny-ok
     *c.*   $suron^y\text{-}t$      $\{su\}\{ron^yt\}$            szurony-t
            $sörn^y\text{-}t$       $\{sör\}\{n^yV_dt\}$          szörny-et
            $torV_dn^y\text{-}t$    $\{tor\}V_d\{n^yV_dt\}$       torny-ot

*Szörny* can syllabify as a CVCC syllable because the final consonant cluster can form a licit coda. There is no difference between the syllabification of *szurony* and *torony* in isolation because the defective vowel in the latter can syllabify in a singly closed syllable (30a). However, the same two stems do not syllabify in the same way when suffixed with a Type A suffix. As can be seen in (30b) the last vowel of *szurony* can syllabify as the nucleus of an open syllable in *szuronyok* (since it is a full vowel). By contrast, the second vowel of *torony* cannot syllabify in the suffixed form because (i) right-to-left template matching and the Onset Principle require the stem-final consonant to syllabify as an onset, and (ii) the vowel preceding the stem-final consonant is a defective one and (23a) disallows $V_d$ in open syllables. Thus, this vowel of the stem remains syllabically unaffiliated (it is skipped by syllabification). Therefore (22) does not apply to it, and consequently, it will not be interpreted phonetically and the form surfaces as [torn$^y$ok]. Consider (30c). The accusative, a Type B suffix, can syllabify as a coda in *szuronyt* because /n$^y$t/ is a well-formed coda. An empty nuclear position preceding it is created by syllabification when it is attached to *szörny* because codas are maximally binary branching. Syllabification will always create an empty nuclear position before a Type B suffix when it is added to an 'epenthetic' stem. If the 'epenthetic' stem ends in a consonant with which the suffixal consonant cannot form a licit coda, then the reason is the same as in the case of similar major stems (compare *ret$V_d$k-et* 'radish' (acc.) and *ének-et* 'song' (acc.)). Overparsing by syllabification occurs even if the 'epenthetic' stem ends in a consonant with which the suffixal consonant could form a licit coda because of (23b) since the stem-final consonant is preceded by a defective vowel (e.g. *bok$V_d$r-ot* 'bush' (acc.), *tor$V_d$ny-ot*—compare *tábort* 'camp' (acc.), *szuronyt*). The $V_d$ of the stem cannot syllabify in a syllable doubly closed by the stem-final consonant and the suffixal consonant, but it cannot syllabify in the syllable preceding the degenerate syllable created by

syllabification either—because that syllable would have to be open
($*\{to\}\{rV_d\}\{n^yV_dt\}$). So it remains unsyllabified and is not interpreted phonetically ($\{tor\}$ $V_d$ $\{n^yV_dt\}$ *torny-ot*).[49]

The syllabification of multiply suffixed 'epenthetic' stems is unproblematic
and follows from the mechanism discussed above:

(31)   $T_A+T_A$     $ba:tV_dr-V_fbb-V_fk$       $\{ba:t\}V_d\{rV_fb\}\{bV_fk\}$       bátrabbak
       $T_A+T_B$     $bokV_dr-V_fk-t$            $\{bok\}V_d\{rV_f\}\{kV_dt\}$          bokrokat

### 8.1.4.3. Syllabification and lowering

We noted in section 8.1.3 above that Lowering (i) determines the quality of
unstable suffix-initial vowels and (ii) interacts with syllabification (cf. (16)).
We also saw that the source of lowering in the representation of a stem or a
suffix may be an unpredictable 'mark' or one which is morphologically predictable (cf. (19)). We claim that, representationally, this mark consists of two
distinct (though always co-ocurring) characteristics. A lowering stem/suffix
has a final *floating* [+open$_1$] feature (cf. section 6.1) and a morpheme-final
defective vowel $V_d$ as shown in (32):

(32)

$$
\begin{array}{c}
N \\
| \\
X \quad X \\
| \qquad\qquad ]_{stem} \\
\text{root} \\
\triangle \\
[+\text{open}_1]
\end{array}
$$

Thus, specially marked nouns and pronouns, all adjectives, and all inflectional suffixes end in a structure shown in (32). This means that some of
the representations discussed above have to be modified because they are,
or they contain, such formatives. For instance, *szörny* must be represented
as /sörn$^y$V$_d$/ instead of /sörn$^y$/ because it is an (unpredictably) lowering noun
and *bor-t* must be /bortV$_d$/ and not /bort/ because the final suffix is inflectional and therefore lowering. Naturally, the syllabification of such forms

---

[49] It is sometimes claimed (e.g. Vago 1980*a*, Törkenczy 1992) that alternative forms exist
in the accusative if the last consonant of the 'epenthetic' stem can form a licit coda with the
following *-t*: e.g. *öböl-t/öbl-öt* 'bay' (acc.). As pointed out in 8.1.2.2, this is not true of
*all* 'epenthetic' stems: only *some* show this variation (cf. Papp 1975). We assume that for
those that do, there are two entries in the lexicon: an 'epenthetic' one and a major one. Given
this assumption, syllabification will yield the alternative forms: $\{\ddot{o}\}\{bV_flt\}$ *öböl-t* vs.
$\{\ddot{o}b\}$ $V_d\{lV_dt\}$ *öbl-öt*. The selection of one or the other entry is often idiosyncratic. Moreover,
different suffixes may select different entries: *öböl-t/öbl-öt* 'bay' (acc.) vs. *öbl-ök* but *\*öböl-ök*
'bay' (pl.).

is also different, but given our assumptions about the syllabification and the interpretation of defective vowels, these modified representations will not change the outcome of the derivations, i.e. the phonetically realized forms. In the two examples above, for instance, the final defective vowel does not syllabify because it cannot occur in an open syllable, so these forms are parsed as /{sörn$^y$}V$_d$/ and /{bort}V$_d$/, and thus the correct surface forms ([sörn$^y$] and [bort]) are derived. We shall discuss some more complex cases below.

We interpret lowering as a process that spreads the floating [+open$_1$] feature locally to a (full or defective) vowel which is *licensed* (i.e. incorporated into a syllable) and is at the edge of a morpheme. Lowering applies regardless of whether the licensed vowel is morpheme-initial or morpheme-final. The spreading process is local and non-iterative, i.e. it targets a single V. If the target is a full vowel, the floating feature can spread to its aperture node. In the case of defective target vowels, we assume that the nodes necessary for preserving well-formedness (e.g. root, vocalic, aperture) are automatically created in the course of the spreading to the empty skeletal position (cf. Sagey 1986, Clements and Hume 1995). This is indicated by parentheses enclosing the relevant nodes.

(33)                                 *Lowering*

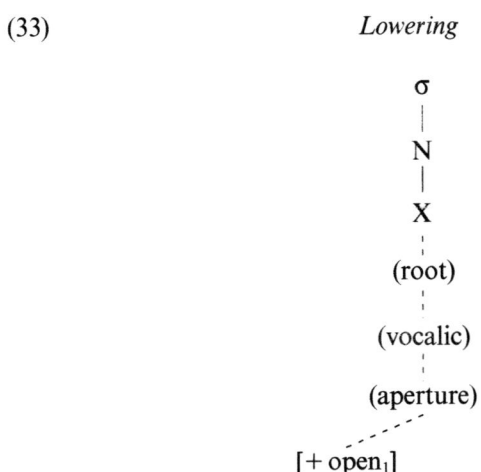

Condition: the target is peripheral in a morpheme.

Whether the target of Lowering is V$_d$ or V$_f$, the output of the process is a structure shown in (34) (where irrelevant nodes are omitted). Crucially, this means that (in addition to its lowering effect) Lowering turns a defective vowel into a full one.

(34)                                          σ
                                              |
                                              N
                                              |
                                              X
                                              |
                                             root
                                              |
                                           vocalic
                                              |
                                           aperture
                                              |
                                          [+open₁]

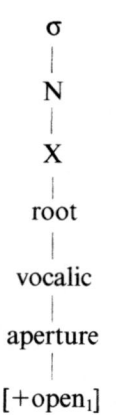

Spreading is a feature filling process, therefore Lowering is blocked if the target vowel has an aperture specification which is incompatible with the feature that is being spread.

Let us now examine how lowering interacts with syllabification.[50] As pointed out above, a word-final defective vowel is not realized phonetically because it cannot be syllabified. Therefore, in this position, a floating [+open₁] feature does not surface since (33) cannot apply to a syllabically unparsed $V_d$. (35) shows this with monomorphemic lowering stems and inflectional suffixes (Type A and Type B) attached to non-lowering stems:[51]

(35)   $fog_{OP}V_d$          →    $\{fog_{OP}\}V_d$          [fog]       fog 'tooth'
       $hal_{OP}V_J$          →    $\{hal_{OP}\}V_d$          [hɔl]       hal 'fish'
       $bor\text{-}V_fk_{OP}V_d$    →    $\{bo\}\{rok_{OP}\}V_d$    [borok]     borok 'wine' (pl.)
       $bor\text{-}t_{OP}V_d$       →    $\{bort_{OP}\}V_d$         [bort]      bort 'wine' (acc.)

Recall that Type B suffixes (such as the accusative) always show up with a (lowered) linking vowel after lowering stems regardless of whether the stem final consonant can or cannot form a licit coda with the suffixal -t. The reason is that lowering stems are vowel-final. In both cases the stem-final defective vowel can syllabify with the -t and, consequently, can be the target of Lowering, which turns it into a low full vowel (36a). This contrasts with the behaviour of -t after major stems where the defective vowel only appears as

---

[50] For expository purposes we shall use the following special symbols in the representations below: Subscripted '$_{OP}$' before $V_d$ stands for the floating [+open₁] feature. $V_{FOP}$ denotes the lowered full vowel that is the result of (33). It must be borne in mind, however, that the linear representations used are just shorthand for the corresponding non-linear ones in the same way as phonetic symbols are for the appropriate feature trees.

[51] The placement of $_{OP}$ relative to a syllable boundary is irrelevant and is not meant to indicate whether the floating feature is inside or outside a syllable.

a result of overparsing by syllabification after stem-final consonants that cannot form a licit coda with the suffixal consonant (36b):

(36) *a.* $n^yak_{OP}V_d\text{-}t_{OP}V_d \rightarrow \{n^ya\}\{kV_{FOP}t_{OP}\}V_d$   [n$^y$ɔkɔt]   nyakat 'neck' (acc.)

$fal_{OP}V_d\text{-}t_{OP}V_d \rightarrow \{fa\}\{lV_{FOP}t_{OP}\}V_d$   [fɔlɔt]   falat 'wall' (acc.)

*b.* $bak\text{-}t_{OP}V_d \rightarrow \{ba\}\{kV_dt_{OP}\}V_d$   [bɔkot]   bakot 'buck' (acc.)

$dal\text{-}t_{OP}V_d \rightarrow \{dalt_{OP}\}V_d$   [dɔlt]   dalt 'song' (acc.)

Thus, the generalization stated in (16) follows from the representation of lowering stems, the syllabification algorithm and the special constraints on the syllabification of defective vowels.

We have seen that Type A suffixes show up with a low linking vowel after lowering stems. Since Type A suffixes underlyingly begin with a full vowel, the stem-final $V_d$ of lowering stems is deleted by Hiatus and the floating [+open$_1$] feature of the stem can spread to the licensed suffix-initial $V_f$.

|  | (37) | fog-ak 'tooth' (pl.). | fog-atok 'your tooth' |
|---|---|---|---|
|  |  | $/fog_{OP}V_d\text{-}V_fk_{OP}V_d/$ | $/fog_{OP}V_d\text{-}V_ftok_{OP}V_d/$ |
| Syllabification |  | $\{fog_{OP}\}V_d\{V_fk_{OP}\}V_d$ | $\{fog_{OP}\}V_d\{V_f\}\{tok_{OP}\}V_d$ |
| Hiatus |  | $\{fog_{OP}\}\{V_fk_{OP}\}V_d$ | $\{fog_{OP}\}\{V_f\}\{tok_{OP}\}V_d$ |
| Syllabification |  | $\{fo\}\{g_{OP}V_fk_{OP}\}V_d$ | $\{fo\}\{g_{OP}V_f\}\{tok_{OP}\}V_d$ |
| Lowering |  | $\{fo\}\{gV_{FOP}k_{OP}\}V_d$ | $\{fo\}\{gV_{FOP}\}\{tok_{OP}\}V_d$ |
|  |  | [fogɔk] | [fogɔtok] |

As pointed out above, a suffix-initial $V_f$ does not lower (i.e. it cannot receive the spreading feature), if it has an aperture feature which is incompatible with the feature spread by Lowering. That is the reason why the suffix-initial vowels of two Type A suffixes, superessive -on/-en/-ön, and possessive 1pl -unk/-ünk do not lower after lowering stems: e.g. *fal-on* 'on the wall' and *fal-unk* 'our wall'. The initial $V_f$s of both these suffixes are underlyingly specified as [−open$_1$] (cf. section 6.1) and thus cannot receive the spreading [+open$_1$] feature.

Lowering may be unordered with respect to Hiatus as both possible orderings yield the correct results. Note that Lowering *can* spread [+open$_1$] past an unlicensed $V_d$ onto the closest potential target (the suffix-initial $V_f$) because defective vowels have no melodic structure. Compare the two ways of ordering Hiatus and Lowering in (37) and (38).

Multiply suffixed forms of lowering stems are also derived in a straightforward manner. (39) shows how the accusative plural of a lowering major stem (*fog* 'tooth') and an 'epenthetic' lowering stem (*sátor* 'tent') is derived.

(38)                     fog-ak 'tooth' (pl.)

$/fog_{OP}V_d\text{-}V_fk_{OP}V_d/$

Syllabification         $\{fog_{OP}\}V_d\{V_fk_{OP}\}V_d$

Lowering                $\{fog\}V_d\{V_{FOP}k_{OP}\}V_d$

Hiatus                  $\{fog\}\{V_{FOP}k_{OP}\}V_d$

Syllabification         $\{fo\}\{gV_{FOP}k_{OP}\}V_d$

$[fogɔk]$

(39)                     fog-ak-at 'tooth' (pl. acc.)        sátr-ak-at 'tent' (pl. acc.)

$/fog_{OP}V_d\text{-}V_fk_{OP}V_d\text{-}tV_d/$        $/\check{s}a{:}tV_dr_{OP}V_d\text{-}V_fk_{OP}V_d\text{-}t_{OP}V_d/$

Syllabification  $\{fog_{OP}\}V_d\{V_f\}\{k_{OP}V_dt_{OP}\}V_d$   $\{\check{s}a{:}\}\{tV_dr_{OP}\}V_d\{V_f\}\{k_{OP}V_dt_{OP}\}V_d$

Hiatus           $\{fog_{OP}\}\{V_f\}\{k_{OP}V_dt_{OP}\}V_d$   $\{\check{s}a{:}\}\{tV_dr_{OP}\}\{V_f\}\{k_{OP}V_dt_{OP}\}V_d$

Syllabification  $\{fo\}\{g_{OP}V_f\}\{k_{OP}V_dt_{OP}\}V_d$   $\{\check{s}a{:}t\}V_d\{r_{OP}V_f\}\{k_{OP}V_dt_{OP}\}V_d$

Lowering         $\{fo\}\{gV_{FOP}\}\{kV_{FOP}t_{OP}\}V_d$   $\{\check{s}a{:}t\}V_d\{rV_{FOP}\}\{kV_{FOP}t_{OP}\}V_d$

$[fogɔkɔt]$        $[\check{s}a{:}trɔkɔt]$

As can be seen in the derivations, syllabification is continuous, i.e. potentially it reapplies after each phonological rule. Both Hiatus and Lowering show derived environment effects: the former can only apply if its target is in another morpheme and the latter at the edge of a morpheme in the environment of another one. Neither applies intramorphemically. However, there is no evidence that their *application* is cyclic: they only ever need to apply once in the course of the derivation (naturally, they may have multiple targets). Rules of this kind challenge the traditional claim in Lexical Phonology that only cyclic rules are subject to the derived environment constraint on rule application.[52] In fact, we know of no phonological rules in Hungarian that must be considered cyclic on grounds other than the derived environment constraint.[53] Therefore—although the phonological rules belong to blocks (Block 1 and Block 2) and each suffix is marked according to whether it is analytic or synthetic—we assume that the derivation proceeds in a non-cyclic way.[54]

---

[52] Similar rules have been identified in a number of other languages, e.g. Finnish and Ondarroan Basque, cf. Hualde (1989) and Cole (1995).

[53] Jensen and Stong-Jensen (1989a) argue for cyclic epenthesis in Hungarian to account for the behaviour of 'epenthetic' stems. However, their arguments do not contradict our claim because (i) essentially, they are based on their epenthesis process blocking in a non-derived context; and (ii) the arguments do not carry over to the present analysis because they are crucially dependent on the assumption that it is epenthesis that is responsible for the vowel ~ zero alternation in 'epenthetic' stems, Type A, and Type B suffixes alike. This is a view that we reject for the reasons discussed in 8.1.1.

[54] It must be pointed out that as long as only synthetic suffixes are attached to the stem, it makes no difference if we assume that the whole 'preassembled' suffixed stem is subjected to the

If lowering stems that are phonetically vowel-final in isolation are represented on a par with the lowering stems discussed above, then they must end in a sequence of a full vowel and a defective vowel underlyingly ($CV_{f\ OP}\ V_d$). The syllabification algorithm and the rules discussed predict that both Type A and Type B suffixes attach to these stems with a phonetically expressed lowered linking vowel. Type A suffixes are underlyingly vowel-initial. When they are added to these stems, an underlying sequence of three vowels is created: $-CV_{f\ OP}\ V_d+V_fC$.[55] Hiatus deletes the stem-final $V_d$, and the two full vowels syllabify in the following way: $-\{CV_{f\ OP}\}\{V_fC\}$.[56] Lowering can apply to the suffix-initial $V_f$ giving $-\{CV_f\}\{V_{FOP}C\}$. Type B suffixes are consonant-initial. Thus, suffixation by a Type B suffix creates the string $-CV_{f\ OP}\ V_d+C$. Note that Hiatus cannot delete the stem-final $V_d$ because the vowel sequence is not derived. Syllabification yields $-\{CV_{f\ OP}\}\{V_dC\}$ and via Lowering the derived representation is the same as in the case of Type A suffixes: $-\{CV_f\}\{V_{FOP}C\}$. Thus, the prediction is that Type A and Type B suffixes behave in the same way when added to surface vowel-final lowering stems: a lowered linking vowel shows up before both types of suffixes.

This prediction is not borne out. Some suffixes never have a linking vowel after a surface vowel-final lowering stem (e.g. accusative, superessive, comparative), others do (e.g. plural, modal).[57] This difference in behaviour only partially correlates with the distinction between Type A and Type B suffixes. In 8.1.3 above we pointed out that the unpredictability of behaviour indicates that lexical marking must be involved. We suggest that the source of this

relevant phonological rules, or that the rules are (re)applied gradually (i.e. 'cyclically') as each suffix is considered, over the suffixed form. This can be seen in the derivation below, in which we adopt the Halle and Vergnaud (1987) approach to cyclicity:

$$fog_{OP}V_d\text{-}V_fk_{OP}V_d\text{-}t_{OP}V_d$$

| | |
|---|---|
| cycle1 | |
| Syllabification | $\{fog_{OP}\}V_d\text{-}V_fk_{OP}V_d\text{-}l_{OP}V_d$ |
| Hiatus | n.a. |
| Lowering | n.a. |
| cycle2 | |
| Syllabification | $\{fog_{OP}\}V_d\{V_fk_{OP}\}V_d\text{-}t_{OP}V_d$ |
| Hiatus | $\{fog_{OP}\}V_fk_{OP}V_d\text{-}t_{OP}V_d$ |
| Syllabification | $\{fo\}\{g_{OP}V_fk_{OP}\}V_d\text{-}t_{OP}V_d$ |
| Lowering | $\{fo\}\{gV_{FOP}k_{OP}\}V_d\text{-}t_{OP}V_d$ |
| cycle3 | |
| Syllabification | $\{fo\}\{gV_{FOP}\}\{k_{OP}V_dt_{OP}\}V_d$ |
| Hiatus | n.a. |
| Lowering | $\{fo\}\{gV_{FOP}\}\{kV_{FOP}t_{OP}\}V_d$ |
| | [fogɔkɔt] |

[55] In the representations that follow, the *irrelevant* suffix-final floating [+open₁] feature and $V_d$ are disregarded.

[56] Note that Hiatus must be non-iterative, because it would delete the suffix-initial $V_f$ if it could apply to its own output.

[57] *Modulo* the OCP effect and the optionality of the linking vowel after mid vowels as discussed in 8.1.3.

idiosyncratic behaviour is allomorphy. Surface vowel-final lowering stems have two lexical allomorphs: a 'normal'[58] one that (like all lowering stems) ends in a defective vowel, and another one whose final defective vowel is missing. By default, suffixes select the 'normal' allomorph. Some suffixes, however, are marked to select the other allomorph. When subjected to phonology, the former concatenations will surface with a lowered linking vowel while the latter ones will not have a linking vowel. Under this interpretation, a linking vowel is *phonologically* required after all lowering stems. Some lowering stems are special in that they have non-lowering lexical allomorphs as well,[59] and some suffixes are *morphologically* irregular because they select the non-lowering allomorphs of these stems.

### 8.1.4.4. The past suffix

We saw in section 8.1.2.2 that the behaviour of the other Type B suffix, the past tense morpheme, is more complex than that of the accusative. This suffix displays vowel ~ zero alternation *as well as* an alternation involving its consonant(s): -*Vtt* ~ -*t* (*lop-ott* '(s)he stole', *fal-t* '(s)he devoured', *lop-t-am* 'I stole'). The length of the suffix-final consonant depends on the presence/absence of the linking vowel: it appears as a geminate after a phonetically expressed linking vowel (*lop-ott*).[60] Recall, however, that the occurrence of the linking vowel depends on (i) the identity of the stem-final consonant (there is no linking vowel if *non-geminate* (!) *t* can form a licit coda[61] with the stem-final consonant (*fal-t*)), and (ii) whether a vowel-initial (non-analytic) suffix follows (there is no linking vowel if it does (*lop-t-am*)).

The interdependence of the length of the suffixal consonant and the conditions on the occurrence of the linking vowel raises some questions about

---

[58] We deliberately avoid using the word 'regular' here, since lowering stems are marked compared to non-lowering ones. 'Normal' is intended to mean 'representing the norm for lowering stems'.

[59] We have no explanation why all surface vowel-final lowering stems belong to this set. It must be pointed out, however, that the set contains some surface consonant-final lowering stems as well, cf. footnote 22. It is an interesting fact that, in contrast to consonant-final inflectional suffixes, vowel-final ones do not lower: *lány-ai-m* 'my daughters' and not *\*lány-ai-am*. The latter is predicted if Hiatus and Lowering apply to the underlying representation *\**/la:n$^y$-ai$_{OP}$V$_d$-V$_f$m$_{OP}$V$_d$ /, whereas the correct output is derived if the UR is /la:n$^y$-ai-V$_f$m$_{OP}$V$_d$ /. It is as if a hiatus consisting of a full vowel and a defective one were disfavoured *within a morpheme*.

[60] In the following discussion we abstract away from the effects of postlexical Degemination (cf. section 9.4), which may shorten a geminate past -*tt*, compare *Eve*[t] *körtét*. '(S)he ate some pears' and *Eve*[t:] *epret*. '(S)he ate some strawberries'.

[61] (i) Recall that only *some* licit codas are available for the past tense suffix to syllabify. We disregard this complication here (cf. 8.1.2.2) and assume that there must be a stipulation specific to the past suffix that disallows its syllabification into a complex coda whose first term is an obstruent. It would be desirable to derive this effect from the representation of the past suffix and/or (more) general conditions on syllabification. At present, we do not see how this could be done. (ii) Note that, similarly to the accusative (cf. 8.1.2.2), after *t*-final verbs a linking vowel appears even though geminate /tt/ is a well-formed coda: *üt-ött* 'hit' (3sg past indef.) and not *\*üt-t*. For a discussion of this problem cf. 8.1.4.6.

the representation of the past suffix. Since it is a Type B suffix, it is conso-nant-initial. When the suffix-initial linking vowel appears, it is a $V_d$ that is the result of overparsing by syllabification. This is doubly problematic if we assume that the suffixal consonant is an underlying geminate.[62] First, it is hard to see how the linking vowel could be absent after (some) consonant-final stems if the suffix is underlyingly -CC. As codas are maximally binary branching, it could not syllabify into the coda of the final syllable of a con-sonant-final stem regardless of the identity of the coda consonant—the expected string that results from syllabification would be C-$V_d$CC. Second, given (23b), it is not even possible to overparse a final CCC string in this way, since a defective vowel is not licensed to occur in a doubly closed syllable (*C-$V_d$CC). Furthermore, the non-occurrence of the linking vowel before vowel-initial suffixes (lop-t-am) would also be a problem. As the geminate could not syllabify as the onset of the syllable whose nucleus is the suffix-initial vowel (*{lop}$V_d${tt-$V_f$m}), the $V_d$ that is the result of overparsing by syllabification preceding the past tense suffix would not be skipped since it could syllabify in a syllable closed by the first half of the geminate ({lo}{p$V_d$t}{t-$V_f$m}). This wrongly predicts that the linking vowel surfaces even before vowel-initial suffixes: *[lopottɔm].

To sum up, the past suffix behaves as a single /t/ in the derivation when the presence/absence of the linking vowel is determined by syllabification, but appears as a geminate if the linking vowel occurs at the surface. We can express this by assuming that the length of the suffixal consonant is the result of gemination. Since the past suffix has to be distinguished from similar suffixes (i.e. the accusative) whose suffixal consonant does not gemi-nate in the same context, we suggest that its underlying representation is the following:

(40)

$$
\begin{array}{ccc}
 & & N \\
 & & | \\
X & X & X \\
| & & \\
t & & [+open_1]
\end{array}
$$

Thus, the past suffix is a /t/ whose root node is associated to a *single* timing slot followed by an empty timing slot (i.e. a timing slot devoid of melodic content). It ends in a floating [+open$_1$] feature and a $V_d$ because it is lowering (since it is an inflectional suffix: lop-t-am). We assume that an empty timing slot is completely invisible to syllabification: it may remain unparsed such

---

[62] This consonant would degeminate postconsonantally later. Since the past suffix must be distinguished from the accusative (whose suffixal consonant never shows up as a geminate), the former cannot be a single consonant underlyingly.

that (i) it may be left 'outside' syllables (41*a*), or (ii) it may be 'inside' a syllable, but unassociated to a subsyllabic constituent (41*b*).[63]

(41)                    *a.*      ]$_\sigma$ X [$_\sigma$

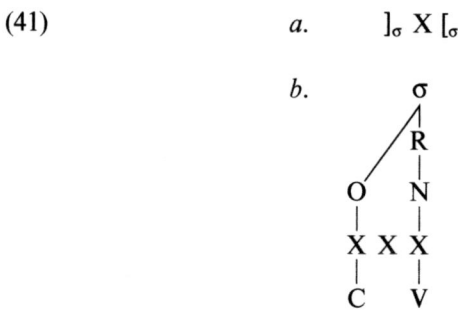

*b.*

Empty timing slots that are unparsed at the end of the derivation are not interpreted phonetically. They become visible to syllabification if they receive content. Then, like other ordinary segments they will be (and must be) parsed. We suggest that this is what happens to the past tense suffix in some contexts. Specifically, its empty timing slot may be filled by spreading from the preceding segment. This process spreads the root node of the /t/ onto a following empty timing slot if the /t/ is preceded by a full vowel:

(42)                    */t/-spread*

$$
\begin{array}{cc}
\text{X} & \text{\textcircled{X}} \\
\text{|} & \\
\text{V}_\text{f} & \text{t}
\end{array}
$$

(42) applies after Default V (22) has applied. Note, however, that it does not have to be ordered with respect to (22). If we assume that (42) applies whenever it can, it will automatically only apply after (22) (if (22) does apply).

Figure (43) below shows the behaviour of the past suffix after stems ending in a single consonant when the stem-final consonant cannot form a licit coda with the suffixal consonant (43*a*), and when it can (43*b*):[64]

---

[63] This presupposes that the syllabification algorithm looks at root nodes when the syllable trees are erected. Then, a timing unit without a root node is skipped (i.e. invisible). Note that defective vowels are different. They may be skipped by syllabification (because of the special constraints they are subject to), but they may not occur unparsed within a syllable because they are prelinked to a nucleus node.

[64] To simplify non-essential features of the derivations that follow, OP stands for floating [+open$_1$] and V$_\text{FOP}$ is the lowered full vowel that results from spreading by Lowering.

(43)  *a.* lop-ott 'steal' (3sg past indef.)  *b.* fal-t 'devour' (3sg past indef.)

UR

Syllabification

Hiatus  n.a.  n.a.

Lowering  n.a.  n.a.

Default  n.a.

Spread /t/  n.a.

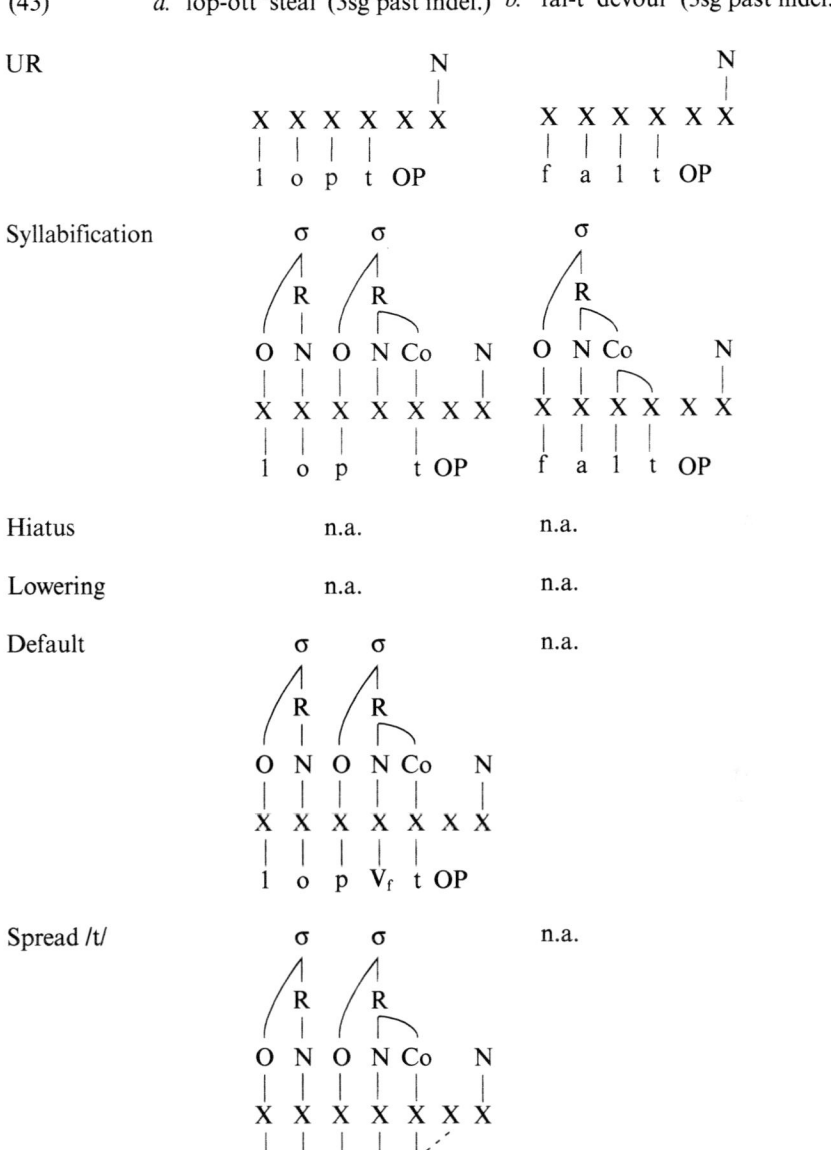

Syllabification

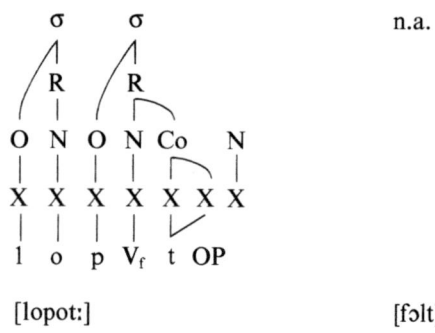

n.a.

[lopot:]　　　　　　　　　　　　　　　　[fɔlt]

As can be seen in (43*a*) the suffixal consonant cannot syllabify into the coda of the stem-final syllable, so a degenerate syllable is created by syllabification. The licensed $V_d$ of this syllable becomes a full vowel by Default V and thus the suffixal /t/ can spread to the empty X slot on its right (*lop-ott*). No degenerate syllable is created, however, if the suffixal consonant *can* form a coda with the stem-final one (43*b*). In this case /t/-spread cannot apply since its structural description is not met, and the past suffix surfaces as a non-geminate [t] (*fal-t*). Comparable forms of cluster-final stems (e.g. *dong-ott* 'buzz' (3sg past indef.), *csukl-ott* 'hiccup' (3sg past indef.)) derive like (43*a*).

The derivation of multiply suffixed forms of the same stems (i.e. when the past suffix is followed by a Type A suffix) is shown in (44).

Note that when Hiatus deletes the degenerate vowel before the full vowel of the Type A suffix, crucially, the /t/ can syllabify 'across' the empty timing slot as the onset of the initial syllable of the following suffix. Thus, /t/-spread cannot apply because its structural description is not met and the past suffix surfaces as a non-geminate [t] (*lop-t-am, fal-t-am*). The difference between the two stems is that in the case of *lop-t-am* the stem-final consonant is followed by an unsyllabified $V_d$ which is the result of overparsing by an earlier round of syllabification. This $V_d$ eventually cannot syllabify (because it is not licensed to occur in an open syllable) and is not interpreted phonetically.

'Epenthetic' stems whose final consonant cannot form a licit coda with the /t/ of the past suffix (e.g. *forog* 'revolve') behave similarly to the comparable major stems in (43*a*) and (44*a*). The only difference in their behaviour is due to the underlying defective vowel in the final syllable of 'epenthetic' stems. (45) shows the (intermediate) representation of *forg-ott* 'revolve' (3sg past indef.) which is the result of syllabification.

When Default V applies to this representation, it turns the licensed $V_d$ into $V_f$, and the /t/ can spread to the available empty position on its right. The unsyllabified $V_d$'s do not receive phonetic interpretation, thus the surface form is [forgot:].

(44)                    *a.*  lop-t-am 'steal' (1st sg. past)      *b.*  fal-t-am 'devour' (1sg past)

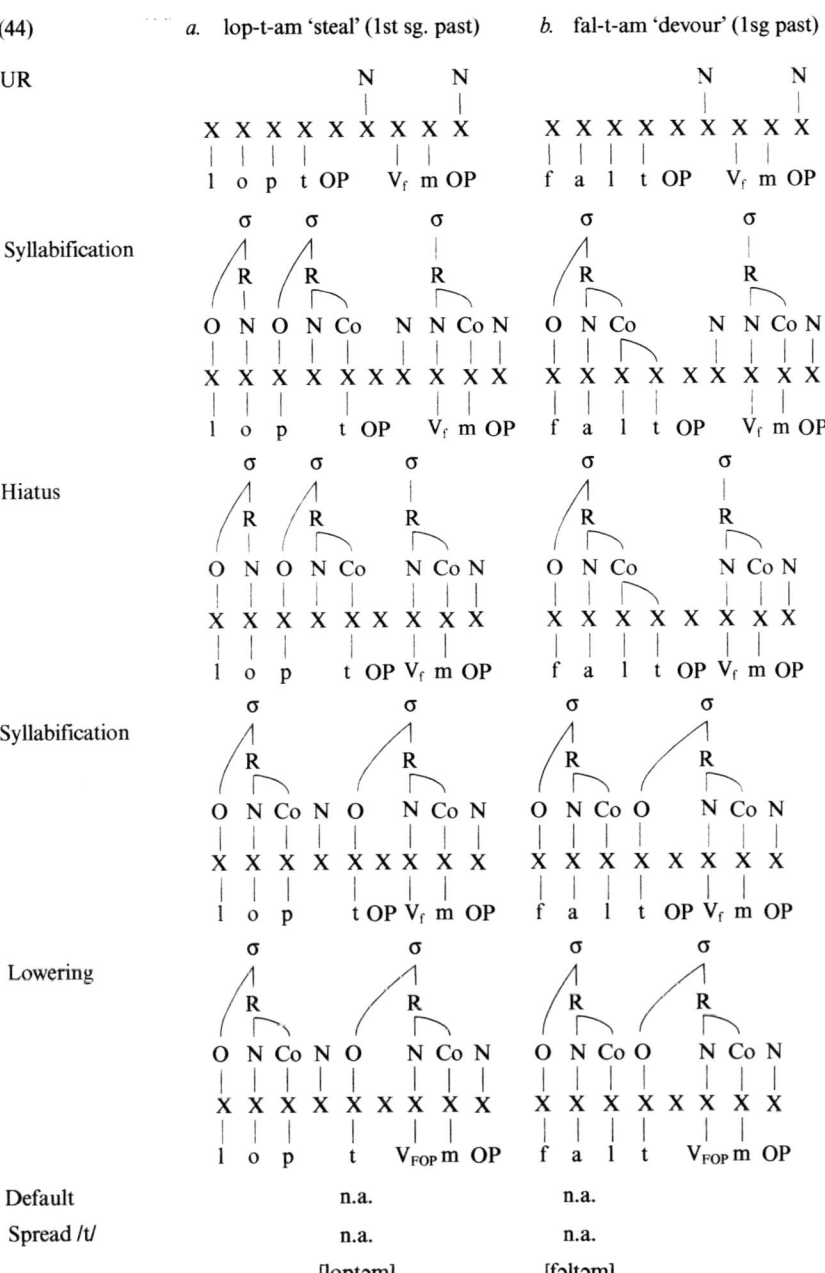

UR

Syllabification

Hiatus

Syllabification

Lowering

Default                                n.a.                              n.a.

Spread /t/                             n.a.                              n.a.

                                     [loptɔm]                          [fɔltɔm]

(45)

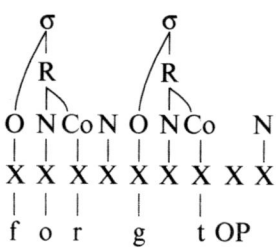

(46) shows a multiply suffixed form of the same stem (*forog-tam* 'revolve' (1sg past)) after Hiatus and syllabification (and Lowering).[65]

(46)

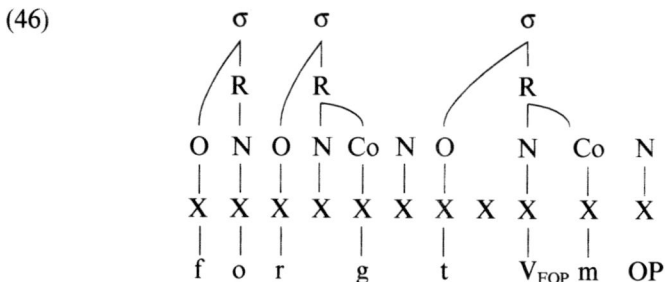

The stem-internal licensed $V_d$ becomes a full vowel when Default V applies to this representation. As /t/-spread cannot apply, the surface form is [foroktɔm].[66]

Given our assumptions about syllabification, the prediction for 'epenthetic' stems that end in a consonant with which /t/ can form a licit coda (e.g. /rabV$_d$l/ 'rob', /šodV$_d$r/ 'roll', /ugV$_d$r-/[67] 'jump', /omV$_d$l-/ 'collapse') is that they should form their singly and multiply suffixed past forms like the 'epenthetic' stems discussed above: -C}V$_d${CV$_f$tt}V$_d$# (like *forgott*) and -{CV$_f$C}V$_d${tV$_{FOP}$- (like *forogtam*).[68] This prediction is only borne out in the case of *some* past forms of *some* of these 'epenthetic' verbs. In (47) below we have charted the possible singly and multiply suffixed past forms of representative 'epenthetic' stems that end in the right consonants for branching

---

[65] Note that the defective vowel between the stem final consonant and the past suffix is the result of a round of syllabification before Hiatus because the two consonants cannot form a licit branching coda.

[66] The [k] is the result of Voicing Assimilation, cf. 7.3.

[67] /ugV$_d$r/ and /omV$_d$l-/ are bound stems (of the -*ik* class, cf. section 2.4): *ugr-ik*, *oml-ik* (3sg pres.) vs. *ugor-j*, *omol-j* (imp.)

[68] The reason is that syllabification will overparse the string consisting of the stem-final consonant and the suffix /t/ in spite of the fact that they could form a branching coda because the $V_d$ that occurs in the last syllable of the stem is disallowed in a doubly closed syllable (cf. the discussion of the accusative of 'epenthetic' nouns in section 8.1.4.2).

codas. The present form and the nominalized one are included for comparison. We have capitalized the forms that are *not predicted* given the representation of 'epenthetic' stems and the syllabification algorithm.

(47)

| stem | 3sg past indef. | 1sg past | 1sg pres. def. | nominalized form |
|------|------|------|------|------|
| rabV$_d$l 'rob' | RABOL-T | rabol-t-am | RABOL-OM | – |
|  | – | – | rabl-om | rabl-ás |
| ugV$_d$r- 'jump' | – | ugor-t-am | – | – |
|  | ugr-ott | UGR-OTT-AM | ugr-om | ugr-ás |
| omV$_d$l- 'collapse' | OMOL-T | omol-tam | – | – |
|  | oml-ott | OML-OTT-AM | oml-om | oml-ás |

It must be pointed out that (i) all these 'epenthetic' stems seem to have unexpected forms, sometimes as the only form at a given point in the paradigm,
sometimes as an alternative to an expected one; (ii) the unexpected forms are
not confined to the past paradigm; (iii) it is unpredictable which forms of
which stems will be unexpected.[69] We suggest that the reason for this complex
state of affairs is that not all forms of these stems derive from the same
underlying representation. Parallel underlying representations exist for these
verbs, one of which is 'epenthetic'.[70] For instance, *rabol* has an underlying
major stem too, which has a full vowel in the last syllable (CV$_f$C), hence
*rabol-t* (and *rabol-om*). It is unpredictable which forms are derived from
which UR(s) and whether only one, or more than one parallel UR is available
for the same form (as in *omol-t/oml-ott*).[71] The parallel UR is not necessarily
CV$_f$C-final. Forms like *ugr-ott-am* and *oml-ott-am* are derivable neither from
a CV$_d$C-final nor from a CV$_f$C-final UR. We propose that these forms derive
from an underlying stem that ends in a CC cluster which is not a possible
coda.[72] Thus, some of the lexemes discussed show allomorphy to such an
extent that they may have as many as three parallel UR variants from which
the different forms are derived.

We noted above that the singly suffixed past forms of cluster-final stems
can be handled in a straightforward way. Multiply suffixed cluster-final stems,
on the other hand, present a problem.

---

[69] It is interesting to note that the nominalized form is always the expected one.

[70] The fact that there is variation among native speakers as to which alternative forms they
find acceptable confirms this interpretation.

[71] Compare the almost identical *boml-ott* 'unfold' (3sg past indef.), which has no alternative
*bomol-t*.

[72] That is, the UR of the stem of these forms is like the bound stem /čukl-/ *csukl-ik* 'hiccup'
(cf. the discussion below), whose stem-final cluster is never separated by a vowel.

Multiply suffixed forms of stems ending in clusters that are not well-formed codas (i.e. defective stems, e.g. /čukl-/ *csukl*-ott-am 'hiccup' (1sg past), /bü:zl-/ *bűzl-ött-em* 'stink' (1sg past), /vedl-/ *vedl-ett-em* 'slough' (1sg past), cf. Károly (1957), Hetzron (1975)) derive in the following way:

(48)     csukl-ott-ak 'hiccup' (3pl past indef.)

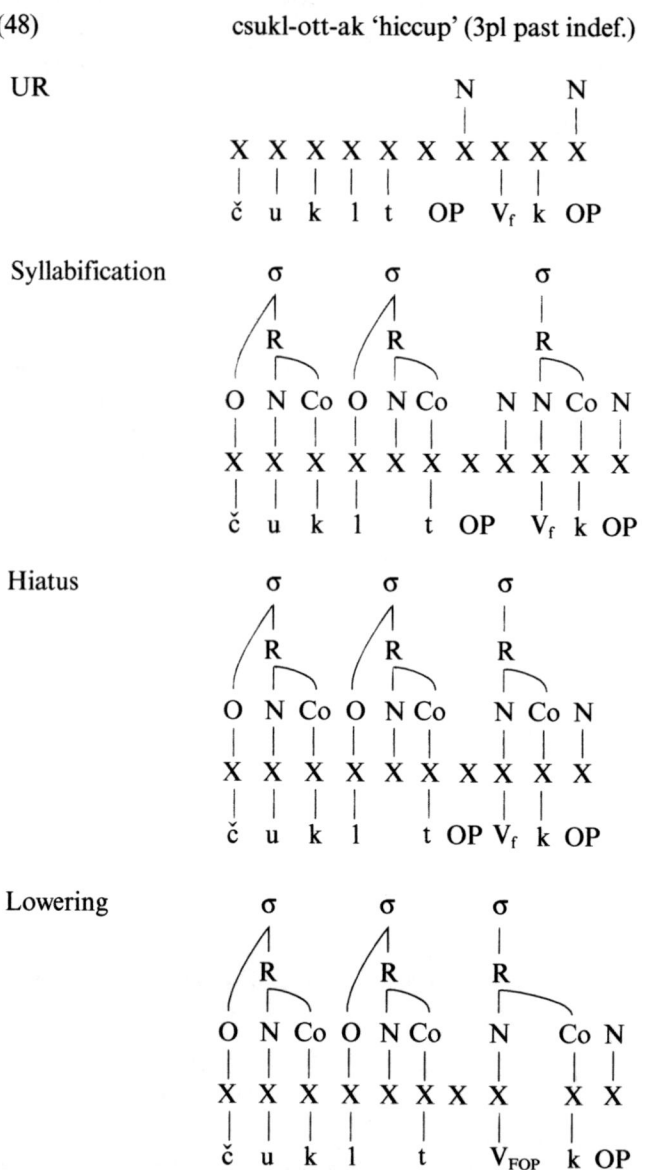

Default

Spread /t/

Syllabification

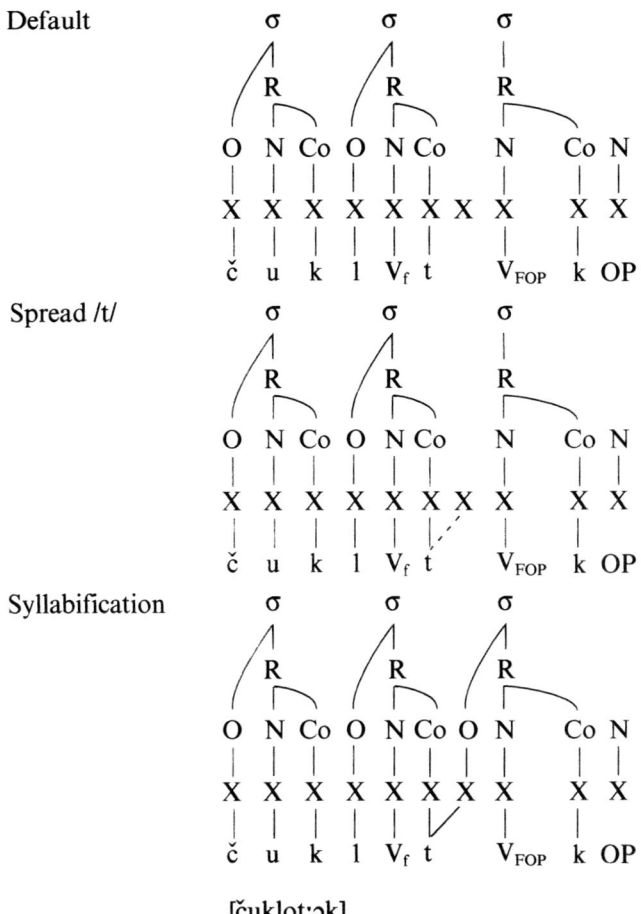

[čuklot:ɔk]

The interesting point in this derivation is the output of Hiatus. If syllabification applied to the output of Hiatus to syllabify the /t/ into the onset of the last syllable of the word, the rest of the word could not be syllabified. The defective vowel preceding the /t/ could remain unparsed, but the consonant before it could not be syllabified into the coda on its left since they do not make up a licit coda (*{čukl}$V_d${t ...}). Thus, the whole $CV_d$ string before the /t/ would have to remain unparsed (*{čuk}l$V_d${t ...}). This is excluded by non-exhaustiveness, which we restate here in a stricter form:[73]

(49)    *Non-exhaustiveness*
        Only defective material (i.e. defective vowels and empty positions)
        may remain unparsed into syllables.

---

[73] Compare (25). For a more precise (re)formulation of non-exhaustiveness, see section 8.1.4.5.

Another option would be for syllabification to overparse the stem-final cluster, but this is not possible either, since overparsing is a structure changing operation and thus can only happen in a derived environment $(*\{\check{c}u\}\{kV_dl\}V_d\{t\ldots\})$.[74] Thus, syllabification cannot apply to the output of Hiatus and the derivation proceeds as shown in (48).

For multiply suffixed forms of cluster-final stems that end in a well-formed coda (e.g. *dong-t-ak* 'buzz' (3pl past indef.)), the syllabification algorithm predicts that that they should follow the derivation of *lop-t-am* (cf. (44*a*)). That is, after Hiatus the past /t/ syllabifies as the onset of the syllable whose nucleus is the full vowel of the suffix following it. The stem-final consonants can syllabify as a coda and the defective vowel following them remains unparsed; /t/-spread cannot apply. This is shown in (50):

(50)                    dong-t-ak 'buzz' (3pl past indef.)

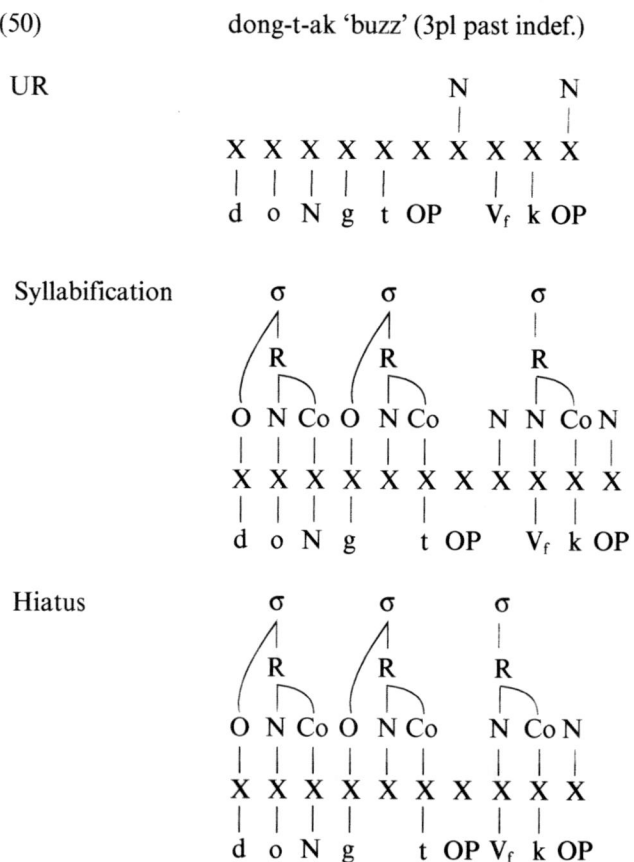

Syllabification

Lowering

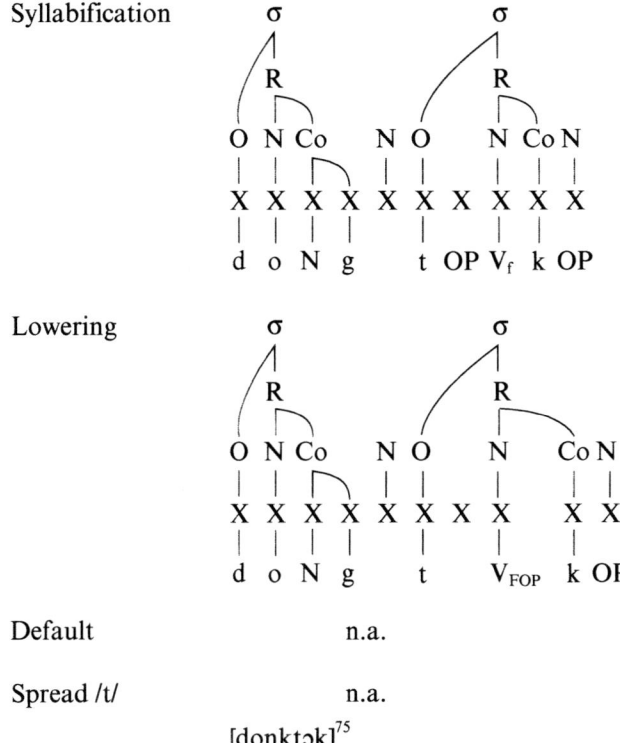

Default                    n.a.

Spread /t/                 n.a.

[doŋktɔk][75]

The prediction is correct for the stem *dong*, but recall that there are other stems ending in a branching coda that (i) either have an alternative multiply suffixed past form alongside the expected one (e.g. *fing* 'fart': *fing-tak*/*FING-OTT-AK*, *told* 'lengthen': *told-t-ak*/*TOLD-OTT-AK*) or (ii) only have a different form (*OLD-OTT-AK*, but *\*old-t-ak*). These unpredicted forms (which are capitalized in the previous sentence) are always of the same shape: they have a linking vowel after the stem (and consequently a geminate /tt/). The unexpectedness of these forms consists in the unmotivated occurrence of the linking vowel after the stem. The $V_d$ (which results from a previous round of syllabification) is eventually unparsed after some stems (e.g. *dong*: ... Ng}$V_d${t ...)—which is the predicted case, after others it *may* be parsed (e.g. *fing*: ... Ng} $V_d${t... / ... N} {g$V_d$t}...) or *must* be parsed (e.g. *old*: ... 1} {d$V_d$t...}). We do not really have an explanation for these forms and can only offer some speculation as to why the defective vowel behaves in this way after these stems. First of all, obviously, lexical marking must be involved since all these stems have well-formed codas, and the defective vowel may be parsed or unparsed after the same coda clusters in different stems (compare *fing* and *dong*, *old* and *told*)

---

[75] The first [k] is the result of Voicing Assimilation, cf. 7.3.

and therefore the occurrence of the linking vowel cannot be predicted on the basis of the melodic content of the coda clusters.[76] It is certainly the stems that must be marked in some way. Second, the reason why this differential behaviour is only observed after cluster-final stems must be related to the status of internal CCC clusters. We have pointed out in section 5.2.2 that, apart from sporadic irregular monomorphemic examples, internal CCC clusters only occur if they are not within the same analytic domain. There is one systematic set of counterexamples to this generalization: multiply suffixed past forms of verb stems that end in a branching coda, such as [dɔŋktɔk] (recall that the past suffix is synthetic). The internal CCC cluster of these forms always consists of a branching coda followed by an onset.[77] However, one could argue that the data above suggest that *internal* branching codas are disfavoured. This would make the unexpected forms above the regular case, and the stems that allow the underparsing of a $V_d$ after a branching coda would have to be lexically marked. In the present treatment we leave this question open.

To conclude, we summarize the different types of (singly and multiply suffixed past forms of) verb stems discussed in this section. Only those forms of the stems are included in (51) that are predicted on the basis of the UR identified. The notation is as follows: $C_\alpha$ is a consonant such that /$C_\alpha$t/ is a well formed coda; $C_\beta$ is a consonant such that /$C_\beta$t/ is not a well formed coda; $C_\gamma C_\delta$ is a well-formed coda; and $C_\kappa C_\lambda$ is not a well-formed coda. The parenthesized question marks are meant to show our indecision about which of the forms syllabification should predict (both forms are attested!). $CV_dC$ final stems are the ones that are traditionally called 'epenthetic' and $C_\kappa C_\lambda$ final stems are 'defective'.

(51)

| stem-final string in UR | | singly suffixed past form | multiply suffixed past form |
|---|---|---|---|
| $CV_fC$ | $CV_fC_\alpha$ | fal-t | fal-t-am |
| | $CV_fC_\beta$ | lop-ott | lop-t-am |
| $CV_dC$ | $CV_dC_\alpha$ | ugr-ott | ugor-t-am |
| | $CV_dC_\beta$ | forg-ott | forog-t-am |
| CC | $C_\gamma C_\delta$ | dong-ott | dong-t-am (?) / old-ott-am (?) |
| | $C_\kappa C_\lambda$ | csukl-ott | csukl-ott-am |

### 8.1.4.5. Analytic suffixes

When Block 1 syllabification happens and the alternations dependent on syllable structure are calculated, material in one (dependent or independent)

---

[76] The stems with /t/-final clusters will be discussed in section 8.1.4.6.

[77] Domain-internal CCC clusters cannot have a different structure (*C.CC) since branching onsets are disallowed in Hungarian. On monomorphemic words with internal CCC clusters cf. section 8.1.4.5.

analytic domain is not visible to that in the other. This can be seen in (52) below where 'epenthetic' stems are shown in isolation, followed by a vowel-initial analytic suffix (terminative -ig), and by a vowel-initial synthetic suffix (plural -$V_f k$):

(52)        _#              _V-initial analytic suffix      _V-initial synthetic suffix
    bokor 'bush'            bokor-ig                        bokr-ok
    retek 'radish'         retek-ig                        retk-ek
    kölyök 'kid'           kölyök-ig                       kölyk-ök

(52) shows that the underlying defective vowel of 'epenthetic' stems is phonetically expressed before terminative -ig (and other vowel-initial analytic suffixes)[78] in spite of the fact that the stem-final consonant syllabifies as the onset of the suffix-initial syllable at the surface. We attribute this to Default V having applied in Block 1 (while syllabification applies in both blocks). This means that all the licensed $V_d$'s are turned into full vowels before Block 2 syllabification applies, which can then syllabify the stem-final consonant as an onset since the syllable which is opened up by this operation no longer contains a defective vowel. Compare the syllabification of bokor-ig and bokr-ok:[79]

(53)                       $[\![[bokV_dr]\!]ig]\!]$              $[\![bokV_dr\text{-}V_fk]\!]$
Block 1
    Syllabification        $[\![[\{bo\}\{kV_dr\}]\!]\{ig\}]\!]$      $[\![\{bok\}V_d\{rV_fk\}]\!]$
    Default V              $[\![[\{bo\}\{kV_fr\}]\!]\{ig\}]\!]$      n.a.
Block 2
    Syllabification        $[\![\{bo\}\{kV_f\}\{rig\}]\!]$           n.a.
                           [bokorig]                        [bokrok]

Consonant-initial analytic suffixes (e.g. inessive -ban/-ben, dative -nak/-nek, ablative -tól/-től, delative -ról/-ről etc.) behave in the same way, except that Block 2 syllabification cannot syllabify the stem-final consonant as (part of) the onset of the suffix-initial syllable (bo.kor.ban 'in the bush', re.tek.ről 'about horseradish').[80]

We have noted (cf. sections 5.3.1 and 8.1.2.2) that hiatus is possible morpheme-internally (kies /kieš/ 'picturesque'), when the two vowels are

---

[78] Other suffixes of this type are causal-final -ért (bokor-ért 'for the bush'), anaphoric possessive -é (bokor-é 'that of the bush'), adverb-forming -ul/ül (bantu-ul 'in Bantu').

[79] Recall that Block 1 rules apply within analytic domains and then the whole word is submitted to the Block 2 rules; cf. section 1.3. The fact that the internal brackets are not shown in the Block 2 stage of the derivation is not meant to imply that they have been erased. It is simply that the derivation interprets the larger domain at this stage.

[80] We assume that appendices are not maximized to the detriment of a preceding coda, hence *re.te.krół.

in different independent and/or dependent analytic domains ($[[[ki]$ $[esik]]$ /kiešik/ 'fall out' (verb), $[[[kapu]$ $ig]]$ /kapuig/ 'up to the gate'),[81] but is not possible when the second vowel is initial in a synthetic suffix. In the last case, Hiatus deletes the suffix-initial vowel (cf. 8.1.4.2). This pattern can be accounted for if we assume that Hiatus is only a Block 1 rule (where it is subject to the derived environment constraint), and does not apply in Block 2:

| (54) | $[kies]$ | $[[[kapu]$ $ig]]$ | $[kapu\text{-}V_fk]$ |
|------|----------|----------|----------|
| Block 1 | | | |
| Syllabification | $[\{ki\}\{eš\}]$ | $[[[\{ka\}\{pu\}]$ $\{ig\}]]$ | $[\{ka\}\{pu\}\{V_fk\}]$ |
| Hiatus | n.a. | n.a. | $[\{ka\}\{pu\}k]$ |
| Syllabification | n.a. | n.a. | $[\{ka\}\{puk\}]$ |
| Block 2 | | | |
| Syllabification | $[\{ki\}\{eš\}]$ | $[\{ka\}\{pu\}\{ig\}]$ | $[\{ka\}\{puk\}]$ |

Vowel-initial analytic suffixes can be used to argue for the stray erasure of defective material (defective vowels and empty skeletal slots) at the end of Block 1. We have seen above that Block 2 syllabification parses the last consonant of underlyingly consonant-final stems (e.g. *pad* 'bench') as an onset when a vowel-initial analytic suffix follows: {pa}{dig} *pad-ig* 'up to the bench'. Lowering stems (e.g. *vad* 'beast'—compare *vad-ak* 'beast' (pl.)) are expected to syllabify in a different way if defective vowels are visible in Block 2 derivation. The reason is that lowering stems end in a defective vowel ($/vad_{OP}V_d/$), and defective vowels may only remain unparsed outside a syllable, i.e. an unparsed $V_d$ cannot occur within a syllable that has a nucleus.[82] Consequently, the consonant preceding the final $V_d$ of a lowering stem cannot syllabify 'across' the $V_d$ to become the onset of the analytic vowel-initial suffix;[83] *vad-ig* 'up to the beast' is predicted to syllabify as $\{vad\}V_d\{ig\}$. However, for native speakers, there is no difference between the syllabification of *padig* and *vadig*—both syllabify the intervocalic consonant into the second syllable. In order to avoid the unnecessary and counterintuitive difference between the syllabification of these items, we shall assume that defective vowels are erased at the end of Block 1 derivation. The two words will then be identical when Block 2 syllabification happens and will syllabify in the same way: {pa}{dig}, {va}{dig}.

Although syllabification applies in both blocks, it is subject to different conditions in them. Block 1 syllabification can build syllable structure on the segmental melody, but it is subject to the derived environment constraint, so

---

[81] Even sequences of identical vowels are possible under these conditions: *kiismer* /kiišmer/ 'learn all about', *taxiig* /taksiig/ 'up to the taxi', *bantuul* /bantuul/ 'in Bantu'.

[82] Compare the different behaviour of an empty skeletal slot, cf. 8.1.4.4.

[83] The stem-final $V_d$ cannot be deleted by Hiatus because Hiatus only applies in Block 1.

it can only overparse it (i.e. insert defective vowels) in a derived environment created by a synthetic suffix. Furthermore, syllabification is based on the core template in Block 1, and appendices (i.e. the extended syllable template) only become available in Block 2.[84] This accounts for the behaviour of subsyllabic analytic suffixes such as the definite imperative -d discussed in section 5.2.4.3. Recall that this suffix always attaches to a stem without a linking vowel, regardless of what the stem-final segment is: nyom-d 'push' (imp. def.)—compare nyom-ot 'trace' (acc.). As -d is analytic, a word in which it occurs has a dependent analytic domain containing the suffix only: $[[[nyom]\ d]]$. When Block 1 syllabification applies in the dependent domain, it cannot create a $V_d$ preceding the suffix because—as the suffix is the only phonological material in the domain—the environment is not derived, and thus overparsing is excluded by the derived environment constraint.[85] Thus, the suffix remains unparsed by Block 1 syllabification and will only syllabify in Block 2. Here, however, overparsing will not happen because the extended syllable template is available and -d can syllabify as an appendix (indicated with an 'A' subscripted to the segment in question):

| (55) | $[[[n^yom]\ d]]$ | $[n^yom\text{-}t]$ |
|---|---|---|
| Block 1 | | |
| Syllabification | $[[[\{n^yom\}]]\ d]$ | $[[\{n^yo\}\ \{mV_dt\}]]$ |
| Default V | n.a. | $[[\{n^yo\}\ \{mV_f t\}]]$ |
| Block 2 | | |
| Syllabification | $[[\{n^yomd_A\}]]$ | n.a. |
| | $[n^yomd]$ | $[n^yomot]$ |

Subsyllabic analytic suffixes do not always have to syllabify as appendices. After vowel-final stems they can syllabify as a coda in Block 2 (e.g. lő-j {lö:j}).[86]

The derived environment constraint on overparsing by syllabification in Block 1 together with the availability of the extended syllable template

---

[84] Block 1 rules (including syllabification) must be allowed to apply to dependent analytic domains as well as non-dependent ones (i.e. the material in a dependent domain cannot 'wait' uninterpreted until Block 2 rules apply to the larger domain) because synthetic suffixes may follow analytic ones, and processes that target material within a domain consisting of a stem and a synthetic suffix also target that within a dependent domain consisting of an analytic suffix and a synthetic suffix. For instance, (i) overparsing by syllabification can take place in the accusative of nouns ending in an analytic suffix (such as -ság/-ség '-hood' or deverbal noun-forming -vány/ -vény): [[[ lány] ság-ot ] 'maidenhood' (acc.) vs. [[ lát ] vány-t ] 'spectacle'; (ii) Lowering applies after analytic lowering suffixes (e.g. -van/-ven): [[ hat ] van-at ]; (iii) Hiatus (which is Block 1 only since it does not delete the initial vowel of vowel-initial analytic suffixes) applies after a vowel-final analytic suffix such as diminutive -ka/ke: [[ malac ] ká-k ] 'piglets' (where the length of the suffix-final vowel is due to Low Vowel Lengthening (cf. section 6.2.1)).

[85] Compare Jensen and Stong-Jensen (1989b).

[86] Verb stems that can appear as vowel-final have /v/-final allomorphs before underlyingly vowel-initial suffixes. We assume that both allomorphs are listed in the lexicon (cf. section 2.4).

in Block 2 can account for the licensing of initial consonant clusters and 'impossible' final ones in monomorphemic words (e.g. /št/ *stoppol* 'hitch-hike', /pr/ *prém* 'pelt', /pš/ *taps* 'clapping', /kt/ *akt* 'nude', /rj/ *férj* 'husband', cf. Chapter 5). The peripheral consonant in these clusters remains unsyllabi-fied by Block 1 syllabification[87] (as they are not derived, overparsing is excluded), and the unparsed consonants can syllabify as an appendix in Block 2:

| (56)            | ⟦preːm⟧           | ⟦feːrj⟧          |
|-----------------|-------------------|------------------|
| Block 1         |                   |                  |
| Syllabification | ⟦p{reːm}⟧         | ⟦{feːr}j⟧        |
| Default V       | n.a.              | n.a.             |
| Block 2         |                   |                  |
| Syllabification | ⟦{p$_A$reːm}⟧     | ⟦{feːrj$_A$}⟧    |
|                 | [preːm]           | [feːrj]          |

It is an advantage of this treatment that clusters containing a subsyllabic ana-lytic suffix and identical monomorphemic clusters receive the same analysis in terms of syllable structure: compare *kér-j* {keːrj$_A$} 'ask' (imp. indef.) and *férj* {feːrj$_A$} 'husband'.

Appendices are thus available for syllabification for consonants peripheral in an (independent or dependent) analytic domain if they are left unparsed by Block 1 syllabification. It is to be noted, however, that the licensing of morpheme-internal clusters consisting of more than two consonants is still unaccounted for. In section 5.3.2.2 we argued that these clusters are

---

[87] This appears to violate non-exhaustiveness as formulated in (49). Note, however, that (49) was designed to prevent the resyllabification of an already syllabified form such as {čuk}{lV$_d$t} ..., while here we have the underparsing of non-defective material in an unsyllabified form. Let us suppose that the vacuous application of the syllabification algorithm is the same as non-appli-cation. Then, the application of syllabification may mean (i) the full parsing/reparsing, (ii) the overparsing, or (iii) the underparsing of a string. In the case of the already syllabified string {čuk}{lV$_d$t} ... application would result in (ii) (*{ču}{kV$_d$l}V$_d${t ...) or (iii): *{čuk}lV$_d${t .... Of these, (ii) is excluded by the derived environment constraint and (49) is intended to exclude (iii). Non-application, however, would still 'yield' a licit syllabified form ({čuk}{lV$_d$t}...). The case of monomorphemic clusters under consideration is different: here, non-application is not possible because it would leave the whole morpheme unsyllabified. Overparsing is excluded for the same reason as above, and thus the minimal underparsing of non-defective material is the only option left ({feːr}j). Thus, the appropriate version of Non-exhaustiveness must ban the underparsing of non-defective material in the first case, but must permit it in the second one. (49′) is a possible formulation:

(49′) *Non-exhaustiveness*
  Syllabification may leave phonological material unparsed. Non-defective material may only be left unparsed as a last resort (where defective material is V$_d$ or an empty timing slot).

This is obviously the kind of problem that could be given an optimality theoretic interpretation.

irregular.[88] Nevertheless, they are not broken up by overparsing and they are not simplified by deletion in the lexical phonology.[89] The fact that they are not overparsed is due to the derived environment constraint, but it is not yet clear how they are licensed, since, morphologically, they are not peripheral in an analytic domain. We suggest here that the reason is a mismatch between purely morphological domains and phonologically relevant ones (cf. Törkenczy and Siptár 1999). Although the words containing these clusters are monomorphemic, phonologically they are treated in Hungarian as if they were compounds, i.e. a morphologically unitary domain is phonologically analysed as if it were two independent domains. The actual point at which the division of the morphological domain is made may vary from speaker to speaker, but is always in the middle of the cluster.[90] Thus, every word containing a cluster longer than two consonants has more than one (re)analysis.

(57)[91]

| templom | [ [ tem ] [ plom ] ] | or | [ [ temp ] [ lom ] ] | | |
|---|---|---|---|---|---|
| export | [ [ ek ] [ sport ] ] | or | [ [ eks ] [ port ] ] | | |
| puzdra | [ [ puz ] [ dra ] ] | or | [ [ puzd ] [ ra ] ] | | |
| asztma | [ [ as ] [ tma ] ] | or | [ [ ast ] [ ma ] ] | | |
| lajstrom | [ [ laj ] [ štrom ] ] | or | [ [ lajš ] [ trom ] ] | or | [ [ lajšt ] [ rom ] ] |

Block 1 syllabification can only partially syllabify the material in each independent analytic domain[92] ([ [ {ek} ] [ s{port} ] ] or [ [ {ek}s ] [ {port} ] ]) and Block 2 syllabification can incorporate the unsyllabified peripheral consonants into extended syllables ([ [ {ek} ] [ {s$_A$port} ] ] or [ [ {eks$_A$} ] [ {port} ] ]).

The fact that different native speakers may syllabify these words differently and even the same native speaker may find more than one syllable division possible shows that they are, or can be, reanalysed as compounds in different ways. With some items, one syllabification is much more likely than the

---

[88] Internal clusters of more than three consonants cannot be analysed as a coda + onset sequence. Internal CCC clusters cannot be syllabified as a simplex coda plus an onset because the onset may not branch. In section 8.1.4.4 we saw that multiply suffixed past forms of cluster-final stems suggest that the well-formedness of domain-internal branching codas is questionable. They are probably ill-formed (or at least marked). Morpheme internally they certainly seem to be ill-formed since morpheme-internal C1C2C3 clusters where C1C2 could be a licit coda are just as irregular/rare as those in which it could not.

[89] On postlexical cluster simplification cf. section 9.5.

[90] A probable scenario is that native speakers try to syllabify these words in the usual way ({laj}št{rom}) and make the division somewhere in the unsyllabifiable portion (the number of ways depends on how many consonants would be left unsyllabified).

[91] Glosses: *templom* 'church', *export* 'id.', *puzdra* 'arrowcase', *asztma* 'asthma', *lajstrom* 'list'.

[92] Overparsing is excluded since the string within the domain is not derived. Note that in some cases both domains of the reanalysed word may be fully syllabified in Block 1, e.g. [ [temp ] [ lom ] ].

alternative one(s): e.g. most (if not all) speakers would syllabify *asztma* as /ast.ma/ rather than /as.tma/. This suggests that everybody analyses this word as ⟦⟦ ast ⟧⟦ ma ⟧⟧, which is unexpected in the present account. In most cases, however, all the predicted syllabifications seem equally possible: laj.štrom = lajš.trom = lajšt.rom.

The above treatment of monomorphemic words containing clusters longer than two consonants is compatible with all other facts of Hungarian phonology.[93] Its weakness is that there is very little internal independent motivation supporting it. In principle, evidence might come from backness/frontness harmony. As compound members do not have to harmonize (cf. 3.2), we would expect that there should be disharmonic stems among those that contain these overlong clusters. This appears to be true (e.g. *angström* 'id.', *ösztrogén* 'oestrogen'). It has to be pointed out, however, that (i) real disharmony is just as rare among these words as in words that do not contain clusters longer than two consonants (e.g. *sofőr* 'driver'), and (ii) most of the words with CCC clusters whose vowels do not agree in backness contain /e/ as a non-harmonizing vowel, which we claim is neutral. If /e/ is neutral, then these words are not disharmonic.[94] Thus, the 'evidence' is inconclusive. It must be pointed out that the 'evidence' would not be better even if /e/ *were* harmonic. The reason is that we would then have a lot of disharmonic words with /e/ that do not contain a CCC cluster (e.g. *betyár* 'highwayman', *haver* 'friend'), i.e. disharmony would be just as frequent in these words as it is in those that have a cluster consisting of more than two consonants (e.g. *export* 'id.', *komplett* 'complete').

### 8.1.4.6. OCP effects, residual problems

In this section we discuss some residual problems concerning the vowel ~ zero alternations analysed above.

#### 8.1.4.6.1. 'Epenthetic' stems

In section 8.1.1 we argued that the phonological mechanism responsible for the vowel ~ zero alternation in 'epenthetic' stems is not epenthesis, i.e. it is not phonotactically motivated. Nevertheless, we pointed out that there are certain phonotactic restrictions that hold between the consonants flanking the defective vowel of these stems. These are static well-formedness constraints that disallow morpheme-shapes that do not conform to them. We repeat them in (58) below:

---

[93] Stress, for instance, is not a problem: both compound and non-compound words have initial stress (cf. 2.3).
[94] The status of /e/ is controversial in the literature (cf. Nádasdy and Siptár 1994). Some authors consider it harmonic rather than neutral (cf. Ringen 1988*a*, *b*, Ringen and Vago 1995, 1998*b*).

(58) *a.* $*C_iV_uC_j$ if $C_i=C_j$
   *b.* $*C_iV_uC_j$, if $C_i$, $C_j$ = [–son], and only one of them has a laryngeal node
   *c.* $*CCV_uC$, $*CV_uCC$

Such constraints may appear unexpected since the consonants involved are non-adjacent. We suggest that all three constraints can be attributed to the transparency of the intervening defective vowel. Although the consonants flanking the defective vowel are not string-adjacent on the skeletal tier, their root nodes *are* adjacent since a $V_d$ has no material below its X slot. Thus, constraints that apply to features and nodes below the skeleton can hold between consonants that are separated by a defective vowel.

Therefore, (58a) can be attributed to the OCP. On the root tier two identical consonants that are separated by a $V_d$ would be a fake geminate (cf. e.g. Perlmutter 1995) and would be banned by the OCP intramorphemically. Thus, the underlying representation of an 'epenthetic' stem could not contain the string in (59a) (where $C_\alpha$ is a consonantal root node dominating a particular feature tree):[95]

(59)

The structure in (59b) could not occur in an 'epenthetic' stem either. Although (59b) conforms to the OCP, the defective vowel 'embedded' in the true geminate could *never* surface since Default V ((22)) could not apply to the $V_d$ of (59b) because of the No Crossing Constraint (NCC)—a geminate integrity effect, cf. e.g. Kenstowicz and Pyle (1973), Schein and Steriade (1986), Yip (1987), Clements and Hume (1995). In order for (22) to specify

---

[95] This presupposes a 'strict' interpretation of the OCP in which it can be determined by inspecting a *single* tier whether a given configuration is an OCP violation or not (e.g. McCarthy 1988: 'Adjacent identical elements are prohibited.') rather than a 'loose' one under which the structural tier (such as the skeleton) to which the features/nodes concerned are anchored has to be examined as well (cf. Hewitt and Prince 1989: 'No melodic element may be structurally adjacent to an identical element'). Under the loose interpretation (59) would not be an OCP violation because the two $C_\alpha$'s are not structurally adjacent (they are tier-adjacent, but their structural anchors, the X-slots in this case, are non-adjacent).

the $V_d$ in (59b) association lines would have to cross, which is banned by the NCC (e.g. Goldsmith 1976). Thus, an 'epenthetic' stem that has identical consonants separated by a $V_d$ cannot be represented.[96]

Although the transparency of $V_d$ plays a role in the other two phonotactic restrictions (58b, c) as well, they cannot be derived from general constraints like the OCP or the NCC. (58b) is identical with the constraint which requires that adjacent obstruents must agree in voicing (cf. Chapter 5), i.e. that they either must share a laryngeal node, or neither should have one. What it shows is that for this constraint, adjacency must be defined on the laryngeal tier. (58c) is more problematic. It is easy to see why 'epenthetic' stems cannot end in more than one consonant (*$CV_uCC$). (i) This string would be unsyllabifiable if the stem is in isolation, or if it is followed by a consonant-initial suffix because of (23b); (ii) in a hypothetical stem ending in this string, the first one of the two stem-final consonants would always syllabify as a coda when a synthetic vowel-initial suffix follows: ... $CV_uC$}{C-V ... (as onsets may not branch). This means that the $V_d$ would *always* surface since it would be parsed in a closed syllable and Default V would apply to it—consequently such a stem would not show vowel ~ zero alternation, which is what 'epenthetic' stems do by definition. The problem is that there is no similar reason why consonant clusters do not *precede* the $V_d$ in 'epenthetic' stems (*$CCV_uC$). Epenthetic stems containing this string could be syllabified in isolation (... C}{$CV_uC$}) and would display vowel ~ zero alternation, i.e. the defective vowel would not surface when a vowel-initial synthetic suffix is attached to the stem (... CC}$V_u${C-V ...).[97] It is not clear how the constraint could be explained.[98] We tentatively suggest that it may derive from the constraint excluding three adjacent consonants in Hungarian (which would then have to be formulated in terms of root nodes rather than X-slots), but here we leave this question open.

### 8.1.4.6.2. /t/-final stems

In the discussion of the accusative and the past tense suffix, both of whose consonantal melody is /t/, and which may receive a defective vowel by overparsing (both are Type B suffixes), we noted that such a linking vowel appears even if the stem ends in /t/ in spite of the fact that geminates are licit codas (cf. sections 8.1.2.2 and 8.1.4.4). Consider the examples below:

[96] It must be pointed out that, apparently, identical place nodes can occur on the two sides of a $V_d$ in an 'epenthetic' stem (cf. 8.1.1 footnote 7). We have no explanation for this.

[97] Unless the cluster is impossible as a coda.

[98] In GP a principled explanation of similar phenomena in French was proposed by Charette (1990, 1991).

(60)  *accusative*                          *past (3sg indef.)*

    rét-et     'field'         vét-ett     'do wrong'

    hat-ot     'six'          hat-ott     'effect'

    szövet-et  'fabric'       szövet-ett  'make weave'

    lapát-ot   'shovel'      matat-ott  'fumble'

The question is why these forms have a linking vowel, i.e. why syllabification overparses the string consisting of the stem-final consonant and the suffix when the suffixal consonant could be syllabified into the stem-final syllable as part of a well-formed branching coda: e.g. *rét-t 'field' (acc.), *vét-t 'do wrong' (3sg past indef.), etc.

The accusative of /t/-final lowering stems (e.g. hát-at 'back' (acc.)) and the accusative forms or the singly suffixed past forms of Ct-final stems (e.g. ezüst-öt 'silver' (acc.), ébreszt-ett 'wake sob. up' (3sg past indef.)) do not require a special explanation. They behave like all the other lowering stems and cluster-final stems (see sections 8.1.4.3 and 8.1.4.4). It is the accusative and the past forms of non-lowering non-cluster-final stems like those in (60) that are problematic.

It would be desirable to avoid stipulating constraints that are specific to these stems and are not directly related to the general syllable template, and to be able to motivate the occurrence of the linking vowel with some general principle. It is tempting to find this principle in the OCP. We shall discuss a possible OCP-based account, but will point out that—because of some arbitrary complexities of the data and certain theoretical difficulties—it is not possible to give an account which connects the OCP and syllabification/overparsing.

The fact that a linking vowel shows up in the accusative and the singly suffixed past forms of /t/-final stems can be interpreted as a repair of an OCP violation (which consists in the juxtaposition of two identical root nodes as a result of suffixation) if we assume that (i) the OCP is not only a constraint on lexical representations, but is effective in the derivation as well; and (ii) the OCP violation created by synthetic suffixation is not repaired by merging the identical root nodes into a true geminate. Given these two assumptions we could motivate why overparsing happens: fake geminates may not be parsed as a branching coda (e.g. reːt-t$_{OP}$V$_d$ → {reː}{tV$_d$t}$_{OP}$V$_d$ [reːtɛt] 'field' (acc.)). This solution is attractive because it is in conformity with the fact that, *typically*,[99] true geminates are not created by concatenation in Block 1: in this block they are either underlying (*cigaretta* /tˢigaretːa/ 'cigarette') or the result of spreading (e.g. *hat-tal* /hat-val/ → [hotːɔl] 'with six', *ad-ott* /adotː/ cf. 8.2.1 and 8.1.4.4).[100]

---

[99] The only exception, some multiply suffixed past forms, will be discussed below.

[100] By contrast, analytic suffixation may freely create (fake) geminates: [[[meg]] [gátol]] [megːaːtol] 'prevent', [[ad]d] [odː] 'give' (imp. indef.), [[bab]ban] [bɔbːɔn] 'in (the) bean'. Note that true geminate consonants and fake ones are phonetically indistinguishable (both have a single

There are two problems, however: the first one concerns some data we have not examined yet and the second one is theoretical.

As shown in (60), the linking vowel is always present in the accusative and the singly suffixed past forms of /t/-final stems. The multiply suffixed past forms of /t/-final verbs (in which the past suffix is followed by another (vowel-initial) suffix[101]) behave in a complex and *ad hoc* way.

The generalization is the following: in these forms the past suffix appears without a linking vowel if the stem ends in the string *atlet*:

(61)

| | *singly suffixed form* *(3sg past indef.)* | *multiply suffixed form* *(1sg past)* | *gloss* |
|---|---|---|---|
| *a.* | ápolgat-**ott** | ápolgat-t-am | 'nurse repeatedly' |
| | emelget-**ett** | emelget-t-em | 'lift repeatedly' |
| | várat-**ott** | várat-t-am | 'make wait' |
| | dolgoztat-**ott** | dolgoztat-t-am | 'make work' |
| *b.* | faggat-**ott** | faggat-t-am | 'interrogate' |
| | dédelget-**ett** | dédelget-t-em | 'pamper' |
| | ugat-**ott** | ugat-t-am | 'bark' |
| | matat-**ott** | matat-t-am | 'rummage' |

Otherwise, the past suffix shows up with a linking vowel:

(62)

| | *singly suffixed form* *(3sg past indef.)* | *multiply suffixed form* *(1sg past)* | *gloss* |
|---|---|---|---|
| *a.* | vakít-**ott** | vakít-ott-am | 'blind' |
| | hűsít-**ett** | hűsít-ett-em | 'cool' |
| | tanít-**ott** | tanít-ott-am | 'teach' |
| *b.* | bocsát-**ott** | bocsát-ott-am | 'allow' |
| | tát-**ott** | tát-ott-am | 'open wide' |
| | fűt-**ött** | fűt-ött-em | 'heat' |
| | fut-**ott** | fut-ott-am | 'run' |
| | köt-**ött** | köt-ött-em | 'tie' |
| | süt-**ött** | süt-ött-em | 'bake' |

There are four exceptions to the generalization above. Two of them have a linking vowel in the multiply suffixed form although the verb stem ends in *atlet* (63*a*), and the other two do not have a linking vowel in the same form although the verb stem does not end in *atlet* (63*b*):

release stage). Affricates are the only exception, because the first half of a fake geminate affricate may (optionally) be released too (e.g. [tˀtˀ]) while the first half of a true one may not (e.g. [tˀ]) but *[tˀtˀ]). Compare (true) *viccel* [vitˀːɛl, *vitˀtˀɛl] 'joke', *gleccser* [glečːɛr, *gleččɛr] 'glacier' with (fake) *bohóc-cipő* [bohoːtˀːipöː, bohoːtˀtˀipöː] 'clown shoe', *apacs csónak* [ɔpɔčːoːnɔk, ɔpɔččoːnɔk] 'apache boat'.

[101] Note that the accusative may not be followed by another suffix.

| (63) | | *singly suffixed form* <br> *(3sg past indef.)* | *multiply suffixed form* <br> *(1sg past)* | *gloss* |
|------|---|------|------|------|
| | *a.* | hat-**ott** | hat-**ott**-am | 'effect' |
| | | vet-**ett** | vet-**ett**-em | 'sow' |
| | *b.* | lát-**ott** | lát-t-am | 'see' |
| | | alkot-**ott** | alkot-t-am | 'create' |

It is difficult to make sense of this pattern. It is not clear why verbs ending in *at/et* should behave differently from other /t/-final verbs. Note that it is not the morphological make-up of the stems that distinguishes those in (61) from those in (62). While in many of the relevant verb stems the *at/et* string is (part of) a derivational suffix (cf. (61*a*); *ápol-gat*, *vár-at*, *dolgoz-tat*, etc.), no such morphological complexity is obvious in others (cf. (61*b*)) and, furthermore, there *are* also stems that end in a suffix among the /t/-final stems that do not end in *at/et* (cf. (62*a*); *vak-ít*, *hűs-ít*, *tan-ít*).[102] Thus, the reason for the differential behaviour is not morphological. There appears to be an arbitrary division in the set of /t/-final verbs.

If we want to keep the OCP as an explanation we have to assume that the merging of identical root nodes juxtaposed by suffixation in Block 1 is possible for the set of stems that end in *at/et*. This merging, however, is only possible if another suffix follows the past suffix. The singly suffixed past forms of all *at/et*-final stems, including the ones that allow merging in their multiply suffixed forms, do have a linking vowel.

Even if we make some provision for the above complications, there are also theoretical problems with the idea that the OCP drives overparsing here.

(i) It is difficult to conceive overparsing, i.e. the insertion of a defective vowel by syllabification, as a process that repairs an OCP violation. The reason is that the 'repair' would not eliminate the OCP violation: as defective vowels do not have phonological material below their skeletal point, the two identical root nodes (that of the stem final /t/ and the suffix-initial one) whose skeletal points the $V_d$ separates as a result of overparsing would remain adjacent. This is a problem for all three forms under consideration (the accusative, the singly and the multiply suffixed past forms of /t/-final stems). A possible way out is to say that here the 'loose' interpretation of the OCP is in force. In this case overparsing *would* be a repair since it would separate the structural anchors (the X slots) of the identical root nodes. Thus, they would no longer be in violation of the OCP (see footnote 95). This, however, would be in contradiction with the way the OCP is supposed to work in 'epenthetic' stems: there, crucially, the 'strict' interpretation was required (cf. 8.1.4.6.1). Allowing different interpretations of the same supposedly general principle

---

[102] *-(V)gat/-(V)get* is the frequentative/diminutive suffix, *-(t)at/-(t)et* is the causative suffix and *-it* is a denominal/deadjectival verb-forming suffix.

within the same language (' "strict" in the lexicon, but "loose" in the derivation') would make the principle so unrestrictive that it would lose much/all of
its explanatory power.

(ii) It might be argued that the 'loose' interpretation is possible in the
derivation because overparsing is crucial in the elimination of the OCP violation. Default V will eliminate the violation even in the strict sense of the
OCP and the $V_d$ created by overparsing is necessary for Default V to apply.
It must be pointed out, however, that not all $V_d$'s created by overparsing can
syllabify. As Default V does not apply to those that do not, the OCP will be
violated in the strict sense if the unsyllabified vowel is flanked by identical
consonants. The syllabification algorithm predicts this state of affairs in the
multiply suffixed past forms of the verb stems discussed, which are supposed
to derive like *loptam* as shown in (44) in section 8.1.4.4. Here the $V_d$ created
by overparsing between the verb stem and the initial /t/ of the past suffix
eventually cannot syllabify. Thus, these forms violate the OCP in the strict
sense even after Default V applies.[103]

Thus, we conclude that OCP-motivated overparsing is not a tenable
account of the behaviour of the multiply suffixed forms of /t/-final stems. The
OCP does play a role, however, but the repair is not overparsing. Let us
assume a strict interpretation of the OCP and that in Hungarian it applies to
underlying forms and in Block 1 (but not in Block 2 and postlexically). OCP
violations can be repaired in two ways: epenthesis (64) and merging (65)
(where non-essential structure is suppressed and root$_V$ is the root node of a
vowel):

(64)

$$\varnothing \;\longrightarrow\; \overset{\displaystyle X}{\underset{\substack{\mid \\ \text{root}_V \\ \mid \\ \text{vocalic} \\ \mid \\ [+\text{open}_2]}}{}} \;\Big/\; \underset{\text{root}_\alpha}{\overset{\mid}{X}} \; \_ \; \underset{\text{root}_\alpha}{\overset{\mid}{X}}$$

(65)

$$\underset{\text{root}_\alpha}{\overset{\mid}{X}} \; \underset{\text{root}_\alpha}{\overset{\mid}{X}} \; \underset{\text{root}_V}{\overset{\mid}{X}} \;\longrightarrow\; \underset{\text{root}_\alpha}{\overset{\diagdown\diagup}{X}} \qquad \underset{\text{root}_V}{X}$$

Both rules are triggered by the OCP. (64) inserts a full vowel, and thereby can
eliminate a violation. (65) achieves the same by creating a true geminate. We
assume that these rules do not apply across an analytic boundary and that

---

[103] The stray erasure of defective material at the end of Block 1 creates an OCP violation in
the loose sense too.

they are 'morphological' in the sense that they apply to the precompiled stem before phonological rules apply. Only (64) can apply to the accusative and the singly suffixed forms of /t/-final stems because (65) requires that there should be an adjacent vowel root node after the root of the second consonant in the input. (64) and (65) are in an 'elsewhere' relationship, with (65) being the more specific rule. This predicts that only (65) applies to the multiply suffixed past forms. This is correct for items like those in (61) but not for those in (62). Therefore the latter stems (and *hat* and *vet*) have to be marked in the lexicon so that (65) may not apply to them, in which case (64) will.

## 8.2. ALTERNATIONS INVOLVING CONSONANTS

### 8.2.1. Alternating v-suffixes: -vall-vel, -vál-vé

There are two suffixes (instrumental *-vall-vel: só-val* 'with salt' and translative *-vál-vé: só-vá* '(turn) into salt') which begin with a [v] after vowel-final stems, but after stems ending in consonants, the segmental content of their initial consonant is identical with that of the stem-final consonant. The stem-final consonant and the initial consonant of the suffix are realized as a geminate only if the stem ends in a single consonant (cf. (66a)). These suffixes will be referred to as 'alternating v-suffixes'. There are 'non-alternating v-suffixes' as well (such as *-van* '-ty': *hat-van* 'sixty', deverbal noun-forming *-vány/-vény: lát-vány* 'sight', deverbal adverb-forming *-val-ve: lop-va* 'stealthily'), which are [v]-initial after vowel-final stems, but whose initial /v/ is unchanged/retained even after consonant-final stems (cf. (66b)).

(66)[104]   a.   *alternating v-suffix*

| V_ | VC_ | CC_ |
|---|---|---|
| nő-vel | csap-pal [pː] | domb-bal [mb] |
| Feri-vel | méz-zel [zː] | vers-sel [rš] |
| falu-val | léc-cel [tːˢ] | lánc-cal [ntˢ] |
| lé-vel | kar-ral [rː] | férj-jel [rj] |

　　　　b.   *non-alternating v-suffix*

| V_ | VC_ | CC_ |
|---|---|---|
| lő-ve | lop-va [pv] | old-va [ldv] |
| ró-va | néz-ve [zv] | zeng-ve [ŋgv] |
| nyű-ve | nyom-va [mv] | tart-va [rtv] |
| rí-va | zár-va [rv] | fest-ve [štv] |

---

[104] Glosses: *csap* 'tap', *domb* 'hill', *falu* 'village', *Feri* <name>, *férj* 'husband', *fest* 'paint', *kar* 'arm', *lánc* 'chain', *lé* 'juice', *lő* 'shoot', *lop* 'steal', *méz* 'honey', *néz* 'watch', *nyom* 'push', *nyű* 'wear down', *old* 'solve', *rí* 'cry', *ró* 'scold', *nő* 'woman', *szem* 'eye', *tart* 'hold', *vers* 'poem', *zár* 'lock', *zeng* 'resound'.

The crucial analytical problem is how to distinguish the alternating v-suffixes from the non-alternating ones. The classical generative analysis was to set up an abstract underlying segment (usually /w/) as the initial consonant of the alternating v-suffixes while the non-alternating ones were considered /v/-initial (cf. Szépe 1969, Vago 1980*a*). Autosegmental representation makes it possible to avoid this excessive abstractness. While the non-alternating suffixes are underlyingly /v/-initial, the alternating ones can be assumed to begin with an empty position that receives melody (by spreading) from the final consonant of the stem to which the suffix is attached. The result of the spreading is a geminate (*csap-pal* [čɔp:ɔl]) which degeminates if the stem is cluster-final by an independently motivated process of general postlexical degemination that applies in the environment of another consonant (*domb-bal* [dombɔl]). After vowel-final stems the position remains empty and is later specified as /v/ by default (*nő-vel* [nö:vɛl]). This idea is pursued by Vago (1989).[105]

Here, we propose a different analysis which obviates the need for the default rule and can explain some asymmetries in the working of the putative general degemination rule. We retain the basic idea of the autosegmental analysis, i.e. that the alternation is due to spreading melody from the stem-final consonant to a suffix-initial empty position, but we claim that the rule that spreads the stem-final melody is the same rule that applies to derive the geminate of the past tense suffix, i.e. we generalize /t/-spread (42) as (67), where C is any consonantal root:

(67)                           *C-spread*

$$
\begin{array}{cc}
\text{X} & \textcircled{\text{X}} \\
| & \nearrow \\
\text{V}_f & \text{C}
\end{array}
$$

As (67) spreads the root node of a consonant to a following empty position only if the consonant is preceded by a full vowel, no rule of degemination is needed to account for cases like *domb-bal* [dombɔl]. In these cases (67) does not apply since its structural description is not met. This would explain why the putative degemination process is compulsory in this case while in other cases it is often optional and/or speech-rate dependent. What is compulsory is really the lack of gemination (i.e. spreading); it is only postlexical degemination, which is necessary for independent reasons and does not apply to the case at hand, which can be optional (cf. section 9.4).

Assuming that (67) is responsible for the 'complete assimilation', we now examine the two related questions: how are alternating and non-alternating v-suffixes to be distinguished, and what is the source of the surface [v] in the

---

[105] Vago uses moraic syllable structure (i.e. there is no skeletal tier); consequently, for him, empty positions are empty root nodes and not empty timing slots.

alternating suffixes after vowel-final stems? In principle the two kinds of
v-suffixes may be distinguished (i) representationally, (ii) in terms of domains
(i.e. the analytic vs. synthetic distinction), or (iii) with reference to a combination of (i) and (ii).

Let us assume (following Szépe 1969, Vago 1980*a*, 1989) that non-alternating v-suffixes have an underlying initial /v/ (68*a*). Suppose that alternating
v-suffixes are different in that they begin with an empty timing slot (68*b*).

(68)            *a.* -va/-ve            *b.* -val/-vel

$$\begin{array}{cc} X & X \\ | & | \\ v & V_f \end{array} \qquad \begin{array}{ccc} X & X & X \\ & | & | \\ & V_f & l \end{array}$$

The non-alternating v-suffixes are certainly analytic, since they may be
attached to any stem, regardless of the identity or the number of the stem-
final consonants (cf. (66*b*)). Alternating v-suffixes, on the other hand, do display phonotactic interaction with the stem-final consonant and the maximum
number of consonants that can arise as a result of the affixation is two. This
suggests that they are synthetic. However, with these assumptions, i.e. that
alternating v-initial suffixes are synthetic and begin with an empty timing
slot, the analysis runs into serious difficulties. Recall that, as opposed to
defective vowels, empty X-slots are invisible to syllabification until they
receive segmental content, and as such they can float 'inside' a syllable, i.e. if
an empty X-slot is preceded by a consonant and followed by a vowel, the
consonant can syllabify 'across' the empty X-slot as an onset (cf. section
8.1.4.4). This invisibility is not a problem when an alternating v-suffix is
attached to major stems like *csap*. The stem-final consonant could first
syllabify as an onset to the suffixal vowel, but after the spreading, the suffix-
initial position would become visible and the resulting geminate would syllabify as a coda + onset sequence (*csap.pal*). The problem arises when the
alternating suffix is attached to an 'epenthetic' stem e.g. *bokor* 'bush'. The
(intermediate) representation of the suffixed form would be the following
after syllabification:

(69)

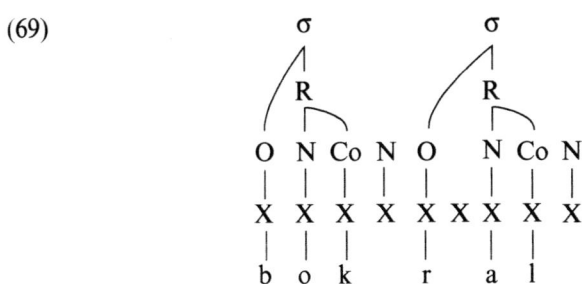

The stem-final consonant would syllabify as an onset ('across' the invisible X) and the stem-internal $V_d$ of the 'epenthetic' stem would remain unparsed (since it cannot syllabify in an open syllable). The problem is that Default V (22) would not target the unparsed stem-internal $V_d$ and consequently (67) could not apply because its structural description is not met (the spreading consonant is not preceded by a $V_f$). Thus, the predicted surface form would be *[bokrɔl] instead of the correct [bokor:ɔl] (*bokorral* 'bush' (instr.)). Assuming that alternating v-initial suffixes are analytic (and allowing (67) to apply both in Block 1 and Block 2) does not help either. The reason is that since Block 1 rules apply in the dependent domain too (cf. sections 1.3 and 8.1.4.5), the suffix-initial empty position would be deleted by the convention that defective material is erased at the end of Block 1 derivation (cf. 8.1.4.5). Thus, (67) would have no chance to apply in Block 2.

In order to avoid these problems we propose that the difference between the two kinds of suffixes is not representational, but simply a difference of domains: both of them begin with an underlying /v/ (whose root node is associated to a timing slot) but alternating v-suffixes are synthetic while non-alternating ones are analytic ([[ hat-val ]] *hat-tal* 'with six' vs. [[ [[ hat ]] van ]] *hat-van* 'sixty'). We maintain that the assimilation is the result of spreading by (67), but claim that the empty timing slot targeted by (67) in the alternating suffixes is not underlying but derived by (70):

(70)                                *v-delink*

$$
\begin{array}{cc}
X & X \\
| & \dagger \\
C & v
\end{array}
$$

v-delink is a Block 1 rule that feeds (67). As it is a Block 1 rule, it is subject to the Derived Environment Constraint. Therefore, it does not apply in monomorphemic words containing postconsonantal /v/, e.g. *tviszt* 'twist', *szvetter* 'sweater', *özvegy* 'widow', *olvas* 'read', *szarv* 'horn', *könyv* 'book'. Neither can it apply to analytic /v/-initial suffixes (or in compounds whose second member is /v/-initial) since their initial /v/ is not postconsonantal within the analytic domain even when they are preceded by a consonant-final stem, e.g. [[ [[ hat ]] van ]], [[ [[ lop]] va ]], [[ [[ ár ]] [[ víz ]] ]] 'flood', [[ [[ át ]] [[ vág ]] ]] 'cut through'. Assuming that the timing slot remains to be linked to the onset node after (70) delinks the /v/, i.e. that the onset 'branch' is only removed if the timing slot is also erased or if the nucleus is deleted (cf. Hayes 1989), (70) need not be ordered with respect to Default V (22). This is a crucial assumption, since if the X becomes dissociated from the syllable as a result of (70), then it becomes invisible to syllabification and we are back to the problem with the 'epenthetic' stems discussed above. This could only be remedied by ordering (70) after (22), because after Default V, 'epenthetic' stems can behave like major stems ending in VC (like *csap* as described above).

If the /v/ is underlying in alternating v-suffixes, there is no need for a default rule to insert it after vowel-final stems.[106] (70) simply does not delink it in these cases and thus C-spread is inapplicable (*nő-vel*).

Lowering stems behave just like non-lowering ones: (70) applies when they are followed by an alternating v-suffix; compare non-lowering *csap* (*csap-ok* 'taps') *csa*[p:]*al* 'with a tap' and lowering *fal* 'wall' (*fal-ak* 'walls') *fa*[l:]*al*. This suggests that (70) must be slightly modified to permit the delinking of the suffix-initial /v/ even if there is an intervening $V_d$ between it and the last consonant of the stem:

(71)                          *v-delink*


$$X \quad (V_d) \quad X$$
$$| \qquad\qquad +$$
$$C \qquad\qquad v$$


(67) must be modified in a similar way to optionally permit a $V_d$ before the source and the target of the spreading. Note that the spreading does not result in line-crossing because defective vowels do not have root nodes.[107]

(72)                          *C-spread*


$$X \quad (V_d) \quad \textcircled{X}$$
$$| \;\text{------}$$
$$V_f \qquad C$$


[106] This is a desirable consequence since a default rule that inserts /v/ would only ever apply to the two alternating v-suffixes. This is hardly the general scope one would expect from a rule which in essence would mean '/v/ is the default consonant in Hungarian'. In Vago (1989) the default rule is somewhat more general as it also applies in 'v-adding' stems too. These stems end in a vowel in isolation, but have a stem-final [v] before vowel-initial synthetic suffixes: *ló* 'horse' ~ *lov-ak* 'horses', *lő* 'shoot' (3sg pres. indef.) ~ *löv-ök* 'shoot' (1sg pres. indef.); cf. section 2.4. In the present treatment this is considered to be suppletive allomorphy, i.e. the phonology does not derive the allomorphs from a single underlying representation. There are only a small number of stems that show the /v/ ~ ø alternation (n=19) which is sometimes (unpredictably) accompanied by changes in the quality and/or the quantity of the vowel in the stem-final syllable: compare *l*[ö:] 'shoot' (3sg pres. indef.) ~ *l*[ö]*v-ök* (1sg pres. indef.) and *f*[ö:] 'cook' (3sg pres. indef.) ~ *f*[ö:]*v-ök* (1sg pres. indef.); *l*[o:] ~ *l*[o]*v-ak* 'horses' and *t*[o:] 'lake' ~ *t*[ɔ]*v-ak* 'lakes'. Several forms have an alternative form in which the stem behaves as a regular vowel-final stem: e.g. *szó* 'word' ~ *szav-ak*/*szó-k* 'words', *falu* 'village' ~ *falv-ak*/*falu-k* 'villages', etc. All this suggests that the alternation these stems display is non-phonological.

[107] In essence, the modified rules say that defective vowels are invisible to the two operations. Note that we could not attribute this invisibility to the convention that erases defective material at the end of Block 1 because (i) the rules show derived environment effects and (ii) Block 2 application of (71) would neutralize the difference between alternating and non-alternating v-suffixes. This raises the question whether the convention should be 'split' in such a way that it could differentiate between defective vowels and empty slots. We do not pursue this option here.

To sum up, v-delink (71) and C-spread (72) apply when an alternating v-suffix is attached to a stem ending in a single consonant, and the resulting geminate syllabifies as a coda + onset sequence in Block 1 (*csap.pal*). v-delink and C-spread do not apply in monomorphemic words like *olvas*, in analytic v-initial suffixes as in *hat-van*, and in alternating v-suffixes after vowel-final stems (*nő-vel*). In these words the /v/ syllabifies as an onset in Block 1: *ol.vas*, *hat.van*, *nő.vel*. v-delink applies but C-spread does not when an alternating v-suffix follows a cluster-final stem (including those ending in a geminate): /domb-val/ → /domb-⊗al/. The stem-final consonant syllabifies as an onset in Block 1 'across' the floating X which is erased at the end of Block 1 derivation. Lowering stems (*fal*, *talp* 'sole') behave analogously, except that the last consonant of those ending in a (surface) cluster can only syllabify as an onset in Block 2 after the stem-final $V_d$ has been erased (by convention).

### 8.2.2. *h*-alternations

There are two alternations involving the sound [h].[108] It can alternate (i) with zero: [h] ~ [Ø] (e.g. *cseh* [čɛ] 'Czech' vs. *cseh-es* [čɛhɛš] 'Czech-like'), or (ii) with a voiceless velar fricative: [h] ~ [x] (*doh* [dox] 'musty smell' vs. *doh-os* [dohoš] 'musty'). It is unpredictable whether a morpheme that has an allomorph with a final [h] ([čɛh-ɛš, doh-oš]) displays alternation (i) or (ii). In these words [h] appears prevocalically / in onset position. Otherwise (preconsonantally and word-finally), we get zero in the former set of items (henceforward *cseh*-type words) and [x] in the latter set (henceforward *doh*-type words). This is shown in (73) below:

(73) *cseh* type

| | | | *doh* type | | |
|---|---|---|---|---|---|
| cseh | [čɛ] | 'Czech' | doh | [dox] | 'musty smell' |
| cseh-től | [čɛtö:l] | 'Czech' (abl.) | doh-tól | [doxto:l] | 'musty smell' (abl.) |
| cseh-es | [čɛhɛš] | 'Czech-like' | doh-os | [dohoš] | 'musty' |

The context beyond the word is irrelevant, i.e. we do not get alternants with [h] preceding vowel-initial words in either type: *cseh asszonyok* [čɛɔs:onʸok, *čɛhɔs:onʸok] 'Czech women', *doh okozta* [doxokostɔ, *dohokostɔ] 'musty smell caused [it]'.

[h] and [x] never contrast in Hungarian. [h] cannot occur as a geminate, only [x] can (e.g. *fach* [fɔx:] 'pigeon-hole', *pech* [pɛx:] 'misfortune', *ahhoz* [ɔx:oz][109] 'to that', *Bachot* [bɔx:ot] 'Bach' (acc.). [x] occurs preconsonantally and word-finally (e.g. *doh* [dox], *doh-tól* [doxto:l] 'from [the] musty smell', *ihlet* [ixlɛt] 'inspiration', *jacht* [jɔxt] 'yacht'), while [h] only occurs in prevo-

---

[108] We are abstracting away from two minor variations on the general pattern: the (postlexical) voicing of [h] between sonorants and vowels (e.g. *konyha* [konʸɦɔ] 'kitchen', *csehes* [čɛɦɛš] 'Czech-like'), and the (postlexical) fronting of [x] after front vowels (e.g. *pech* [pɛҫ:] 'misfortune'). See Siptár (1994*b*).

[109] *Ahhoz* can also be pronounced [ɔhoz].

calic position (e.g. *doh-os* [dohoš], *cseh-es* [čɛhɛš], *hol* [hol] 'where', *néha* [neːhɔ] 'sometimes', *nátha* [naːthɔ] 'flu').

Traditionally, the *cseh* type is considered to be the native pattern (cf. Deme 1961). This assumption was taken over by most generative accounts of the phenomenon (e.g. Vago 1980*a*, Siptár 1994*b*, Törkenczy 1994*a*). As there are no systematic constraints on the occurrence of vowels preceding the [h] in *cseh*-type words, and stems can end in the same set of vowels that can occur before a [h], *h*-deletion (rather than *h*-insertion) was assumed to apply in these words (preconsonantally and finally, or (equivalently) in the coda). Because of the complementary distribution between [h] and [x], *doh*-type morphemes were assumed to end in the same underlying segment as *cseh*-type morphemes and therefore had to be marked in the lexicon so as not to undergo the deletion rule (i.e. the *doh* type was considered exceptional). In these treatments, typically, /h/ is assumed to be underlying and all instances of surface [x] are derived by rule (but, in principle, this could be the other way round).

We claim that in present-day ECH it is the *doh* type that is the systematic pattern (cf. Siptár 1998*b*) rather than the *cseh* type, which we suggest is not phonological (anymore) and is best considered as suppletive allomorphy. The principal reason is that it is the *doh* type that is productive in the sense that (i) in ECH there is a tendency for *cseh*-type morphemes to be reclassified as *doh*-type items while the reverse is unattested (e.g. *méh* [meː]/[meːx] 'bee', but *eunuch* 'id.' [ɛunux], *[ɛunu]); and (ii) new *h*-final items (loans and acronyms) are always of the *doh* type (e.g. *Hezbollah* [hɛd²bolːɔx] 'id.', *APEH* [ɔpɛx] <name of the tax office>, *BAH* [bɔx] <name of an intersection in Budapest>, etc.). Thus, while *doh*-type morphemes are an open class, there is only a single lexical item (*cseh*) that consistently represents the *cseh* type for all ECH speakers. The rest of the morphemes that are traditionally considered to belong to the *cseh* class show variation across ECH speakers or even within the speech of one and the same speaker (*juh* 'sheep', *méh* 'bee', *céh* 'guild', *düh* 'anger', *rüh* 'scabies', *éh-* 'hunger', *oláh* 'Wallachian'), or have been reclassified as *doh*-type morphemes (*méh* 'womb', *?keh* 'wheeziness') or as vowel-final ones (*pléh* 'tin').[110]

---

[110] Of the items that show variation some may be *doh* type only or *cseh* type only for a particular ECH speaker. In our own speech most of them clearly belong to the *doh* class (e.g. *méh* 'bee' [meːx], *méh-ek* [meːhɛk] (pl.), *méh-et* [meːhɛt] (acc.), *méh-től* [meːxtöːl] (abl.)). It can also happen that for the same speaker some forms of a given morpheme show *doh*-like behaviour while other forms of the same morpheme are *cseh*-like: for many ECH speakers *düh* behaves in this way: *düh* [düx], *düh-től* [düxtöːl, dütöːl] (abl.), *dühroham* [dürohɔm, *düxrohɔm] 'a fit of anger'. Note that *éh-* is a bound stem that only occurs with (some) derivational suffixes and in compounds (e.g. *éh-es* [eːhɛš] 'hungry', *éh-ség* [eːxšeːg, eːšeːg] 'hunger', *éhkopp* [eːxkopː] '[go] hungry'). Inasmuch as *keh* occurs in ECH at all in isolation and before analytic suffixes, it is a *doh*-type stem (*?keh* [kex], *?keh-től* [kɛxtöːl] (abl.), *keh-es* [kɛhɛš] 'wheezy'). *Pléh* behaves exactly like vowel-final *vécé* 'loo': compare *pléh* [pleː], *pléh-k* [pleːk] (pl.), *pléh-t* [pleːt] (acc.), *pléh-től* [pleːtöːl] (abl.) vs. *vécé* [veːtᵉeː], *vécé-k* [veːtᵉeːk] (pl.), *vécé-t* [veːtᵉeːt] (acc.), *vécé-től* [veːtᵉeːtöːl] (abl.). For most speakers *oláh* is vowel-final, but, exceptionally, Type A suffixes attach to it with a linking vowel: *oláh* [olaː], *oláh-t* [olaːt] (acc.), but *oláh-ok* [olaːok] (pl.) (see Papp 1975).

We conclude that only the *doh*-type alternation is phonological synchronically, and *cseh*-type morphemes have two underlying allomorphs, a consonant-final and a vowel-final one, whose selection is morphological. *Doh*-type stems, on the other hand, are always consonant-final. Given the complementary distribution of [h] and [x], a decision has to be made as to which of the two segments is underlying in these stems. We suggest that /x/ is the underlying segment since in this case the rule can be formulated as the delinking of the C-place node of /x/ in onset position. This is less arbitrary than the strengthening of /h/ into [x] since, if /h/ were taken to be underlying, and it were assumed to acquire DOR (and a C-place node) in the coda, then it would be impossible to identify the source of the the DOR feature assigned by the rule as it is not (necessarily) present in the environment of the putative /h/'s in the relevant stems (cf. Siptár 1998*b*).[111]

The 'weakening' rule can be formulated as follows:

(74)

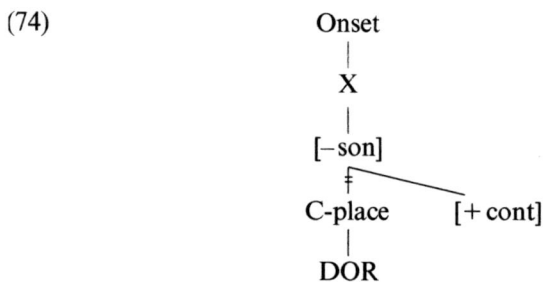

The output of (74) is a placeless fricative. In order for [h] to be derived from the underlying /x/, we have to assume that a placeless [+cons, −son, +cont] segment is phonetically implemented as [−cons, −son, +cont], i.e. an obstruent glide. This means that implementation in this case should be feature changing. In order to avoid this, we can assume that the segment underlying [h] and [x] is unspecified for [cons]. Given this assumption, (74) derives a placeless non-sonorant continuant (which is not specified for [cons]). The correct surface realizations are derived if we assume two implementation rules: (i) [−son, +cont, DOR] segments have to be implemented as [+cons] and (ii) placeless non-sonorants as [−cons].

The non-application of (74) to a geminate /xx/[112] (i.e. the impossibility of surface [h:]) is an instance of geminate inalterability (cf. Kenstowicz and Pyle 1973, Perlmutter 1995). As the rule explicitly refers to the timing tier, it is to

---

[111] It is to be pointed out that putative /h/ → [x] is also unnatural in the sense that it is 'strengthening' in the *coda*, i.e. in a lenition site. /x/ → [h] is somewhat better because here the 'weakening' often happens in intervocalic position, which is a typical lenition site (cf. Harris 1990, 1997). Note, however, that in this account, the latter event also happens in *initial onsets* where lenition typically does not take place. This is just as problematic as the strengthening in the coda in the alternative analysis.

[112] This notation is just shorthand for a single root node associated to two timing slots.

be interpreted exhaustively, i.e. it does not apply to an input in which the segmental content is multiply linked to two timing slots (and is in coda and onset position at the same time), cf. Hayes (1986), Schein and Steriade (1986). This behaviour of geminate [x:] reveals another advantage of the /x/-based account over the /h/-based one. If /h/ were the underlying segment, surface [x:] would have to be derived from an underlying geminate /hh/. The rule that derives surface [x] from a coda /h/ would not apply to a geminate /hh/ because of geminate inalterability (the same reason as above). Note, however, that while non-application yields the correct surface result in the /x/-based account (since the unchanged underlying segmental melody is the attested surface melody), in the /h/-based account an extra rule (specific to /hh/) is needed to make sure that geminate /hh/ surfaces as [x:].

Thus, we conclude that /h/ is not an underlying segment in Hungarian. All instances of [h] are derived from /x/ (or more precisely /X/, which is unspecified for [cons]) by (74).

The rule is not postlexical because it does not apply across a word boundary (*doh okozta* [doxokostɔ, *dohokostɔ] 'musty smell caused [it]'). Given our assumptions about Block 1 and Block 2 derivation, this means that (74) is a Block 2 rule since it has to apply in non-derived environments (as well), e.g. in *holló* [hol:oː] 'raven', *tehén* [tɛheːn] 'cow' and it is not a structure building rule. Assuming that (74) is a Block 2 rule makes the prediction that [h] (not [x]) occurs preceding vowel-initial analytic (i.e. Block 2) suffixes (like -*ig*, -*ért*, -*é*, -*ul/ül*, cf. section 8.1.4.5) in *doh*-type words since the stem-final /x/ is syllabified as an onset before a vowel-initial suffix in Block 2. If (74) were a Block 1 rule, [x] would be expected before vowel-initial analytic suffixes in these words because at the point when it applies, the stem-final consonant is still in the coda. It must be pointed out that native speaker intuitions/judgements (including our own) are uncertain on this point. While there are speakers who (claim to) pronounce *doh-ért* 'for musty smell' as [doxeːrt], others feel/make no difference between the pronunciation of the *h* in *bohém* 'bohemian' (monomorphemic), *doh-ot* (stem + synthetic suffix) and *doh-ért* (stem + analytic suffix). As no experimental evidence or large scale survey is available, we merely point out that the former pronunciation would be problematic since it would involve the application of a Block 1 rule in a non-derived environment.

# 9

**SURFACE PROCESSES**

In this chapter, various issues in the surface (postlexical) phonology of
Hungarian will be briefly discussed. Section 9.1 summarizes data about
the surface vacillation of vowel length (shortening, lengthening); section
9.2 gives an overview of compensatory lengthening processes. Hiatus filling
is the topic of section 9.3; the issue of degemination is discussed in 9.4.
Finally, the simplification of CCC clusters in fast speech is considered in
section 9.5.

## 9.1. THE SURFACE VACILLATION OF VOWEL DURATION

The term 'vacillation' is normally used in two different senses in phonology.
One is inter-speaker variability, i.e. the case where, with respect to some
phonological phenomenon, some speakers behave in one way, whereas other
speakers (consistently) behave in some other way. In such cases, the two
groups of speakers have different grammars: we could say that they speak different—though very similar—languages (or dialects). Ideally, the linguist
describes a single coherent linguistic system, and refrains from taking
glimpses at 'neighbouring' systems; however, if the difference is observed
within the same (sociolinguistically defined) language, in our case, standard
Hungarian, the description usually has to take data from several systems into
consideration.

The other type of vacillation is based on speakers' inconsistent behaviour
(e.g. when the same speaker sometimes says *dzsungelben* 'in the jungle' and
sometimes *dzsungelban* 'id.'). This type of vacillation has to be accounted for
even if what is described is the idiolect of a single speaker (the limiting case
of a homogeneous 'speech community'). Usually, however, the two types of
vacillation occur in conjunction. For instance, if some speakers always say
*dzsungelben*, others always say *dzsungelban*, and yet others (probably the
majority) use both forms indiscriminately, we have a mixture of both types of
vacillation.

Cases of vacillation can be classified in another, less superficial manner,
too. In some cases, the indeterminacy is located within the lexicon, in the
form of alternative underlying forms (e.g. *tejfel/tejföl* 'sour cream',

*vakond/vakondok* 'mole', as well as *dzsungel*, if its ambiguous behaviour is ascribed to two alternative underlying representations, one in which the *e* is opaque, and another one in which it is transparent, cf. section 6.1). In other cases, the rule concerned may be optional (or rate/style-dependent), or the rules may be applied in several different orders, giving rise to surface vacillation. For instance, *analízis-ben/ban* 'in analysis' can be described by an optional rule turning sequences of neutral vowels into front-harmonic (the degree of optionality depending on vowel height, see section 3.2.3).

Turning to the topic of the present section, the surface variability of vowel duration, the said types are found here, too. Inter-speaker variability based on alternative underlying forms is found, for instance, in *sz*[i]*nész/sz*[i:]*nész* 'actor', *h*[u]*ga/h*[u:]*ga* 'his sister', *gy*[ü]*jt/gy*[ü:]*jt* 'collect', *arr*[o]*l/arr*[o:]*l* 'about that', *egyb*[ö]*l/egyb*[ö:]*l* 'at once', *p*[o]*sta/p*[o:]*sta* 'post office', *k*[ö]*r-*[u:]*t/k*[ö:]*r*[u]*t* 'boulevard', *k*[ɛ]*l/k*[e:]*l* 'rise', *h*[ɔ]*nyas/h*[a:]*nyas* 'which number', as well as the vacillation between [a:] and short unrounded [a] in words like *spájz* 'larder', *Svájc* 'Switzerland', *Mozart*. Inter-speaker variability based on optional rule application is found e.g. with respect to final high vowel shortening (see Nádasdy and Siptár 1998): *szomor*[u]*/szomor*[u:] 'sad', *men*[ü]*/men*[ü:] 'set dinner' (for speakers who do not have invariable short vowels in all such items). Similarly, but this time resulting in variability within the speech of a single speaker, non-high long vowels may optionally shorten in non-final closed syllables in colloquial speech: *általános* %[altɔla:noš] 'general', *vásárváros* %[va:šarva:roš] 'market town', *érthetetlen* %[ert(h)ɛtɛtlɛn] 'unintelligible', *keménység* %[kɛmenʸše:g] 'hardness', *szórványos* %[sorva:nʸoš] 'sporadic', *őrmester* %[örmɛštɛr] 'sergeant', etc. Being a postlexical process, such shortening does not change the quality of /a: e: o: ö:/ (as opposed to lexical shortening of the *nyár* → *nyarat* 'summer'/(acc.), *szél* → *szeles* 'wind(y)', *ló* → *lovam* '(my) horse', *cső* → *csövek* 'pipe(s)' type (FSVS, cf. sections 3.1.2.1 and 6.2.2) whose result—in a structure-preserving manner—takes up the quality of the corresponding short vowel). Thus, shortened [a] is unrounded and central (like [a:]), and shortened [e o ö] are closer (tenser) than the realization of dialectal mid /e/ and that of short /o/ or /ö/ (and of course much closer than standard low [ɛ]).

With respect to high vowels, it is hard to tell if the above type of shortening (that in non-final closed syllables) does or does not apply, since their duration is highly variable to begin with (cf. footnote 2 in Chapter 3). Also, the quality difference between corresponding short and long vowels is very slight (hence, lexical and postlexical shortening cannot be told apart on the basis of vowel quality). It is nevertheless noteworthy that in the paradigms of high-vowelled FSVS stems like *út/utat* 'road'/(acc.), *tűz/tüzet* 'fire'/(acc.), *víz/vizet* 'water'/(acc.), the short vowel seems to be gaining ground outside FSVS environments, too. In examples like *úttörö* 'pioneer', *tűzhely* 'fireplace', *vízcsap* 'water-tap'; *útnak* 'to the road', *tűzben* 'in the fire', *víztől* 'from water', the shortening may be ascribed to the above process ([u]*ttörő* like [ö]*rmester*). But this time, short vowels crop up in open syllables (*úton* %[uton] 'on the road', *tűzoltó* %[tüzolto:] 'fireman', *vízierőmű* %[viziɛrö:mü:] 'hydroelectric power station') and in final closed syllables (*gyalogút* %[dʲɔlogut] 'footpath', *erdőtűz* %[ɛrdö:tüz] 'forest-fire', *kölnivíz* %[kölniviz] 'eau de cologne', too. For many speakers, stable use of the long alternants is restricted to unsuffixed, uncompounded instances of *út* 'road', *tűz* 'fire', *víz* 'water'.

Along with the surface shortening rules reviewed so far, there are surface lengthening rules as well. 'Pause-substituting' (i.e. hesitational or phrase-final) lengthening, just like compensatory lengthening as discussed in section 9.2, does not convert the short vowels into their long counterparts but only increases their physical duration. Emphatic lengthening either keeps the vowel quality or changes it in the 'wrong' direction (e.g. emphatic *ooolyan* 'so much' with an *o* opener than usual, whereas long /o:/ is closer/tenser than /o/). Other types of surface lengthening will produce [i:] out of /i/, (tense) [o:] out of /o/, etc. For instance, names of letters and sounds are usually quoted in a lengthened version as in *Ezt rövid* [i:]-*vel kell írni* 'This is spelt with short I', *A magyarban nincs rövid* [o:]-*ra végződő szó* 'There are no word-final short O's in Hungarian', etc. On the other hand, the names of the letters/sounds *a* and *e* exhibit a curiously intricate pattern. The basic case can be observed in contexts like *Nagy* [ɔ:]-*val írjuk* 'It is spelt with capital A', *Az* [ɛ:] *alsó nyelvállású magánhangzó* 'E is a low vowel'. But the musical notes *A* and *E* are called [a:] and [e:], and the word *ábécé* [a:be:tˢe:] 'alphabet' makes it likely that the name of the letter *A* used to be pronounced [a:], perhaps due to some latinate influence.[1]

Abbreviations, if they are pronounced as a sequence of letters, contain [a:] and [e:] (or [ɛ:]) if A and E are initial (e.g. *AB* [a:be:] 'abortion committee', *EKG* [e:ka:ge:] 'electrocardiogram', *EGK* [ɛ:ge:ka:] 'European Economic Community') but always [ɔ:] and [ɛ:] if final (e.g. *MTA* [ɛmte:ɔ:] 'Hungarian Academy of Sciences', *BSE* [be:ɛše:] 'Budapest Sports Club'). Abbreviations that are read out as words (e.g. *USA* [ušɔ] 'United States', *ELTE* [ɛltɛ] 'Eötvös Loránd University') behave as normal words do: they end in short [ɔ]/[ɛ] which reguarly undergoes LVL ([uša:bɔn] 'in the US', [ɛlte:röl] 'from ELTE', cf. section 6.2.1), hence they are uninteresting for our present purposes. What is interesting is that [ɔ:] and [ɛ:] never undergo LVL (in the sense that they never shift into [a:] and [e:]): [ɛmte:ɔ:vɔl], not \*[ɛmte:a:vɔl] 'with MTA', [be:ɛše:bɛ], not \*[be:ɛše:bɛ] 'into BSE'; cf. also [ɔ:ɦoz] 'to A', [ɛ:nɛk] 'for E', etc.

---

[1] Letters used for identification exhibit a mixed pattern. The bus *7A* is [he:tɔ:] but a school class *7a* is [he:ta:] or [hetɛdika:] (where the variation is in how the numeral is read out but the letter is invariably [a:]); *A épület* 'building A' can be either [a:] or [ɔ:]; in geometry, *A pont* 'point A' is either [ɔ:] or [a:]; but the bus *7E*, the class *7e*, building E, and point E are all invariably [ɛ:].

If the underlying representation of the name of the letter *E* is a short /e/, how can its surface (postlexical) lengthening block the application of a lexical rule like LVL? Such bleeding interaction (between a postlexical and a lexical rule) undoubtedly runs counter to all current assumptions concerning the way phonological systems are organized. However, the phenomena discussed in this section are both peripheral and variable: therefore, the alternative approach (positing underlying /ɔː/, /ɛː/) will be discarded here and it will be assumed that either some exception device takes care of the offending cases (e.g. the names of letters/sounds are marked in the lexicon as exceptions to LVL), or else the formulation of LVL must be modified so that a segment (a consonant or a vowel) is given as left environment. In the latter case, the difference between e.g. *fa* + *t* → *fát* 'tree' (acc.) and *a* + *t* → [ɔːt] 'the letter *a*' (acc.) is accounted for, but at the cost of restricting the generality (increasing the complexity) of the rule. It is not obvious if the gain is worth the cost.

## 9.2. COMPENSATORY LENGTHENING

In standard Hungarian, compensatory lengthening is exclusively postlexical (a casual-speech phenomenon). In addition to sporadic cases of glide deletion (*autó* %[ɔːtoː] 'car', *Európa* %[ɛːroːpɔ] 'Europe'), there are two major cases of deletion with compensatory lengthening. The deletion of liquids (*l*, *r*, *j*, cf. section 7.4.2) leaves no trace other than the lengthening of the vowel (e.g. *elront* %[ɛːront] 'spoil'); that of nasals (*m*, *n*, *ny*, cf. section 7.4.1) leaves nasality behind on the (lengthened) vowel (e.g. *színház* [sĩːɦiaːz] 'theatre').

From among liquids, it is /l/ that gets deleted the most easily, e.g. *balra* %[bɔːrɔ] 'to the left', *elvisz* %[ɛːvis] 'take away', *el kell menni* %[ɛːkɛːmɛnːi] 'one must leave'. As can be seen, compensatory lengthening does not affect vowel quality (*[baːrɔ] etc.). For long vowels, it applies vacuously, since Hungarian has no 'overlong' (three-mora) vowels (*féltem* %[feːtɛm] 'I was frightened', *leszállt* %[lɛsaːt] 'he got off'). Mid vowels (*zöld* %[zöːd] 'green', *bolt* %[boːt] 'shop', *polc* %[poːtˢ] 'shelf', *tölt* %[töːt] 'fill', *olvas* %[oːvaš] 'read') tend to preserve their quality in standard casual speech (i.e. they do not get tensed into the realization of /öː/ and /oː/); in substandard, 'village-flavour' speech (also in some dialects), however, they switch quality, too: '*ződ*', '*bót*', '*póc*', '*tőt*', '*óvas*' (where non-standard spelling is meant to suggest that the [öː] and [oː] involved are realized exactly like underlying long /öː/ and /oː/ are, rather than simply as physically lengthened [ö] and [o]). With high vowels, this difference is hardly noticeable (*küld* %[küːd] 'send', *kulcs* %[kuːč] 'key'), given that members of such long/short pairs exhibit practically no quality difference (cf. (1) in Chapter 3).

The slight difference in the mid vowels is a telling example of the difference between productive (surface level) and lexicalized instances of compensatory lengthening. Speakers who say '*ződ*' etc. have /zö:d/ etc. as the underlying representation of such words, i.e. compensatory lengthening has become part of their lexical forms. Just like any underlying /ö:/, these are realized as the appropriate tense quality. On the other hand, standard speakers have underlying /zöld/ etc., and compensatory lengthening is an on-line phonological process for them. Being postlexical (non-structure-preserving), this process does not change the quality of /ö/, it just adds (physical) duration to it: hence we get a lax (though long) realization.

The deletion of /r/ (*egyszer csak* %[ɛtˢ:ɛ:čɔk] 'after a while') is usually observed in casual speech only but in *arra* 'that way', *erre* 'this way', *merre* 'which way' it applies in colloquial (and even in moderately formal) speech: [ɔ:rɔ], [ɛ:rɛ], [mɛ:rɛ]. /j/ is primarily dropped after front vowels: *gyűjt* [dʸü:t] 'collect', *szíjra* [si:rɔ] 'on a leash', *mélység* [me:še:g] 'depth', *felejthetetlen* [fɛlɛ:t(h)ɛtɛtlɛn] 'unforgettable'. The deletion of liquids was discussed in section 7.4.2; the rule of liquid deletion was formalized there as (51).

Vowels followed by a nasal are phonetically always (more or less) nasalized, especially their latter portion (that nearest to the nasal). If, however, that nasal is deleted (this is practically restricted to /n/), the nasality of the vowel becomes a lot stronger on the one hand, and phonologically relevant on the other since this is now the only surface trace of the underlying nasal consonant (apart from compensatory lengthening, but the latter only shows that there was a consonant there, not the fact that it was nasal). For instance, in the minimal pair *szánhat* 'may pity' vs. *szállhat* 'may fly', the /n/ and the /l/ may both get deleted, neither resulting in observable compensatory lengthening, as the /a:/ is long to begin with; in this case, it is exclusively the nasality of the [a:] in the first word that carries the distinction between [sã:fiɔt] and [sa:fiɔt].

In Hungarian, nasalization of a vowel does not result in quality change, not even in this latter, phonologized, form. Contrast this with e.g. French where nasalized vowels are all opener than their oral counterparts.

The rule of /n/-deletion—with the concomitant compensatory lengthening and vowel nasalization—was given as (46) in section 7.4.1.

Note finally that intervocalic consonant deletion in fast speech (e.g. *egyedül* %[ɛɛdül] 'alone') does not involve compensatory lengthening, not even where liquids or nasals are deleted (*valódi* %[vɔɔ(:)di], *[vɔ:o:di] 'real', *minek* %[miɛk], *[mĩ:ɛk] 'what for'); in the case of nasals, it does not result in vowel nasalization, either (*menetrend* %[mɛɛtrɛnd], *[mɛ̃ɛtrɛnd] 'timetable').

## 9.3. HIATUS FILLING

As we saw in sections 8.1.4.2–3, Hungarian has a Hiatus rule eliminating either a defective vowel that occurs next to a full vowel (on either side), or else the right member of a vowel cluster. This rule shows derived environment effects and is only in force in Block 1; that is, monomorphemic hiatuses survive and additional hiatuses are freely created by analytic suffixation, compounding, and across word boundaries. The issue of monomorphemic hiatuses was discussed at length in section 5.3.1 but the process of hiatus filling was ignored there. In this section, we look at the surface implementation of all vowel clusters that survive the Hiatus rule or are created in contexts where that rule is not in force.

In section 2.2.1 we saw that one—exceptional—way of implementing an underlying hiatus is to pronounce it as a phonetic diphthong (e.g. *autó* 'car', *Európa* 'Europe'). Otherwise, some hiatuses are filled in (*hiány* [hiʲaːnʸ] 'lack' (noun)), whereas others are not (*leány* *[lɛʲaːnʸ] 'girl'). Whether the vowel cluster is monomorphemic or arises at an analytic boundary (or even at a word boundary) is irrelevant: *kiáltás* 'a cry' and *kiállítás* 'exhibition' (preverb + verb (+ nominalizing suffix)) both exhibit hiatus filling (just like, say, *ki áll itt* 'who's standing here?'), whereas *Bea* <name> and *beadás* 'handing in' (preverb + verb (+ nominalizing suffix)) both surface with unfilled hiatus (as does *be akarok menni* 'I want to go in').

Consider the following data (where hiatus filling is indicated by a superscript *j*).[2] (1*a*) lists combinations of /i/+V, (1*b*) combinations of V+/i/; similarly, (1*c*) gives examples of /iː/+V and (1*d*) examples of V+/iː/:

(1) *a.* kiʲír 'write out', kiʲiktat 'eliminate', diʲéta 'diet', siʲet 'make haste', kiʲűz 'expel', enniʲük 'eat' (3pl inf.), éjjeliʲőr 'night watchman', kiʲönt 'pour out', fiʲú 'boy', adniʲuk 'give' (3pl inf.), piʲóca 'leech', viʲola 'violet', piʲac 'market', hiʲába 'in vain'
   *b.* odáʲig 'as far as that', maʲi 'today' (adj.), utóʲirat 'postscript', kapuʲig 'as far as the gate', nőʲi 'female', műʲintézet 'institution', esküʲig 'as far as the oath', övéʲi 'his/her family', összeʲillik 'fit together'
   *c.* síʲel 'ski' (verb), síʲugrás 'ski jumping', híʲa haza 'your country needs you'
   *d.* ráʲír 'write on', adóʲív 'tax form', dicsőʲít 'praise', színműʲíró 'playwright'

As the data in (1) show, hiatuses are obligatorily filled if (at least) one of the vowels is high and coronal, i.e. /i/ or /iː/. Otherwise, if one of the vowels is /eː/, the pattern is a bit less clear. If the /eː/ comes first, *j*-insertion is optional, see

---

[2] As was briefly mentioned in section 4.2, the hiatus filler may be weaker, more transient than the realization of underlying /j/. For instance, *mi ez* 'what's this' (with hiatus filler) and *milyen* 'like what' (with underlying /j/) or *kiönt* 'pour out' and *kijön* 'come out' do not sound quite the same. The difference is clearly noticeable in careful speech, although it may get blurred in colloquial or casual styles. In the analysis we propose here, that phonetic dissimilarity reflects the distinction between a glide (the hiatus filler) and a liquid (the realization of an underlying /j/).

(2*a*); if it comes second, some vowel combinations involve hiatus filling, whereas others do not (2*b*):

(2)  *a.* %kefé<sup>j</sup>é 'that of a brush', %ketté<sup>j</sup>oszt 'divide', %mellé<sup>j</sup>áll 'stand beside him/her', %elé<sup>j</sup>ül 'sit in front of him/her'

*b.* rá<sup>j</sup>ér 'have time', oda<sup>j</sup>ég 'burn', po<sup>j</sup>én 'punchline', %áru<sup>j</sup>ért 'for goods', *le<sup>j</sup>ég 'burn down', *szökő<sup>j</sup>év 'leap year'

On the other hand, if both vowels are rounded, back, or low, or any combination of these, the hiatus surfaces as it is, without a hiatus filler:[3]

(3)  *a.* LAB + LAB:   nőügy 'affair with a woman' (*nő<sup>j</sup>ügy), mezőőr 'rural constable'

*b.* LAB + DOR:   díszműáru 'fancy goods' (*díszmű<sup>j</sup>áru), nüansz 'nuance', nőalak 'female figure', előáll 'present itself', főúr 'head waiter'

*c.* DOR + LAB:   pályaőr 'signalman' (*pálya<sup>j</sup>őr), aláönt 'pour under', kapuügyelet 'function of doorman'

*d.* DOR + DOR:   ráadás 'encore' (*rá<sup>j</sup>adás), fáraó 'pharaoh', kalauz 'conductor', hozzáolvas 'read alongside', ráun 'get bored with', aktuális 'timely', kapualj 'doorway', oázis 'oasis', oboa 'oboe', lóugrás 'knight's move', hajóút 'voyage', fluor 'id.', duó 'duo', faarc 'wooden face', aláás 'undermine', kooperál 'cooperate', állóóra 'grandfather clock', vákuum 'vacuum', hosszú út 'long trip'

*e.* [ɛ] + LAB/DOR:   beleőrül 'get mad' (*bele<sup>j</sup>őrül), beönt 'pour into', leüt 'strike' (verb), leány 'girl', belead 'give into', neon 'id.', sztereó 'stereo', beleun 'get bored with'

*f.* LAB/DOR + [ɛ]:   betűejtés 'spelling pronunciation' (*betű<sup>j</sup>ejtés), menüett 'minuet', esőember 'rain man', aláesik 'fall under', hazaenged 'allow to go home', adóellenőr 'tax inspector', poentíroz 'embellish with jokes', búcsúest 'farewell party', influenza 'flu'

In sum, the rule of hiatus filling must have an obligatory branch, before/after *i/í*, and an optional branch (before/after *é*). We will abstract away from the minor asymmetries noted in (2*b*) and simply formulate this second branch of the rule as optional both ways. In (4) the spreading process we will assume to operate is illustrated in two representative examples:

---

[3] In non-standard speech, there is another type of phenomenon resembling hiatus filling in examples like *mondta<sup>j</sup>az anyjának* 'he said to his mother'. However, this is not based on a phonological process, given that such inserted glides only occur before the definite article *a/az* (*mondta <sup>j</sup>aztán* 'he said then', **olcsó<sup>j</sup>alma* 'cheap apples'). Hence, in this substandard variety of Hungarian, the definite article has the allomorphs *a/az/a<sup>j</sup>a/a<sup>j</sup>az* (this observation is due to Ádám Nádasdy, personal communication).

(4)    *a. fiú* 'boy'                              *b. mai* 'of today'

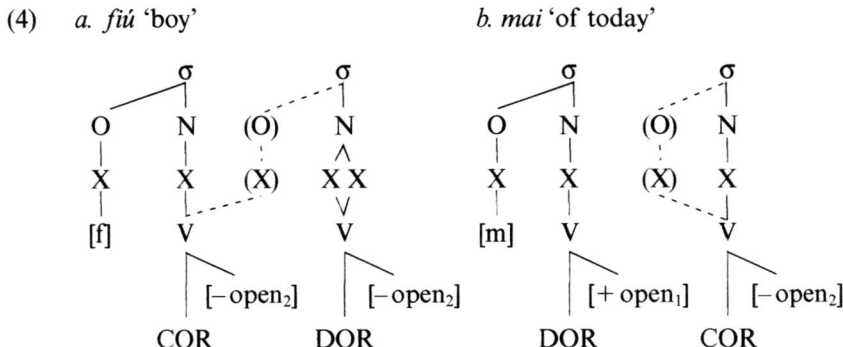

In (4a), the V node of the *i* spreads to the right, creating an X and an O node along the way, to the σ node of the following syllable. In (4b), on the other hand, the V node of the *i* spreads to the left, again creating an X and an O node, but this time the non-nuclear slot thus created will be syllabified as the onset of the right-hand syllable, rather than as the coda of the left-hand syllable. Examples like *síel* 'ski' (verb) and *ráír* 'write on' will work as (4a) and (4b), respectively, except that the source of spreading is a long vowel (has two X slots) here. Finally, examples like *kettéoszt* 'divide' and *ráér* 'have time' differ from the above types in two respects: the rule is optional in their case and the source of spreading is a mid, rather than high coronal vowel. We propose the following rules to account for all these cases:

(5)  *Hiatus Filling*

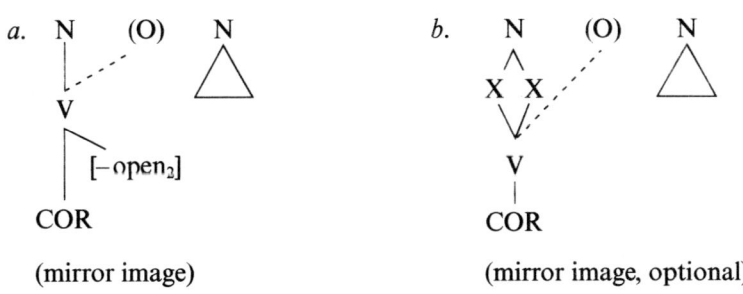

According to (5a), a high coronal (front unrounded) vowel, whether it is short or long,[4] spreads in the direction of an adjacent nucleus (i.e. one that is not divided from it by an onset) and creates an Onset node in the process.

---

[4] The timing tier is suppressed in (5a) to suggest that the rule applies for short and long vowels alike. In (5b), on the other hand, the timing tier information (that the vowel is long) is crucial, therefore it is part of the way the rule is written.

The rule is of the mirror image type which means that it does not care in what order the source of spreading and its target are situated in the input string. Irrespective of whether spreading goes left-to-right, as depicted in (5a) and illustrated in (4a), or right-to-left, the mirror image of (5a) as illustrated in (4b), the onset thus created will end up as the onset of the right-hand syllable (the target of spreading or the source of spreading, as the case may be).

The other rule, (5b), works in the same manner except that it is optional and it applies to all long coronal vowels irrespective of their height. This means that /iː/ is included in both rules but this is the simplest (most general) way of stating the rules and it does not cause any problem in their application.

## 9.4. DEGEMINATION

The traditional insight concerning degemination is that geminates do not occur in Hungarian (i) word initially, or (ii) flanked by another consonant on either side. In other words, the occurrence of geminates is only possible (i) intervocalically (e.g. *állat* 'animal', *áll-ok* 'I stand', *áll Attila* 'Attila stands') and (ii) utterance finally (i.e. before a pause) if preceded by a vowel (e.g. *áll* 'stand'). The latter type is degeminated, however, if a consonant follows, irrespective of whether that consonant comes from synthetic suffixation (e.g. *áll-t* 'stand' (3sg past), *áll-tam* 'stand' (1sg past)), analytic suffixation (e.g. *áll-hat* 'may stand'), compounding (*áll-kapocs* 'jawbone'), or even from a different word (*áll Tamás* 'Tom stands'). However, this traditional view is oversimplified and has to be revised in various ways.[5]

In a detailed study of degemination in Hungarian, Nádasdy (1989a) distinguishes underlying vs. derived geminates and left-flanked vs. right-flanked geminates. Within the class of derived geminates he further distinguishes what we have referred to as 'true' vs. 'fake' geminates in section 4.1.2 (see especially footnote 11 there) and in section 7.2.2 (see footnote 16 there). The following discussion is based on Nádasdy's data and classification but the actual analysis differs from his in some respects.

Right-flanked underlying geminates behave roughly in the way described above, except that across word boundary degemination is optional and varies in terms of speech style and boundary strength (cf. Dressler and Siptár 1989): the 'stronger' the boundary and/or the more formal the register, the less likely degemination is to apply.

---

[5] A point of minor significance concerns the examples in this paragraph rather than the issue of degemination. In a number of lexical items there is free variation between short and geminate consonants; one of the most characteristic combinations where this holds is /aː/ followed by /lː/ as in *áll* 'chin', *áll* 'stand', *állat* 'animal', *istálló* 'stable', *szakáll* 'beard', *száll* 'fly' (verb), *váll* 'shoulder', *vállal* 'undertake', etc. On the small functional load of geminate consonants in Hungarian, cf. Obendorfer (1975).

Left-flanked underlying geminates do not normally occur since no morpheme *begins* with a geminate consonant. There are two possible candidates for morphemes *consisting of* a geminate consonant: comparative *-bb* and past tense *-tt*. The former hardly ever occurs in a degemination context; it is a Type A suffix (e.g. *nagy-obb* 'bigger', *csúnyá-bb* 'uglier'; cf. section 8.1.2.2) that, however, exceptionally 'loses' its unstable vowel in a handful of lexicalized forms: *különb* 'superior', *idősb* 'elder', and *nemesb* 'nobler'; also in some forms containing the verbalizing suffix *-ít*: *helyesbít* 'rectify', *öregbít* 'enhance', *súlyosbít* 'aggravate'. With respect to the past tense suffix, we suggested in section 8.1.4.4 that it exhibits degemination effects without actually undergoing degemination. In particular, we suggested that this suffix is a /t/ whose root node is underlyingly associated to a *single* timing slot followed by an *empty* timing slot. We further assumed that a rule of *t*-spread applies to this configuration *if a full vowel precedes it*. Thus, in a form like *fal-t* 'devour' (3sg past indef.), a geminate never arises in the first place, hence there is nothing to degeminate.

Another type of suffix showing degemination effects without actually undergoing degemination is the set of 'alternating *v*-suffixes' discussed in section 8.2.1. This case (if it did involve degemination) would be that of a left-flanked derived true geminate: *domb-bal* [mb] 'hill' (instr.), *vers-sel* [rš] 'poem' (instr.), *lánc-cal* [ntˢ] 'chain' (instr.), *férj-jel* [rj] 'husband' (instr.). The analysis we offered in section 8.2.1 involved the generalization of *t*-spread into a rule of C-spread (see (72) there; the rule is repeated in (7*b*) below) that applies in e.g. *csap-pal* 'tap' (instr.) but not in *domb-bal* etc., giving the desired degemination effect. Let us now consider if this treatment can be extended to other instances of left-flanked derived true geminates as well.

Recall that in section 7.2.1 we formulated a rule of full assimilation whereby a sequence of strident consonant + /j/ emerged as a long strident consonant, e.g. *hozzon* /hoz-j-on/ → [hoz:on] 'bring' (3sg imp. indef.). We repeat that rule here for convenience:

(6)     *Strident j-Assimilation*

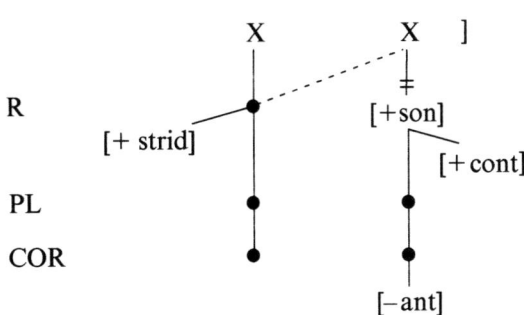

Suppose we simplify this rule so that it is just a delinking rule and let the rule of C-spread apply to its output in a case like *hozzon*. In addition to the improvement that this move represents with respect to the form of the rule, it has the side effect that in a case like *rajzzon* /rajz-j-on/ → [rɔjzon] 'swarm' (3sg imp.) we get the degemination effect for free:[6]

(7)                    *Strident j-Assimilation (revised)*

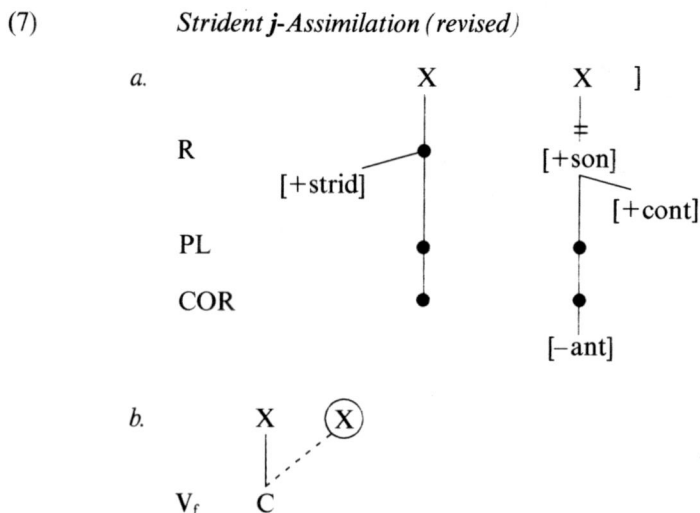

Another rule that can be reformulated in the same manner is Palatal *j*-Assimilation as in *bátyja* [tʲː] 'his brother', *hagyja* [dʲː] 'leave' (3sg ind./imp. def.), *hányja* [nʲː] 'throw' (3sg ind./imp. def.):

(8)                    *Palatal j-Assimilation*

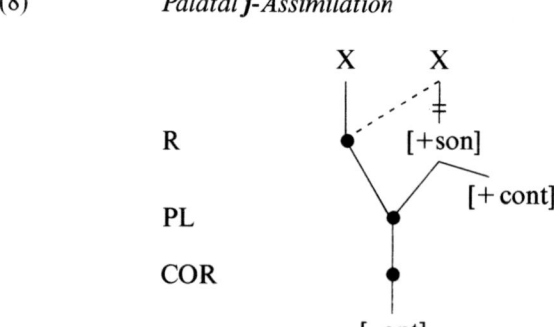

---

[6] Given that (7a) and the other rules mentioned below are Block 2 rules, this solution also means that we allow C-spread to apply in Block 2 as well as in Block 1. Note further that the target empty X's are also created in Block 2, therefore the claim we made earlier that all empty material is erased at the end of Block 1 does not undermine the analysis given here.

This rule also applies to the output of palatalization in cases like *látja* 'see' (3sg ind. def.), *adja* 'give' (3sg ind./imp. def.). Again, if we omit the spreading part, we get the degemination effect in *tartja* 'hold' (3sg ind. def.), *hordja* 'carry' (3sg ind./imp. def.):

(9)           *Palatal j-Assimilation (revised)*

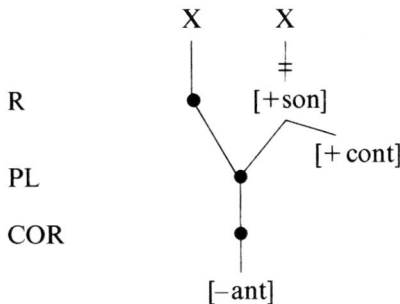

Similarly, it would be a good idea to let the difference between *füts* [č:] 'heat' (2sg imp. indef.) and *önts* [č] 'pour' (2sg imp. indef.) fall out automatically. In principle, this could be done as follows:

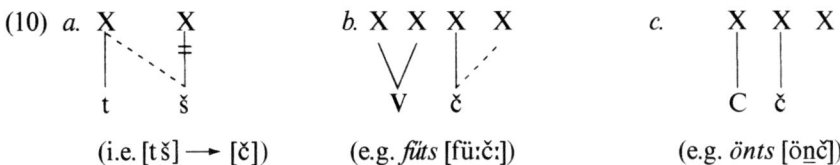

(10)  *a.*   (i.e. [tš] ⟶ [č])          *b.* (e.g. *füts* [fü:č:])          *c.* (e.g. *önts* [önč])

However, we argued in section 7.2.3 that the rule that merges [tš]—the output of rule (15) there—into an affricate can be subsumed under the general rule of Palatalization (28). In order to implement the idea sketched in (10), we would have to undo this latter generalization. Whether it would or would not be worth that loss of generality depends on the rest of the cases falling under the rubric 'left-flanked derived true geminate'. If all of them can be made to show automatic degemination effects without an actual rule of degemination, in order to achieve that state of affairs we might just as well give up the generalization about the affrication process considered here.

Unfortunately, there is at least one further type of case that can by no means be analysed without a degemination rule. The output of voice assimilation[7] may or may not come out as a geminate (it does if the two seg-

---

[7] This problem is by no means particular to voice assimilation. It arises in all cases where a spreading operation involving a single feature or a single class node leads to complete identity (hence to a derived true geminate) by accident, as it were.

ments only differed in terms of voicing; it does not otherwise). The merger of all class nodes dominating identical material that is involved here is an automatic OCP-effect, not a rule—hence we cannot manipulate it in a way similar to what we did above. Thus, we need a degemination rule for these cases. That rule can be informally written as in (11):

(11)                    *Degemination I*

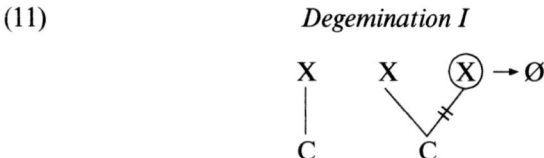

Once we need this rule anyway, we can derive the *önts*-type cases as we suggested in section 7.2.3 and apply (11) to the output where appropriate.

Turning to right-flanked derived true geminates as in *üsd* /üt-j-d/ [üžd] 'hit' (2sg imp. def.), one way to try to let the degemination effect fall out automatically would be, again, to simplify the full assimilation rule that produces the intermediate [š:] by omitting the spreading part from it. The original form of the rule (see (17) in section 7.2.1) is shown in (12*a*); the modified version appears in (12*b*):

(12)        *t-Assimilation*

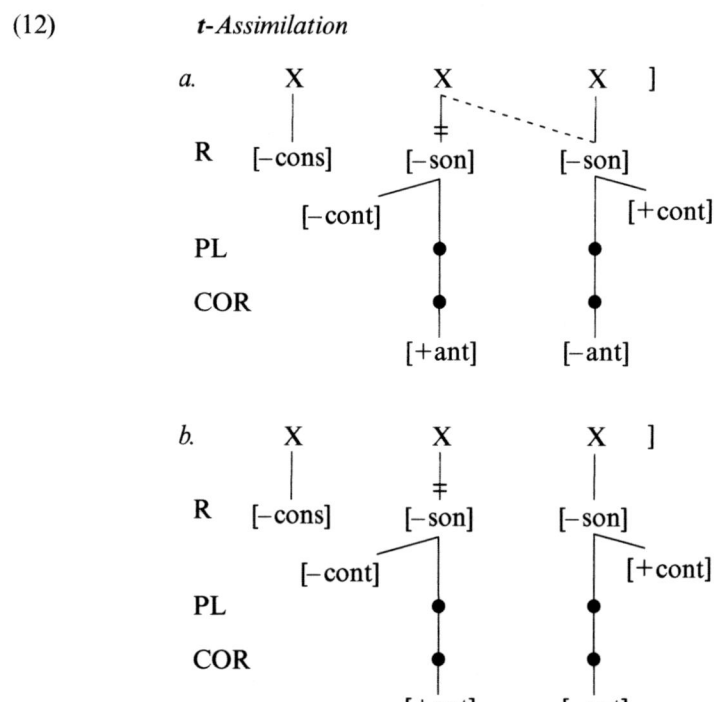

However, for this idea to work, a separate C-spread rule would be required that—unlike (7*b*)—(i) spreads the C *leftwards*, (ii) has a *negative* condition ('unless another consonant follows') rather than a positive environment, and (iii) is *not* independently motivated. But the whole attempt is superfluous anyway since the mirror image of (11) will be independently needed to handle right-flanked underlying geminates as in *hall-gat* 'listen' etc. Therefore, we leave (12*a*) as it is and formulate the following rule for all right-flanked true geminates, whether underlying or derived:

(13)                          *Degemination  II*

The next question is whether we should collapse (11) with (13) into a single mirror-image rule schema or not. The answer is in the negative since (13) applies both lexically and postlexically and—as was mentioned above—it shows optionality effects across a word boundary, whereas (11) is always strictly obligatory (as long as true geminates are concerned).

Let us now consider fake geminates (i.e. sequences of identical consonants arising across analytic morpheme boundaries). Note first of all that—with the possible exception of geminate affricates as in *kulcscsomó* 'bunch of keys' and unlike geminate vowels as in *kiirt* 'exterminate'—fake geminate consonants surface phonetically as if they were true geminates. This means that at some point they will undergo merger (one which is either OCP-driven or rule-based, depending on one's general assumptions). That merger can take place either too early or too late: if it takes place before (postlexical) degemination is considered for application, the difference between the behaviour of true and fake geminates may become inexpressible; if, on the other hand, merger is later than degemination, it may be difficult to refer to adjacent identical consonants that do not form a linked structure (co-indexing is one possibility but not a very pleasant one).

Consider the following data (partly based on Nádasdy 1989*a*):

(14)  *a. Left-flanked fake geminates*:

'Obligatory' degemination if the flanking consonant is an obstruent:
koszt-tól    [kostol]       'from food'              (analytic suffix)
direkttermő [dirɛktɛrmő:] 'a type of vine'        (compound)
lakj jól     [lɔkjo:l]      'eat enough' (2sg imp.)  (phrase)

'Optional' degemination if the flanking consonant is a nasal:

| comb-ból | [tˢomb(ː)ol] | 'from thigh' | (analytic suffix) |
|---|---|---|---|
| csonttányér | [čont(ː)aːnʸeːr] | 'bone plate' | (compound) |
| tank körül | [tɔŋk(ː)örül] | 'around tank' | (phrase) |

'No degemination' if the flanking consonant is a liquid:

| sztrájk-kor | [strajkːor] | 'during a strike' | (analytic suffix) |
|---|---|---|---|
| talppont | [tɔlpːont] | 'foot-end' | (compound) |
| szerb bor | [sɛrbːor] | 'Serbian wine' | (phrase) |

b.  *Right-flanked fake geminates*:

'Obligatory' degemination if the flanking consonant is an obstruent:

| kisstílű | [kištiːlüː] | 'petty' | (compound) |
|---|---|---|---|
| olasz sztár | [olɔstaːr] | 'Italian (film) star' | (phrase) |

'Optional' degemination if the flanking consonant is a nasal:

| őssmink | [öːš(ː)miŋk] | 'proto-make-up' | (compound) |
|---|---|---|---|
| kész sznob | [keːs(ː)nob] | 'a perfect snob' | (phrase) |

'No degemination' if the flanking consonant is a liquid:

| széppróza | [seːpːroːzɔ] | 'prose fiction' | (compound) |
|---|---|---|---|
| ügyes srác | [üdʸɛšːraːtˢ] | 'smart boy' | (phrase) |

The expressions 'obligatory', 'optional', and 'no degemination' appear in quotation marks in (14) since we want to claim that there is a continuous gradient of optionality here in which 'most likely', 'less likely', and 'least likely' would be more appropriate labels. The type of degemination we are considering is simply an optional process whose likelihood co-varies with the type of the flanking consonant as indicated.

The question is whether the phenomenon displayed in (14) is a postlexical phonological process or rather part of phonetic interpretation. An argument that supports the latter option is that the merger of fake geminates into true ones is most probably a phonetic issue and—unless we want to formulate a deletion rule referring to (co-indexed) identical segments—the earliest point where this simplification process can be stated in terms of linked structures is after that merger has taken place. Therefore, we will assume the following two statements as part of the phonetic implementation module of the grammar of Hungarian:

(15) a. *Long Consonant Formation*
      Merge a sequence of two identical short consonants into a single long consonant (applies in all speech styles/tempos with respect to consonants other than affricates; applies to affricates in fast/casual speech only).

*b. Degemination III*

Optionally realize a long consonant as short if it is flanked by
another consonant (applies with decreasing likelihood when the
flanking consonant is (i) an obstruent, (ii) a nasal, or (iii) a liquid).

In this section, we have proposed three different degemination rules, applying
at word level, postlexically, and in the phonetic implementation module,
respectively. (11) is the word level rule that applies obligatorily to all left-
flanked true geminates that emerge from the lexical phonology as such, irrespective of the identity of the flanking consonant and of the morphological
make-up (underlying vs. derived) of the geminate itself. Instances of this
process are cases like *önts* [önč] 'pour' (2sg imp. indef.),[8] *hordtam* [hortɔm]
'carry' (1sg past def.) where degemination is directly fed by palatalization in
the first example and by voice assimilation in the second.[9]

The postlexical rule is (13) that applies obligatorily within words
and optionally in phrasal domains (with decreasing likelihood across increasingly 'stronger' syntactic boundaries and in increasingly formal speech
styles). However, the rule is insensitive to the identity of the flanking consonant and to whether the geminate is underlying or derived. Instances of this
process include *hallgat* [hɔlgɔt] 'listen', *üsd* /üt-j-d/ [üžd] 'hit' (2sg imp. def.),
*adj neki* [ɔdʲnɛki] 'give him', *evett banánt* [ɛvɛdbɔnaːnt] 'he ate some bananas'.

Finally, the phonetic rule is (15*b*) that applies optionally and targets—
primarily—long consonants that are (phonologically) fake geminates.
The gradience of optionality is as stated in (15*b*); examples appear in (14)
above.

## 9.5. FAST CLUSTER SIMPLIFICATION

Clusters consisting of more than two consonants may be simplified in fast
speech (cf. Dressler and Siptár 1989, Siptár 1991*a*). Fast cluster simplification
(FCS) is an optional deletion process that targets consonants flanked by consonants on both sides, i.e. it deletes the middle one of a sequence of three
consonants, as the examples show:

---

[8] Note that cases like *öntse* [önče] 'pour' (3sg imp. def.) and *öntsd* [önjd] ~ [önžd] ~ [õːžd]
'pour' (2sg imp. def.) also belong here, i.e. it does not matter whether the geminate is followed by
nothing, a vowel, or a consonant; what is important is the left-hand consonant, *n* in this case,
that is the necessary and sufficient condition for degemination to apply.

[9] Recall that a number of cases that are traditionally analysed as degemination are reinter-
preted here as lack of gemination. The major cases include (i) past-tense verb forms like *kap-t-a*
'get' (3sg past def.) and *fal-t* 'devour' (3sg past indef.), (ii) noun forms involving 'alternating *v*-
suffixes' like *domb-bal* [dombɔl] 'hill' (instr.), (iii) imperatives of sibilant-final verbs like *rajzzon*
/rajz-j-on/ → [rɔjzon] 'swarm' (3sg imp.), and (iv) verb forms, both indicative and imperative
ones, involving palatal *j*-assimilation as in *tart-ja* [tɔrtʲɔ] 'hold' (3sg ind. def.).

(16)  lambda        [lɔmbdɔ], [lɔmdɔ]              'id.'
      asztma        [ɔstmɔ], [ɔsmɔ]               'asthma'
      röntgen       [röndgɛn], [röŋgɛn]           'X-ray'
      dombtető      [domptɛtö:], [domtɛtö:]       'hilltop'

The process can be informally stated as (17):

(17)  *Fast Cluster Simplification*
      $C \rightarrow \emptyset / C \_ C$

FCS is postlexical as it can apply in monomorphemic words (e.g. *asztma*)
and across any boundary including that between words in compounds
(18*a*) and phrases (18*b*) (in (18*a*) below the hyphens indicate the boundary
between the constituents of compounds and do not appear in normal
spelling):

(18)  *a.* lomb-korona  [lompkoronɔ], [lomkoronɔ]     'foliage of a tree'
          test-nevelés  [tɛštnɛvɛle:š], [tɛšnɛvɛle:š]  'PE'
      *b.* dobd ki      [doptki], [dopki]              'throw (it) out'
          most pedig    [moštpɛdig], [mošpɛdig]        'and now'

Fast cluster simplification does not apply to all CCC clusters. For instance, it
does not apply to the clusters shown in (19) below:

(19)  ámbra         [a:mbrɔ], *[a:mrɔ]            'ambergris'
      eszpresszó    [ɛspres:o:], *[ɛsres:o:]     'espresso'
      centrum       [tˢɛntrum], *[tˢɛnrum]       'centre'
      templom       [tɛmplom], *[tɛmlom]         'church'

The differential behaviour of words like those in (16) and (19) has been used
to suggest that FCS is a syllable structure conditioned process. It has been
claimed that it applies if C2C3 of a C1C2C3 cluster is not a well-formed onset
(e.g. *lambda*), but it does not if C2C3 is a well-formed onset (e.g. *centrum*). In
order to account for this pattern one could assume that there is an optional
postlexical resyllabification process that moves the last consonant of a branching coda into the onset of the following syllable. This process would be
subject to the general well-formedness conditions and would be expected to
block if the resulting onset is ill-formed—hence the FCS effect (cf. Dressler
and Siptár 1989, Siptár 1991*a*, Ács and Siptár 1994). This interpretation
would be problematic for the present analysis since we claim that onsets
may not branch in Hungarian (see section 5.2.2). It is to be pointed out, however, that this position can be shown to be untenable (cf. Törkenczy and
Siptár 1999): contrary to what is predicted by the above interpretation, FCS is
not possible if C3 is a continuant even if C2C3 is *not* a possible branching

onset (granting—for the sake of argument—that branching onsets exist in Hungarian and assuming that (most) occurring word-initial clusters are well-formed onsets):

(20)  handlé          [hɔndle:], *[hɔnle:]             'second-hand dealer'
      pántlika        [pa:ntlikɔ], *[pa:nlikɔ]          'ribbon'
      kompjúter       [kompju:tɛr], *[komju:tɛr]        'computer'
      aktfotó         [ɔktfoto:], *[ɔkfoto:]            'nude photograph'
      pemzli          [pɛmzli], *[pɛmli]                'brush'(noun)
      hangsor         [hɔŋkšor], *[hɔŋšor]              'sound sequence'

(20) shows that FCS does not apply if C3 is [+cont] irrespective of the syllabic affiliation of the consonants in the cluster. Therefore, we conclude that FCS is not sensitive to syllable structure (and is not a problem for our claim that onsets may not branch in Hungarian).

Thus, the reason why FCS does apply to the relevant clusters in (16), but does not apply to those in (19) and (20) is that in the former set of words the C3 of the C1C2C3 clusters is [–cont] while in the latter two it is [+cont].[10]

There are two further conditions on the application of FCS. It does not apply if C1 is a continuant sonorant:

(21)  talpnyaló       [tɔlpnʸɔlo:], *[tɔlnʸɔlo:]        'lackey'
      bazaltkő        [bɔzɔltkö:], *[bɔzɔlkö:]          'basalt stone'
      partner         [pɔrtnɛr], *[pɔrnɛr]              'id.'
      szerbtől        [sɛrptö:l], *[sɛrtö:l]            'from (a) Serb'
      sejtmag         [šɛjtmɔg], *[šɛjmɔg]              'cell nucleus'
      fajdkakas       [fɔjtkɔkɔš], *[fɔjkɔkɔš]          'blackcock'

It is not just the continuancy of C1 that matters here: note that FCS *can* apply if C1 is [+son, –cont] (e.g. *röntgen* [röndgɛn, röŋgɛn]) or if it is [–son, +cont] (e.g. *asztma* [ɔstmɔ, ɔsmɔ]).[11]

---

[10] There are sporadic examples in which FCS seems to apply although C3 is a continuant: e.g. *szoftver* [softvɛr], [sofvɛr] 'software', *szendvics* [sɛndvič], [sɛɱvič] 'sandwich', *testvér* [tɛštve:r], [tɛšve:r] 'brother', *mumpsz* [mumps], [mums] 'mumps'. There are two things to be noted here: (i) Some of the forms that appear to show FCS are actually lexicalized and are not the result of deletion at all. For instance, for many speakers *mumpsz* is [mums] regardless of the tempo. Note that the same cluster cannot be simplified in other items: *kolompszó* [kolompso:], *[kolomso:] 'sound of the cattle bell'; (ii) Most of the problematic examples that we have found have [v] as C3. One could use this fact to suggest that FCS is gradient rather than absolute: it is more likely to apply if C3 is [v] than with other continuants. Compare *dombtető* 'hilltop', *dombvidék* 'hilly region', and *dombról* 'from the hill'. In *dombtető* FCS is definitely possible [domtɛtö:], in *dombról* it is definitely not *[domro:l]; *dombvidék* [domvide:k] is intuitively somewhere in between. We leave this problem for further research.

[11] Naturally, it can also apply if C1 is [–son, –cont]: *receptkönyv* [rɛtʲɛptkönʸv], [rɛtʲɛpkönʸv] 'book of recipes'.

FCS cannot apply either if C2 is a fricative or an affricate:[12]

(22)

| szenvtelen | [sɛnftɛlɛn], *[sɛntɛlɛn] | 'indifferent' |
| könyvtár | [könʸftaːr], *[könʸtaːr] | 'library' |
| eksztázis | [ɛkstaːziš], *[ɛktaːziš] | 'ecstasy' |
| Amszterdam | [ɔmstɛrdɔm], *[ɔmtɛrdɔm] | 'Amsterdam' |
| inspekció | [inšpɛktˢioː], *[inpɛktˢioː] | 'inspection' |
| obskurus | [opškuruš], *[opkuruš] | 'obscure' |
| lánctalp | [laːntˢtɔlp], *[laːntɔlp] | 'caterpillar track' |
| táncdal | [taːndʲdɔl], *[taːndɔl] | 'popular song' |
| parancsnok | [pɔrɔnčnok], *[pɔrɔnnok] | 'commander' |
| narancsból | [nɔrɔn͡jboːl], *[nɔrɔnboːl] | 'from (an) orange' |

To sum up, FCS (17) is subject to three conditions: (i) C2, the target consonant, cannot contain the feature specification [–son, +cont], (ii) C3 must be [–cont] and (iii) C1 cannot be [+son, +cont]. All three conditions must be satisfied in order for FCS to apply.

Substrings of clusters longer than three consonants behave in the same way as clusters containing exactly three consonants. For instance, FCS cannot apply to the four term cluster in *foxtrott* [fokstrotː] 'foxtrot' because of the two potential targets (C2 and C3 of C1C2C3C4), the first one cannot delete as it is [–son, +cont] (*[foktrotː]), and the second one cannot delete since it is followed by a [+cont] segment (*[foksrotː]). By contrast, FCS can apply to C3 of the C1C2C3C4 cluster in *karsztból* 'from (a) karst formation' [kɔrzdboːl], [kɔrzboːl] since all the three conditions are met (C2 cannot delete because it is [–son, +cont] and is preceded by a [+son, +cont] segment *[kɔrdboːl]).

Like other fast-speech processes, the conditions are relaxed gradually and FCS generalizes to other CCC clusters as the tempo of speech increases (cf. Siptár 1991*a*).[13]

---

[12] In essence, this means that the target must be a plosive. It is not possible to test the behaviour of the various sonorants as C2 because they are either (i) unattested as C2, or (ii) if attested, are preceded by [+cont] sonorants, which in itself blocks FCS (e.g. *modernkedik* [modɛrnkɛdik], *[modɛrkɛdik] 'act modern' (3sg pres.), *szörnyben* [sörnʸbɛn], *[sörbɛn] 'in (a) monster', *filmtől* [filmtöːl], *[filtöːl] 'from (a) film'). The segments [ç] and [j] (which are—with a handful of exceptions—always the post-consonantal surface reflexes of imperative -*j* (cf. section 7.3)) present a further complication: they also seem to be omissible in environments where FCS otherwise blocks. Compare *Dobj neki egy törölközőt!* 'Throw her/him a towel!' and *Dobj rá egy törölközőt!* 'Throw a towel on her/him!'. In fast speech deletion of C2 is possible in both cases ([dobjnɛki]/[dobnɛki], [dobjraː]/[dobraː]) in spite of the fact that in the second FCS cannot apply since the target is followed by a continuant ([r]). We suggest that [ç] and [j] are deleted by a separate optional process that targets [ç] and [j] exclusively, between any two consonants.

[13] Because of the lack of solid evidence, the exact way in which this happens must be the subject of future research.

# SUGGESTED READING

## GENERAL

ABONDOLO, DANIEL (1988), *Hungarian Inflectional Morphology* (Budapest: Akadémiai Kiadó).

É. KISS, KATALIN (2002), *The Syntax of Hungarian* (Cambridge: Cambridge University Press).

—— (2005), 'Hungarian', in Keith Brown (ed.), *The Encyclopedia of Language and Linguistics*, Second Edition, Vol. 5 (Oxford: Elsevier), 429–31.

—— and VAN RIEMSDIJK, HENK (eds.) (2004), *Verb Clusters. A Study of Hungarian, German and Dutch* (Amsterdam/Philadelphia: John Benjamins).

FENYVESI, ANNA (ed.) (2005), *Hungarian Language Contact Outside Hungary. Studies on Hungarian as a Minority Language* (Amsterdam/Philadelphia: John Benjamins).

KENESEI, ISTVÁN (2005), 'Hungarian in Focus', *Journal of Linguistics*, 41: 409–35.

—— ; VAGO, ROBERT M.; and FENYVESI, ANNA (1998), *Hungarian* (Descriptive Grammar Series, London/New York: Routledge).

REBRUS, PÉTER, and TÖRKENCZY, MIKLÓS (2004), 'Uniformity and Contrast in the Hungarian Verbal Paradigm', in Laura J. Downing, T. Alan Hall and Renate Raffelsiefen (eds.), *Paradigms in Phonological Theory* (Oxford: Oxford University Press), 263–95.

RITTER, NANCY A. (2002), 'The Hungarian Personal Possessive Suffix Revisited', in István Kenesei and Péter Siptár (eds.), *Approaches to Hungarian 8: Papers from the Budapest Conference* (Budapest: Akadémiai Kiadó), 283–307.

TRÓN, VIKTOR, and REBRUS, PÉTER (2001), 'Morphophonology and the Hierarchical Lexicon', *Acta Linguistica Hungarica*, 48: 101–35.

—— —— (2005), 'Re-presenting the Past: Contrast and Uniformity in Hungarian Past Tense Suffixation', in Christopher Piñón and Péter Siptár (eds.), *Approaches to Hungarian 9: Papers from the Düsseldorf Conference* (Budapest: Akadémiai Kiadó), 303–27.

VAGO, ROBERT M. (2005), 'Hungarian: Phonology', in Keith Brown (ed.), *The Encyclopedia of Language and Linguistics*. Second Edition, Vol. 5. (Oxford: Elsevier), 433–40.

## VOWEL SYSTEM, VOWEL PROCESSES

DIENES, PÉTER (1997), 'Hungarian Neutral Vowels', *The Odd Yearbook 1997*, 151–80.

GÓSY, MÁRIA (1989), 'Vowel Harmony: Interrelations of Speech Production, Speech Perception, and the Phonological Rules', *Acta Linguistica Hungarica*, 39: 93–118.

HARE, MARY (1990), 'The Role of Similarity in Hungarian Vowel Harmony: A Connectionist Account', *Connection Science*, 2: 123–50.

HAYES, BRUCE, and CZIRÁKY LONDE, ZSUZSA (2006), 'Stochastic Phonological Knowledge: The Case of Hungarian Vowel Harmony', *Phonology*, 23: 59–104.

KERTÉSZ, ZSUZSA (2003), 'Vowel Harmony and the Stratified Lexicon of Hungarian', *The Odd Yearbook*, 7: 62–77.

POLGÁRDI, KRISZTINA, and REBRUS, PÉTER (1998), 'There is No Labial Harmony in Hungarian: a Government Phonology Analysis', in Casper de Groot and István Kenesei (eds.), *Approaches to Hungarian 6: Papers from the Amsterdam Conference.* (Szeged: JATEPress), 3–20.

REBRUS, PÉTER, and POLGÁRDI, KRISZTINA (1997), 'Two Default Vowels in Hungarian?', in Geert Booij and Jeroen van de Weijer (eds.), *Phonology in Progress – Progress in Phonology. HILP Phonology Papers III* (The Hague: Holland Academic Graphics), 257–75.

REISS, CHARLES (2003), 'Deriving the Feature-Filling/Feature-Changing Contrast: an Application to Hungarian Vowel Harmony', *Linguistic Inquiry*, 34: 199–224.

## CONSONANT SYSTEM, CONSONANT PROCESSES

BLAHO, SYLVIA (2005), 'Another Look at the Misbehaving Segments of Hungarian Voice Assimilation', in Christopher Piñón and Péter Siptár (eds.), *Approaches to Hungarian 9: Papers from the Düsseldorf Conference* (Budapest: Akadémiai Kiadó), 35–55.

CÔTÉ, MARIE-HÉLÈNE (2004), 'Syntagmatic Distinctness in Consonant Deletion', *Phonology*, 21: 1–41.

CSER, ANDRÁS (2003), *The Typology and Modelling of Obstruent Lenition and Fortition Processes* (Budapest: Akadémiai Kiadó).

—— and SZENDE, TAMÁS (2002), 'The Question of [j]: Systemic Aspects, Phonotactic Position and Diachrony', *Sprachtheorie und germanistische Linguistik*, 12: 27–42.

GÓSY, MÁRIA (2002), 'Temporal Coding of Voicing Assimilation in Speech Production', *Acta Linguistica Hungarica*, 49: 257–76.

JANSEN, WOUTER, and TOFT, ZOË (2002), 'On Sounds that Like to be Paired (After All): An Acoustic Investigation of Hungarian Voicing Assimilation', *SOAS Working Papers in Linguistics*, 12: 19–52.

KENSTOWICZ, MICHAEL; ABU-MANSOUR, MAHASEN; and TÖRKENCZY, MIKLÓS (2003), 'Two Notes on Laryngeal Licensing', in Angela C. Carpenter, Andries W. Coetzee and Paul de Lacy (eds.), *Papers in Optimality Theory II. University of Massachusetts Occasional Papers 26*, 121–40. Also in Stefan Ploch (ed.) (2003), *Living on the Edge: 27 Papers in Honour of Jonathan Kaye* (Berlin and New York: Mouton de Gruyter), 259–82.

KISS, ZOLTÁN, and BÁRKÁNYI, ZSUZSANNA (2006), 'A Phonetically-Based Approach to the Phonology of [v] in Hungarian', *Acta Linguistica Hungarica*, 53: 175–226.

PETROVA, OLGA, and SZENTGYÖRGYI, SZILÁRD (2004), '/v/ and Voice Assimilation in Hungarian and Russian', *Folia Linguistica*, 38: 87–116.

RITTER, NANCY A. (2000), 'Hungarian Voicing Assimilation Revisited in Head-Driven Phonology', in Gábor Alberti and István Kenesei (eds.), *Approaches to Hungarian 7: Papers from the Pécs Conference* (Szeged: JATEPress), 23–49.

SIPTÁR, PÉTER (2003), 'Hungarian Yod', *Acta Linguistica Hungarica*, 50: 457–73.

—— and SZENTGYÖRGYI, SZILÁRD (2002), '*H* as in Hungarian', *Acta Linguistica Hungarica*, 49: 427–56.

SZENTGYÖRGYI, SZILÁRD, and SIPTÁR, PÉTER (2005), 'Hungarian *H*-type Segments in Optimality Theory' in Christopher Piñón and Siptár Péter (eds.), *Approaches to Hungarian 9: Papers from the Düsseldorf Conference* (Budapest: Akadémiai Kiadó), 261–81.

ZSIGRI, GYULA (1998), '/h/ and /v/ in Hungarian Voicing Assimilation', in Casper de Groot and István Kenesei (eds.), *Approaches to Hungarian 6: Papers from the Amsterdam Conference* (Szeged, JATEPress), 87–100.

## PHONOTACTICS, SYLLABLE STRUCTURE, SYLLABLE-RELATED PROCESSES

HUME, ELIZABETH V. (2004), 'The Indeterminacy/Attestation Model of Metathesis', *Language*, 80: 203–37.

KISS, ZOLTÁN (2005), 'Graduality and Closedness in Consonantal Phonotactics: A Perceptually Grounded Approach', in Sylvia Blaho, Luis Vicente and Erik Schoorlemmer (eds.), *Proceedings of Console XIII* (Leiden: Student Organization of Linguistics in Europe), 171–95.

POLGÁRDI, KRISZTINA (2002), 'Hungarian Superheavy Syllables and the Strict CV Approach', in István Kenesei and Péter Siptár (eds.), *Approaches to Hungarian 8: Papers from the Budapest Conference* (Budapest: Akadémiai Kiadó), 263–82.

—— (2003), 'Hungarian as a Strict CV Language', in Jeroen van de Weijer, V. J. van Heuven and Harry van der Hulst (eds.), *The Phonological Spectrum II. Suprasegmental Structure* (Amsterdam: John Benjamins), 59–79.

SIPTÁR, PÉTER (2000), 'Degemination in Hungarian', in László Varga (ed.), *The Even Year-book 4. ELTE SEAS Working Papers in Linguistics* (Budapest: ELTE), 107–15.

—— (2007a),'How to Get Rid of Hiatuses', in Ravi Sheorey and Judit Kiss-Gulyás (eds.), *Studies in Applied and Theoretical Linguistics* (Debrecen: Kossuth Egyetemi Kiadó).

—— (2007b), 'Hiatus Resolution in Hungarian: an Optimality Theoretic Account', in Christopher Piñón and Szilárd Szentgyörgyi (eds.), *Approaches to Hungarian 10: Papers from the Veszprém Conference* (Budapest: Akadémiai Kiadó).

SZIGETVÁRI, PÉTER (2001), 'Dismantling Syllable Structure', *Acta Linguistica Hungarica*, 48: 155–81.

—— (2006), 'The Markedness of the Unmarked', *Acta Linguistica Hungarica*, 53: 433–47.

TÓRKENCZY, MIKLÓS (2001), 'Phonotactic Grammaticality and the Lexicon', *Acta Linguistica Hungarica*, 48: 137–53.

—— (2002), 'Absolute Phonological Ungrammaticality in Output-Biased Phonology', in István Kenesei and Péter Siptár (eds.), *Approaches to Hungarian 8: Papers from the Budapest Conference* (Budapest: Akadémiai Kiadó), 309–24.

—— and SIPTÁR, PÉTER (1999), 'Hungarian Syllable Structure: Arguments for/against Complex Constituents', in Harry van der Hulst and Nancy Ritter (eds.), *The Syllable: Views and Facts* (Berlin/New York: Mouton de Gruyter), 249–84.

## STRESS AND INTONATION

FÓNAGY, IVÁN (1998) 'Intonation in Hungarian', in Daniel Hirst and Albert Di Cristo (eds.), *Intonation Systems. A Survey of Twenty Languages* (Cambridge: Cambridge University Press), 328–45.

HUNYADI, LÁSZLÓ (2002), *Hungarian Sentence Prosody and Universal Grammar* (Frankfurt am Main: Peter Lang).

OLASZY, GÁBOR (2002), 'The Most Important Prosodic Patterns of Hungarian', *Acta Linguistica Hungarica*, 49: 277–306.

VARGA, LÁSZLÓ (2002a), *Intonation and Stress. Evidence from Hungarian* (Basingstoke: Palgrave Macmillan).

—— (2002b), 'The Intonation of Monosyllabic Hungarian Yes/No Questions', *Acta Linguistica Hungarica*, 49: 307–20.

# REFERENCES

## ABBREVIATIONS

*CLS*     Papers from the *n*th Regional Meeting of the Chicago Linguistic Society
CNRS    Centre National de la Recherche Scientifique
ELTE    Eötvös Loránd Tudományegyetem (E.L. University, Budapest)
HAS     Hungarian Academy of Sciences
HIL      Holland Institute of Generative Linguistics
JATE    József Attila Tudományegyetem (J. A. University, Szeged)
MIT     Massachusetts Institute of Technology
*NELS*   Proceedings of the *n*th Annual Meeting of the North Eastern Linguistic
          Society
SEAS    School of English and American Studies
SLA      Stanford Linguistics Archives
SLE      Societas Linguistica Europaea
SOAS    School of Oriental and African Studies
*WCCFL* Proceedings of the *n*th West Coast Conference on Formal Linguistics

ABONDOLO, DANIEL (1988), *Hungarian Inflectional Morphology* (Budapest: Akadémiai Kiadó).
—— (1990), 'Hungarian', in Comrie (1990), 577–92.
ÁCS, PÉTER, and SIPTÁR, PÉTER (1994), 'Túl a gondozott beszéden [Beyond Guarded Speech]', in Kiefer (1994), 550–80.
ALBERTI, GÁBOR (1997), *Argument Selection* (Frankfurt: Peter Lang).
ANDERSON, JOHN, and EWEN, COLIN J. (1987), *Principles of Dependency Phonology* (Cambridge: Cambridge University Press).
ANDERSON, STEPHEN R. (1982), 'The Analysis of French Schwa: Or How to Get Something from Nothing', *Language* 58: 534–73.
ANTAL, LÁSZLÓ (1961), *A magyar esetrendszer [The Hungarian Case System]* (Budapest: Akadémiai Kiadó).
—— (1963), 'The Possessive Form of the Hungarian Noun', *Linguistics*, 3: 50–61.
—— (1977), *Egy új magyar nyelvtan felé [Towards a New Hungarian Grammar]* (Budapest: Magvető Kiadó).
—— CSONGOR, BARNABÁS, and FODOR, ISTVÁN (1970), *A világ nyelvei [Languages of the World]* (Budapest: Gondolat Kiadó).
AOUN, YOUSSEF (1979), 'Is the Syllable or the Supersyllable a Constituent?', *MIT Working Papers in Linguistics*, 1: 140–8.
ARONOFF, MARK (1976), *Word Formation in Generative Grammar* (Cambridge, Mass.: MIT Press).
AUSTERLITZ, ROBERT PAUL (1950), 'Phonemic Analysis of Hungarian', Columbia University M.A. thesis.
—— (1990), 'Uralic Languages', in Comrie (1990), 567–76.

BARKAÏ, MALACHI, and HORVATH, JULIA (1978), 'Voicing Assimilation and the Sonority Hierarchy: Evidence from Russian, Hebrew and Hungarian', *Linguistics*, 212: 77–88.

BARTOS, HUBA (1997), 'On "Subjective" and "Objective" Agreement in Hungarian', *Acta Linguistica Hungarica*, 44: 363–84.

BATTISTELLA, ED (1982), 'More on Hungarian Vowel Harmony', *Linguistic Analysis*, 9: 95–119.

BECKER MAKKAI, VALERIE (1970*a*), 'Vowel Harmony in Hungarian Reexamined in the Light of Recent Developments in Phonological Theory', in Becker Makkai (1970*b*), 634–48.

—— (ed.) (1970*b*), *Phonological Theory: Evolution and Current Practice* (New York: Holt, Rinehart and Winston).

BENKŐ, LORÁND, and IMRE, SAMU (eds.) (1972), *The Hungarian Language* (The Hague: Mouton).

BEŐTHY, ERZSÉBET, and SZENDE, TAMÁS (1985), 'On the Sound Pattern of Hungarian: Research History and Inventory', *Ural-Altaische Jahrbücher*, 57: 1–32.

BLEVINS, JULIETTE (1995), 'The Syllable in Phonological Theory', in J. Goldsmith (ed.), *The Handbook of Phonological Theory* (Oxford: Blackwell), 206–44.

BOOIJ, GEERT E. (1984), 'Neutral Vowels in Hungarian Vowel Harmony', *Linguistics*, 22: 629–41.

—— (1992), 'Lexical Phonology and Prosodic Phonology', in Dressler *et al.* (1992), 49–62.

—— (1995), *The Phonology of Dutch* (The Phonology of the World's Languages. Series editor: Jacques Durand. Oxford: Clarendon Press/Oxford University Press).

—— (1999), 'Morpheme Structure Constraints and the Phonotactics of Dutch', in H. van der Hulst and N. Ritter (eds.), *The Syllable: Views and Facts* (Berlin/New York: Mouton de Gruyter), 53–68.

—— and RUBACH, JERZY (1984), 'Morphological and Prosodic Domains in Lexical Phonology', *Phonology Yearbook*, 1: 1–28.

—— —— (1987), 'Postcyclic versus Postlexical Rules in Lexical Phonology', *Linguistic Inquiry*, 18: 1–44.

BOROWSKY, TONI (1986), 'Topics in the Lexical Phonology of English', University of Massachusetts Ph.D. dissertation.

—— (1989), 'Structure Preservation and the Syllable Coda in English', *Natural Language and Linguistic Theory*, 7: 145–66.

BRENTARI, DIANE, and BOSCH, ANNA (1990), 'The Mora: Autosegment or Syllable Constituent?', in M. Ziolkowski, M. Noske, and K. Deaton (eds.), *CLS 26: Parasession on the Syllable in Phonetics and Phonology* (Chicago: Chicago Linguistic Society), 1–16.

BURZIO, LUIGI (1994), *Principles of English Stress* (Cambridge: Cambridge University Press).

CHARETTE, MONIK (1990), 'Licence to Govern', *Phonology*, 7: 233–53.

—— (1991), *Conditions on Phonological Government* (Cambridge: Cambridge University Press).

CHOMSKY, NOAM (1964), *Current Issues in Linguistic Theory* (The Hague: Mouton).

—— and HALLE, MORRIS (1968), *The Sound Pattern of English* (New York: Harper and Row).

CLEMENTS, GEORGE N. (1976), 'Neutral Vowels in Hungarian Vowel Harmony: an Autosegmental Interpretation', in J. Kegl, D. Nash, and A. Zaenen (eds.), *Proceedings of the 7th Annual Meeting of the North Eastern Linguistic Society* (Cambridge, Mass.: MIT).

—— (1985), 'The Geometry of Phonological Features', *Phonology Yearbook*, 2: 225–52.

—— (1986), 'Syllabification and Epenthesis in the Barra Dialect of Gaelic' in: K. Bogers, H. van der Hulst, and M. Mous (eds.), *The Phonological Representation of Suprasegmentals* (Dordrecht: Foris), 317–36.

—— (1987), 'Phonological Feature Representation and the Description of Intrusive Stops', *CLS* 23/2: 29–50.

—— (1988), 'Toward a Substantive Theory of Feature Specification', *Proceedings of The North Eastern Linguistic Society*, 18: 79–93.

—— (1990), 'The Role of the Sonority Cycle in Core Syllabification', in J. Kingston and M. E. Beckman (eds.), *Papers in Laboratory Phonology 1: Between the Grammar and Physics of Speech* (Cambridge: Cambridge University Press), 283–333.

—— (1992), 'The Sonority Cycle and Syllable Organization', in Dressler *et al.* (1992), 63–76.

—— and HUME, ELISABETH V. (1995), 'The Internal Organization of Speech Sounds', in Goldsmith (1995), 245–306.

—— and KEYSER, SAMUEL JAY (1983), *CV Phonology: A Generative Theory of the Syllable* (Cambridge, Mass.: MIT).

COLE, JENNIFER (1995), 'The Cycle in Phonology', in Goldsmith (1995), 70–113.

COLLINDER, BJÖRN (1960), *Comparative Grammar of the Uralic Languages* (Stockholm: Almqvist & Wiksell).

COMRIE, BERNARD (1981), 'Uralic Languages', in B. Comrie (ed.), *Languages of the Soviet Union* (Cambridge: Cambridge University Press), 92–141.

—— (ed.) (1990), *The World's Major Languages* (New York and Oxford: Oxford University Press).

CRYSTAL, DAVID (1995), 'Documenting Rhythmical Change', in J. W. Lewis (ed.), *Studies in General and English Phonetics: Essays in Honour of Professor J. D. O'Connor* (London: Routledge), 174–9.

CSERESNYÉSI, LÁSZLÓ (1992), 'An Outline of Hungarian Phonology', *The Journal of Intercultural Studies* (Kansai Gaidai University, Japan), Extra Series 2: 79–104.

CSÚRI, PIROSKA (1990), 'Moraic Theory and the Syllable Inventory of Hungarian', manuscript (Waltham: Brandeis University).

DAVIS, STUART (1984), 'Some Implications of Onset–Coda Constraints for Syllable Phonology', *CLS*, 20: 46–51.

—— (1985), 'Topics in Syllable Geometry', University of Arizona Ph.D. dissertation.

—— (1990), 'The Onset as a Constituent of the Syllable', *CLS* 26: 71–81.

—— (1991), 'Coronals and the Phonotactics of Nonadjacent Consonants in English', in C. Paradis and J.-F. Prunet (eds.), *Phonetics and Phonology, Vol. 2. The Special Status of Coronals* (New York: Academic Press), 49–60.

DÉCSY, GYULA (1965), *Einführung in die finnisch-ugrische Sprachwissenschaft* (Wiesbaden: Otto Harrassowitz).

DEME, LÁSZLÓ (1961), 'Hangtan [Phonetics and Phonology]', in J. Tompa (ed.), *A mai magyar nyelv rendszere I. [The System of Present-day Hungarian, Vol. 1]* (Budapest: Akadémiai Kiadó), 57–119.

DEZSŐ, LÁSZLÓ (1965), 'Notes on the Word Order of Simple Sentences in Hungarian', *Computational Linguistics*, 4: 3–60.

—— (1980), 'Word Order, Theme and Rheme in Hungarian, and the Problem of Word Order Acquisition', in L. Dezső and W. Nemser (eds.), *Studies in English and Hungarian Contrastive Linguistics* (Budapest: Akadémiai Kiadó).

DRESSLER, WOLFGANG U. (1985), *Morphonology: The Dynamics of Derivation* (Ann Arbor: Karoma).

—— (1990), *Spoken Language: A Major Challenge to Linguistic Theory and Methodology* (Budapest: Linguistics Institute of HAS).

—— LUSCHÜTZKY, HANS C.; PFEIFFER, OSKAR E.; and RENNISON JOHN R. (eds.) (1992), *Phonologica 1988. Proceedings of the 6th International Phonology Meeting* (Cambridge: Cambridge University Press).

—— and SIPTÁR, PÉTER (1989), 'Towards a Natural Phonology of Hungarian', *Acta Linguistica Hungarica* 39: 29–51.

DUNN, CHRISTIAN (1995), 'Aspects du gouvernement harmonique', University of Montreal Ph.D. dissertation.

É. KISS, KATALIN (1981), 'Structural Relations in Hungarian, a "Free" Word Order Language', *Linguistic Inquiry*, 12: 185–213.

—— (1987), *Configurationality in Hungarian* (Dordrecht: Reidel, and Budapest: Akadémia Kiadó).

—— (1998), 'Multiple Topic, One Focus?', *Acta Linguistica Hungarica*, 45: 3–29.

—— and PAPP, FERENC (1984), 'A *dz* és a *dzs* státusához a mai magyar fonémarendszerben [On the Status of [dᶻ] and [ǰ] in the Phoneme System of Present-day Hungarian]', *Általános Nyelvészeti Tanulmányok*, 15: 151–60.

FABB, NIGEL (1988), 'English Suffixation is Constrained by Selectional Restrictions', *Natural Language and Linguistic Theory*, 6: 527–39.

FARKAS, DONKA F., and BEDDOR, PATRICE S. (1987), 'Privative and Equipollent Backness in Hungarian', *CLS*, 23: 91–105.

—— and SADOCK, JERROLD M. (1989), 'Preverb Climbing in Hungarian', *Language*, 65: 318–38.

FÓNAGY, IVÁN (1989), 'On the Status and Functions of Intonation', *Acta Linguistica Hungarica*, 39: 53–92.

—— and MAGDICS, KLÁRA (1967), *A magyar beszéd dallama [The Melody of Hungarian Speech]* (Budapest: Akadémiai Kiadó).

FUDGE, ERIK (1969), 'Syllables', *Journal of Linguistics*, 5: 193–320.

—— (1987), 'Branching Structure within the Syllable', *Journal of Linguistics*, 23: 359–77.

FUJIMURA, OSAMU (1979), 'English Syllables as Core and Affixes', *Zeitschrift für Phonetik, Sprachwissenschaft und Kommunikationsforschung*, 32: 471–6.

GOLDSMITH, JOHN A. (1976), 'Autosegmental Phonology', MIT Ph.D. dissertation. (Published by Garland Press, New York, 1979.)

—— (1985), 'Vowel Harmony in Khalkha Mongolian, Yaka, Finnish and Hungarian', *Phonology Yearbook*, 2: 253–75.

—— (1990), *Autosegmental and Metrical Phonology* (Oxford: Blackwell).

—— (ed.) (1995), *The Handbook of Phonological Theory* (Cambridge, Mass., and Oxford: Blackwell).

GÓSY, MÁRIA (1989), 'Vowel Harmony: Interrelations of Speech Production, Speech Perception, and the Phonological Rules', *Acta Linguistica Hungarica*, 39: 93–118.

—— (1993), *Speech Perception* (Frankfurt am Main: Hector).

—— and TERKEN, JACQUES (1994), 'Question Marking in Hungarian: Timing and Height of Pitch Peaks', *Journal of Phonetics*, 22: 269–81.

HAJDÚ, PÉTER (1975), *Finno-Ugrian Languages and Peoples*, translated and adapted by G. F. Cushing (London: André Deutsch).

HALL, ROBERT A. (1944), *Hungarian Grammar* (Language Monograph No. 21. Baltimore: Linguistic Society of America).

HALLE, MORRIS (1962), 'Phonology in Generative Grammar', *Word*, 18: 54–72.

—— and KENSTOWICZ, MICHAEL (1991), 'The Free Element Condition and Cyclic vs. Noncyclic Stress', *Linguistic Inquiry*, 22: 457–501.

—— and MOHANAN, K. P. (1985), 'Segmental Phonology of Modern English', *Linguistic Inquiry*, 16: 57–116.

—— and STEVENS, KENNETH N. (1971), 'A Note on Laryngeal Features', *MIT Quarterly Progress Report*, 101: 198–213.

—— and VERGNAUD, JEAN-ROGER (1987), *An Essay on Stress* (Cambridge, Mass.: MIT Press).

HAMMOND, MICHAEL (1987), 'Hungarian Cola', *Phonology Yearbook*, 4: 267–9.

—— (1997), 'Underlying Representations in Optimality Theory', in I. Roca (ed.), *Derivations and Constraints in Phonology* (Oxford: Clarendon Press), 349–67.

HARRIS, JOHN (1990), 'Segmental Complexity and Phonological Government', *Phonology*, 7: 255–300.

—— (1994), *English Sound Structure* (Oxford: Blackwell).

—— (1997), 'Licensing Inheritance: An Integrated Theory of Neutralisation', *Phonology*, 14: 315–70.

HAYES, BRUCE (1980), 'A Metrical Theory of Stress Rules', MIT Ph.D. dissertation. (Published by Garland Press, New York, 1985.)

—— (1982), 'Extrametricality and English Stress', *Linguistic Inquiry*, 13: 227–76.

—— (1986), 'Inalterability in CV Phonology', *Language*, 62: 321–51.

—— (1989), 'Compensatory Lengthening in Moraic Phonology', *Linguistic Inquiry*, 20: 253–306.

—— (1995), *Metrical Stress Theory: Principles and Case Studies* (Chicago: University of Chicago Press).

HETZRON, ROBERT (1972), 'Studies in Hungarian Morphophonology', *Ural-Altaische Jahrbücher*, 44: 79–106.

—— (1975), 'Where Grammar Fails', *Language*, 51: 859–72.

—— (1992), 'Prosodic Morphemes in Hungarian', in I. Kenesei and Cs. Pléh (eds.), *Approaches to Hungarian 4: The Structure of Hungarian* (Szeged: JATE), 141–56.

HEWITT, MARK, and PRINCE, ALAN (1989), 'OCP, Locality and Linking: the N. Karanga Verb', in E. J. Fee and K. Hunt (eds.), *Proceedings of the WCCFL 8* (Stanford: SLA), 176–91.

HOOPER, JOAN B. (1978), 'Constraints on Schwa-deletion in American English', in J. Fisiak (ed.), *Recent Developments in Historical Phonology* (The Hague: Mouton), 183–207.

HORVATH, JULIA (1986), *Focus in the Theory of Grammar and the Syntax of Hungarian* (Dordrecht: Foris).

—— (1998), 'Multiple WH-Phrases and the WH-Scope-marker Strategy in Hungarian Interrogatives', *Acta Linguistica Hungarica*, 45: 31–60.

HUALDE, JOSÉ (1989), 'The Strict Cycle Condition and Noncyclic Rules', *Linguistic Inquiry*, 20: 675–80.

HULST, HARRY VAN DER (1984), *Syllable Structure and Stress in Dutch* (Dordrecht: Foris).

—— (1985), 'Vowel Harmony in Hungarian: A Comparison of Segmental and Autosegmental Analyses', in H. van der Hulst and N. Smith (eds.), *Advances in Nonlinear Phonology* (Dordrecht: Foris), 267–303.

—— (1992), 'The Phonetic and Phonological Basis of the Simplex Feature Hypothesis' in Dressler *et al.* (1992), 119–31.

HUME, ELISABETH V. (1992), 'Front Vowels, Coronal Consonants, and their Interaction in Non-linear Phonology', Cornell University Ph.D. dissertation.

HYMAN, LARRY M. (1985), *A Theory of Phonological Weight* (Dordrecht: Foris).

—— (1990), 'Nonexhaustive Syllabification: Evidence from Nigeria and Cameroon', in M. Ziolkowsky, M. Noske, and K. Deaton (eds.), *CLS 26: Parasession on The Syllable in Phonetics and Phonology*, 175–96.

ITÔ, JUNKO (1986), 'Syllable Theory in Prosodic Phonology', University of Massachusetts Ph.D. dissertation. (Published by Garland Press, New York, 1988.)

—— (1989), 'A Prosodic Theory of Epenthesis', *Natural Language and Linguistic Theory*, 7: 217–59.

JENSEN, JOHN T. (1978), 'Reply to "Theoretical Implications of Hungarian Vowel Harmony"', *Linguistic Inquiry*, 9: 89–97.

—— (1984), 'A Lexical Phonology Treatment of Hungarian Vowel Harmony', *Linguistic Analysis*, 14: 231–53.

—— (1993), *English Phonology* (Amsterdam: John Benjamins).

—— and STONG-JENSEN, MARGARET (1988), 'Syllabification and Epenthesis in Hungarian', paper read at the Institute of Linguistics, Hungarian Academy of Sciences, Budapest.

—— —— (1989a), 'Vowel Length Alternations in Hungarian', *Acta Linguistica Hungarica*, 39: 119–31.

—— —— (1989b), 'The Strict Cycle Condition and Epenthesis in Hungarian', *NELS*, 19: 223–35.

KAGER, RENÉ (1995), 'The Metrical Theory of Word Stress', in Goldsmith (1995), 367–402.

KAHN, DANIEL (1980), *Syllable-based Generalizations in English Phonology* (New York: Garland).

KAISSE, ELLEN M. (1985), *Connected Speech. The Interaction of Syntax and Phonology* (New York: Academic Press).

—— and SHAW, PATRICIA A. (1985), 'On the Theory of Lexical Phonology', *Phonology Yearbook*, 2: 1–30.

KÁLMÁN, C. GYÖRGY; KÁLMÁN, LÁSZLÓ; NÁDASDY, ÁDÁM; and PRÓSZÉKY, GÁBOR (1984), 'Topic, Focus, and Auxiliaries in Hungarian', *Groninger Arbeiten zur Germanistischen Linguistik*, 24: 162–77.

—————— (1989), 'A magyar segédigék rendszere [The System of Auxiliaries in Hungarian]', *Általános Nyelvészeti Tanulmányok*, 17: 49–103.

KÁLMÁN, LÁSZLÓ (1985a), 'Word Order in Neutral Sentences', in I. Kenesei (ed.), *Approaches to Hungarian 1: Data and Descriptions* (Szeged: JATE), 13–24.

—— (1985b), 'Word Order in Non-Neutral Sentences', in I. Kenesei (ed.), *Approaches to Hungarian 1: Data and Descriptions* (Szeged: JATE), 25–38.

—— (1985c), 'Inflectional Morphology', in I. Kenesei (ed.), *Approaches to Hungarian 1: Data and Descriptions* (Szeged: JATE), 247–62.

—— and NÁDASDY, ÁDÁM (1994), 'A hangsúly [Stress]', in Kiefer (1994), 393–467.

KÁROLY, SÁNDOR (1957), 'A *csuklik*-féle ikes igék ragozása, képzése [The Suffixation of verbs of the *csuklik* type]', *Magyar Nyelvőr*, 81: 275–81.

KASSAI, ILONA (1981), 'A magyar beszéd hangsorépítési szabályszerűségei [Phonotactic Regularities in Hungarian]', *Magyar Fonetikai Füzetek*, 8: 63–86.

KAYE, JONATHAN (1974), 'Morpheme Structure Constraints Live!', *Recherches linguistiques à Montréal / Montreal Working Papers in Linguistics*, 3: 55–62.

—— (1989), *Phonology: A Cognitive View* (Hillsdale, New Jersey: Lawrence Erlbaum Associates).

—— (1990), '"Coda" Licensing', *Phonology*, 7: 301–30.

—— (1992a), 'Do You Believe in Magic? The Story of *s+C* Sequences', *SOAS Working Papers in Linguistics and Phonetics*, 2: 293–314.

—— (1992b), 'On the Interaction of Theories of Lexical Phonology and Theories of Phonological Phenomena', in Dressler *et al.* (1992), 141–55.

—— (1995), 'Derivations and Interfaces', in J. Durand and F. Katamba (eds.), *Frontiers of Phonology: Atoms, Structures, Derivations* (London: Longman), 289–332.

—— and LOWENSTAMM, JEAN (1981), 'Syllable Structure and Markedness Theory', in A. Belletti *et al.* (eds.), *Theory of Markedness in Generative Grammar* (Pisa: Scuola Normale Superiore), 287–315.

—— —— and VERGNAUD, JEAN-ROGER (1985), 'The Internal Structure of Phonological Representations: a Theory of Charm and Government', *Phonology Yearbook*, 2: 305–28.

—— —— —— (1990), 'Constituent Structure and Government in Phonology', *Phonology*, 7: 193–231.

KENESEI, ISTVÁN (1995), 'On Bracketing Paradoxes in Hungarian', *Acta Linguistica Hungarica*, 43: 153–74.

—— (1998), 'Adjuncts and Arguments in VP-Focus', *Acta Linguistica Hungarica*, 45: 61–88.

—— VAGO, ROBERT M., and FENYVESI, ANNA (1998), *Hungarian* (Descriptive Grammar Series, London and New York: Routledge).

—— and VOGEL, IRENE (1989), 'Prosodic Phonology in Hungarian', *Acta Linguistica Hungarica*, 39: 149–93.

KENSTOWICZ, MICHAEL (1994), *Phonology in Generative Grammar* (Oxford: Basil Blackwell).

—— and KISSEBERTH, CHARLES (1977), *Topics in Phonological Theory* (New York: Academic Press).

—— and PYLE, CHARLES (1973), 'On the Phonological Integrity of Geminate Clusters', in M. Kenstowicz and C. Kisseberth (eds.), *Issues in Phonological Theory* (The Hague: Mouton), 27–43.

KENSTOWICZ, MICHAEL, and RUBACH, JERZY (1987), 'The Phonology of Syllabic Nuclei in Slovak', *Language*, 63: 463–97.

KEREK, ANDREW (1977), 'Consonant Elision in Hungarian Casual Speech', in D. Sinor (ed.), *Studies in Finno-Ugric Linguistics in Honor of Alo Raun* (Bloomington: Indiana University Press), 115–30.

KIEFER, FERENC (1967), *On Emphasis and Word Order in Hungarian* (The Hague: Mouton).

—— (ed.) (1982), *Hungarian Linguistics* (Amsterdam: John Benjamins).

—— (1985), 'The Possessive in Hungarian: A Problem for Natural Morphology', *Acta Linguistica Hungarica*, 35: 85–116.

—— (1992), 'Compounds and Argument Structure in Hungarian', in I. Kenesei and Cs. Pléh (eds.), *Approaches to Hungarian 4: The Structure of Hungarian* (Szeged: JATE), 51–66.

—— (ed.) (1994), *Strukturális magyar nyelvtan II. Fonológia [A Structural Grammar of Hungarian. Vol. 2: Phonology]* (Budapest: Akadémiai Kiadó).

—— and É. KISS, KATALIN (eds.) (1994), *The Syntactic Structure of Hungarian* (Syntax and Semantics 27. San Diego and New York, Academic Press).

KIPARSKY, PAUL (1982a), 'From Cyclic Phonology to Lexical Phonology', in H. van der Hulst and N. Smith (eds.), *The Structure of Phonological Representations* (Dordrecht: Foris), I, 131–75.

—— (1982b), 'Lexical Morphology and Phonology', in I. S. Yang (ed.), *Linguistics in the Morning Calm* (Seoul: Hanshin), 3–91.

—— (1985), 'Some Consequences of Lexical Phonology', *Phonology Yearbook*, 2: 85–138.

KISSEBERTH, CHARLES (1970), 'On the Functional Unity of Phonological Rules', *Linguistic Inquiry*, 1: 291–306.

KONTRA, MIKLÓS (1992), 'Fonológiai általánosítás és szociolingvisztikai realitás [Phonological Generalization and Sociolinguistic Reality]', in M. Kontra (ed.), *Társadalmi és területi változatok a magyar nyelvben [Sociolectal and Dialectal Variation in Hungarian]* (Linguistica, Series A: Studia et Dissertationes, 9. Budapest: Linguistics Institute of HAS), 87–95.

—— (1995), 'On Current Research into Spoken Hungarian', *International Journal of the Sociology of Language*, 111: 9–20.

—— and RINGEN, CATHERINE O. (1986), 'Hungarian Vowel Harmony: the Evidence from Loanwords', *Ural-Altaische Jahrbücher*, 58: 1–14.

—— —— (1987), 'Stress and Harmony in Hungarian Loanwords', in K. Rédei (ed.), *Studien zur Phonologie und Morphologie der Uralischen Sprachen* (Vienna: Verband der Wissenschaftliche Gesellschafte Österreichs), 81–96.

—— —— and STEMBERGER, JOSEPH PAUL (1991), 'The Effect of Context on Suffix Vowel Choice in Hungarian Vowel Harmony', in W. Bahner, J. Schildt, and D. Viehweger (eds.), *Proceedings of the Fourteenth International Congress of Linguists* (Berlin: Akademie-Verlag) 1, 450–3.

KORNAI, ANDRÁS (1986a), 'Szótári adatbázis az akadémiai nagyszámítógépen [A Computerized Dictionary Data Base at the Hungarian Academy of Sciences]', *Műhelymunkák a nyelvészet és társtudományai köréből*, 2: 30–40.

—— (1986b), 'On Hungarian Morphology', Hungarian Academy of Sciences Ph.D. dissertation, Budapest.

—— (1987), 'Hungarian Vowel Harmony', in M. Crowhurst (ed.), *Proceedings of the 6th West Coast Conference on Formal Linguistics* (Stanford Linguistics Association), 147–61.

—— (1990*a*), 'The Sonority Hierarchy in Hungarian', *Nyelvtudományi Közlemények*, 91: 139–46.

—— (1990*b*), 'Hungarian Vowel Harmony', in I. Kenesei (ed.), *Approaches to Hungarian 3: Structures and Arguments* (Szeged: JATE), 183–240.

—— (1993), 'The Generative Power of Feature Geometry', *Annals of Mathematics and Artificial Intelligence*, 8: 37–46.

—— (1994), *On Hungarian Morphology* (Linguistica, Series A: Studia et Dissertationes, 14. Budapest: Linguistics Institute of HAS).

—— and KÁLMÁN, LÁSZLÓ (1989), 'Hungarian Sentence Intonation', in H. van der Hulst and N. Smith (eds.), *Autosegmental Studies on Pitch Accent* (Dordrecht: Foris), 183–95.

KYLSTRA, ANDRES DIRK (1984), 'Még egyszer a magánhangzó + *j* kapcsolatról a magyarban [Once More on Hungarian Vowel + *j* Combinations]', *Nyelvtudományi Közlemények*, 86: 148–51.

—— and GRAAF, TJEERD DE (1980), 'Vannak-e diftongusok a magyar köznyelvben? [Are there Diphthongs in Standard Hungarian?]', *Nyelvtudományi Közlemények*, 82: 313–17.

LACZKÓ, TIBOR (1997), 'Action Nominalizations and the Possessor Function within Hungarian and English Noun Phrases', *Acta Linguistica Hungarica*, 44: 413–75.

LADEFOGED, PETER (1993), *A Course in Phonetics*, 3rd edition (New York: Harcourt Brace Jovanovich).

LAVER, JOHN (1994), *Principles of Phonetics* (Cambridge: Cambridge University Press).

LAZICZIUS, GYULA (1932), *Bevezetés a fonológiába [Introduction to Phonology]* (A Magyar Nyelvtudományi Társaság Kiadványai 33. Budapest).

—— (1948), 'Phonétique et phonologie', *Lingua*, 1: 293–302; reprinted in Sebeok (1966), 95–105.

LEBEN, WILLAM R. (1973), 'Suprasegmental Phonology', MIT Ph.D. dissertation.

LEVIN, JULIETTE (1985), 'A Metrical Theory of Syllabicity', MIT Ph.D. dissertation.

LOMBARDI, LINDA (1995*a*), 'Laryngeal Features and Privativity', *The Linguistic Review*, 12: 35–59.

—— (1995*b*), 'Laryngeal Neutralization and Syllable Wellformedness', *Natural Language and Linguistic Theory*, 13: 39–74.

—— (1996), 'Postlexical Rules and the Status of Privative Features', *Phonology*, 13: 1–38.

LOTZ, JOHN (1939), *Das ungarische Sprachsystem* (Stockholm: Ungarisches Institut).

MCCARTHY, JOHN J. (1986), 'OCP Effects: Gemination and Antigemination', *Linguistic Inquiry*, 17: 207–63.

—— (1988), 'Feature Geometry and Dependency: A Review', *Phonetica*, 43: 84–108.

MESTER, R. ARMIN, and ITÔ, JUNKO (1989), 'Feature Predictability and Underspecification: Palatal Prosody in Japanese Mimetics', *Language*, 65: 258–93.

MOHANAN, KARUVANNUR PUTHANVEETTIL (1986), *The Theory of Lexical Phonology* (Dordrecht: Reidel).

MOLNÁR, VALÉRIA (1998), 'Topic in Focus', *Acta Linguistica Hungarica*, 45: 89–166.

MYERS, SCOTT (1987), 'Vowel Shortening in English', *Natural Language and Linguistic Theory*, 5: 485–518.

NÁDASDY, ÁDÁM (1985), 'Segmental Phonology and Morphophonology', in I. Kenesei (ed.), *Approaches to Hungarian 1: Data and Descriptions* (Szeged: JATE), 225–46.

—— (1989a), 'The Exact Domain of Consonant Degemination in Hungarian', in T. Szende (ed.), *Proceedings of the Speech Research '89 International Conference* (Hungarian Papers in Phonetics 20), 104–7.

—— (1989b), 'Consonant Length in Recent Borrowings into Hungarian', *Acta Linguistica Hungarica*, 39: 195–213.

—— and SIPTÁR, PÉTER (1989), 'Issues in Hungarian Phonology: Preliminary Queries to a New Project', *Acta Linguistica Hungarica*, 39: 3–27.

—— —— (1994), 'A magánhangzók [The Vowels]', in Kiefer (1994), 42–182.

—— —— (1998), 'Vowel Length in Present-day Spoken Hungarian', in L. Varga (ed.), *The Even Yearbook 3* (ELTE SEAS Working Papers in Linguistics), 149–72.

NESPOR, MARINA (1987), 'Vowel Degemination and Fast Speech Rules', *Phonology Yearbook* 4: 61–85.

OBENDORFER, RUDOLF (1975), 'The Ambiguous Status of Hungarian Long Consonants', *Lingua*, 36: 325–36.

ODDEN, DAVID (1986), 'On the Role of the Obligatory Contour Principle in Phonological Theory', *Language*, 62: 353–83.

—— (1988), 'Anti AntiGemination and the OCP', *Linguistic Inquiry*, 19: 451–75.

OLSSON, MAGNUS (1992), *Hungarian Phonology and Morphology* (Travaux de l'Institut de Linguistique de Lund 26. Lund University Press).

—— (1993), 'The Basic Hungarian Allophone System: Structure and Rules', *Lund University, Dept. of Linguistics Working Papers*, 40: 157–84.

PAPP, FERENC (1975), *A magyar főnév paradigmatikus rendszere [The Paradigmatic System of the Hungarian Noun]* (Budapest: Akadémiai Kiadó).

PERLMUTTER, DAVID (1995), 'Phonological Quantity and Multiple Association', in J. Goldsmith (ed.), *The Handbook of Phonological Theory* (Oxford: Blackwell), 307–17.

PHELPS, ELAINE (1978), 'Exceptions and Vowel Harmony in Hungarian', *Linguistic Inquiry*, 9: 98–105.

PINTZUK, SUSAN; KONTRA, MIKLÓS; SÁNDOR, KLÁRA; and BORBÉLY, ANNA (1995), *The Effect of the Typewriter on Hungarian Reading Style* (Working Papers in Hungarian Sociolinguistics 1, Budapest: Linguistics Institute of HAS).

POLGÁRDI, KRISZTINA (1997), 'Hungarian is Strict CV', paper presented at the CV Workshop, Leiden, 10 June 1997.

—— (1998), 'Vowel Harmony: An Account in Terms of Government and Optimality', HIL/Leiden University Ph.D. dissertation.

—— and REBRUS, PÉTER (1996), *There is No Labial Harmony in Hungarian: A Government Phonology Approach* (Working Papers in the Theory of Grammar 3/3, Budapest: Linguistics Institute of HAS).

—— —— (1998), 'There is No Labial Harmony in Hungarian: A Government Phonology Analysis', in C. de Groot and I. Kenesei (eds.), *Approaches to Hungarian 6: Papers from the Amsterdam Conference* (Szeged: JATE), 3–20.

PRINCE, ALAN, and SMOLENSKY, PAUL (1993), *Optimality Theory: Constraint Interaction in Generative Grammar*, manuscript (Rutgers University, New Brunswick, and University of Colorado, Boulder).

PULLUM, GEOFFREY (1976), 'The Duke of York Gambit', *Journal of Linguistics*, 12: 93–103.

PUSKÁS, GENOVÉVA (1998), 'On the NEG Criterion in Hungarian', *Acta Linguistica Hungarica*, 45: 167–213.

REBRUS, PÉTER (2000), 'Morfofonológiai jelenségek [Morphophonological phenomena]', in F. Kiefer (ed.), *Strukturális magyar nyelvtan III. Morfológia [A Structural Grammar of Hungarian. Vol. 3: Morphology]* (Budapest: Akadémiai Kiadó), 763–947.

—— and POLGÁRDI, KRISZTINA (1997), 'Two Default Vowels in Hungarian?', in G. Booij and J. van de Weijer (eds.), *Phonology in Progress–Progress in Phonology. HILP Phonology Papers III* (The Hague: Holland Academic Graphics), 257–75.

RIALLAND, ANNIE (1993), 'L'Allongement compensatoire: Nature et modéles', in B. Laks and A. Rialland (eds.), *Architecture des représentations phonologiques* (Paris: CNRS).

RICE, KEREN D. (1992), 'On Deriving Sonority: a Structural Account of Sonority Relationships', *Phonology*, 9: 61–99.

RINGEN, CATHERINE O. (1977), 'Vowel Harmony: Implications for the Alternation Condition', in W. U. Dressler and O. E. Pfeiffer (eds.), *Phonologica 1976. Proceedings of the 3rd International Phonology Meeting* (Innsbruck: Innsbrucker Beiträge zur Sprachwissenschaft), 127–32.

—— (1978), 'Another View of the Theoretical Implications of Hungarian Vowel Harmony', *Linguistic Inquiry*, 9: 105–15.

—— (1980), 'A Concrete Analysis of Hungarian Vowel Harmony', in R. M. Vago (ed.), *Issues in Vowel Harmony* (Amsterdam: John Benjamins B. V.), 135–54.

—— (1982), 'Abstractness and the Theory of Exceptions', *Linguistic Analysis*, 10: 191–202.

—— (1988*a*), 'Transparency in Hungarian Vowel Harmony', *Phonology*, 5: 327–42.

—— (1988*b*), *Vowel Harmony: Theoretical Implications* (New York and London: Garland).

—— and KONTRA, MIKLÓS (1989), 'Hungarian Neutral Vowels', *Lingua*, 78: 181–91.

—— and VAGO, ROBERT M. (1995), 'A Constraint Based Analysis of Hungarian Vowel Harmony', in I. Kenesei (ed.), *Approaches to Hungarian 5: Levels and Structures* (Szeged: JATE), 307–19.

—— —— (1998*a*), 'Hungarian Roundness Harmony in Optimality Theory', in C. de Groot and I. Kenesei (eds.), *Approaches to Hungarian 6: Papers from the Amsterdam Conference* (Szeged: JATE), 21–39.

—— —— (1998*b*), 'Hungarian Vowel Harmony in Optimality Theory', *Phonology*, 15: 393–416.

RITTER, NANCY (1995), 'The Role of Universal Grammar in Phonology: A Government Phonology Approach to Hungarian', New York University Ph.D. dissertation.

—— (1998), 'The Unifying Effects of Proper Government', in C. de Groot and I. Kenesei (eds.), *Approaches to Hungarian 6: Papers from the Amsterdam Conference* (Szeged: JATE), 41–60.

ROACH, PETER (1982), 'On the Distinction between Stress-timed and Syllable-timed Languages', in D. Crystal (ed.), *Linguistic Controversies* (London: Edward Arnold), 73–9.

RUBACH, JERZY (1993), *The Lexical Phonology of Slovak* (The Phonology of the World's Languages. Series editor: Jacques Durand. Oxford: Clarendon Press/Oxford University Press).

—— (1996) 'Shortening and ambisyllabicity in English', *Phonology*, 13: 197–237.

—— and BOOIJ, GEERT E. (1990), 'Edge of Constituent Effects in Polish', *Natural Language and Linguistic Theory*, 8: 427–63.

SAGEY, ELIZABETH (1986), 'The Representation of Features and Relations in Nonlinear Phonology', MIT Ph.D. dissertation.

SAUVAGEOT, AURÉLIEN (1971), *L'Édification de la langue hongroise* (Paris: Klincksieck).

SCHEIN, BARRY, and STERIADE, DONCA (1986), 'On Geminates', *Linguistic Inquiry* 17: 691–744.

SEBEOK, THOMAS A. (ed.) (1966), *Selected Writings of Gyula Laziczius* (The Hague: Mouton).

SELKIRK, ELISABETH O. (1982), 'The Syllable', in H. van der Hulst and N. Smith (eds.), *The Structure of Phonological Representations, Part II* (Dordrecht: Foris), 337–83.

SIMONYI, ZSIGMOND (1907), *Die ungarische Sprache* (Strassburg: Karl J. Trübner).

SIPTÁR, PÉTER (1980), 'A Note on Initial Clusters in English and Hungarian', *Acta Linguistica Hungarica*, 30: 327–43.

—— (1983), 'Robert M. Vago: The Sound Pattern of Hungarian', *Journal of the International Phonetic Association*, 13: 52–9.

—— (1984), 'Robert M. Vago: The Sound Pattern of Hungarian' [review article], *Acta Linguistica Hungarica*, 34: 138–45.

—— (1989*a*), 'How Many Affricates are There in Hungarian?', in T. Szende (ed.), *Proceedings of the Speech Research '89 International Conference.* Hungarian Papers in Phonetics 20, 123–6.

—— (1989*b*), 'On Fast Speech', *Acta Linguistica Hungarica*, 39: 215–24.

—— (1990), 'Issues in Hungarian Phonology', in *Natural Phonology Workshop at the Annual Meeting of the Societas Linguistica Europaea* (Bern: SLE), 113–16.

—— (1991*a*), 'Fast-speech Processes in Hungarian', in M. Gósy (ed.), *Temporal Factors in Speech: A Collection of Papers* (Budapest: Linguistics Institute of HAS), 27–61.

—— (1991*b*), 'Marginal Vowels in Hungarian', *Proceedings of the XIIth International Congress of Phonetic Sciences* (Aix-en-Provence: Institut de Phonétique, Université de Provence), 3: 214–17.

—— (1993*a*), 'Marginalia in Hungarian Phonology', *Eurasian Studies Yearbook* 65: 73–84.

—— (1993*b*), 'Hungarian Consonants: Rules and Representations', in Z. Kövecses (ed.), *Voices of Friendship. Linguistic Essays in Honor of László T. András* (Budapest: Eötvös Loránd University), 27–50.

—— (1993*c*), 'A magyar mássalhangzók fonológiája [The Phonology of Hungarian Consonants]', Hungarian Academy of Sciences Ph.D. dissertation, Budapest.

—— (1994*a*), 'Palatalization Rules in Hungarian', *Acta Linguistica Hungarica* 42: 5–32.

—— (1994b), 'A mássalhangzók [The Consonants]', in Kiefer (1994), 183–272.

—— (1994c), 'Monotonicity and Deletion', in W. U. Dressler, M. Prinzhorn, and J. Rennison (eds.), *Phonologica 1992. Proceedings of the 7th International Phonology Meeting* (Torino: Rosenberg & Sellier), 257–63.

—— (1994d), 'The Vowel Inventory of Hungarian: Its Size and Structure', in L. Varga (ed.), *The Even Yearbook 1994* (Budapest: ELTE School of English and American Studies), 175–84.

—— (1995), *A magyar mássalhangzók fonológiája [The Phonology of Hungarian Consonants]* (Linguistica, Series A: Studia et Dissertationes, 18. Budapest: Linguistics Institute of HAS).

—— (1996), 'A Janus-faced Hungarian Consonant', in L. Varga (ed.), *The Even Yearbook 2* (ELTE SEAS Working Papers in Linguistics), 83–96.

—— (1998a), 'A magyar magánhangzók lexikális fonológiájából [On the Lexical Phonology of Hungarian Vowels]', in M. Gósy (ed.), *Szófonetikai vizsgálatok [Investigations in the Phonetics of Words]* (Budapest: Linguistics Institute of HAS), 188–230.

—— (1998b), 'Hangtan [Phonology]', in K. É. Kiss, F. Kiefer, and P. Siptár, *Új magyar nyelvtan [A New Hungarian Grammar]* (Budapest: Osiris), 291–390.

SLOAN, KELLY (1991), 'Syllables and Templates: Evidence from Southern Sierra Miwok', MIT Ph.D. dissertation.

SOMMER, BRUCE (1981), 'The Shape of Kunjen Syllables', in D. L. Goyvaerts (ed.), *Phonology in the 1980s* (Ghent: Story-Scientia), 231–44.

SPENCER, ANDREW (1986), 'A Non-linear Analysis of Vowel Zero Alternations in Polish', *Journal of Linguistics*, 22: 249–80.

STERIADE, DONCA (1982), 'Greek Prosodies and the Nature of Syllabification', MIT Ph.D. dissertation.

STIEBELS, BARBARA, and WUNDERLICH, DIETER (1997), 'Second Stems in Hungarian Nouns', *Theorie des Lexikons* (Arbeiten des Sonderforschungsbereichs 282, Nr. 94, Düsseldorf: Heinrich-Heine Universität).

—— —— (1999), 'Second Stems in Hungarian Nouns', *The Linguistic Review*, 16: 253–94.

SZENDE, TAMÁS (1989), 'On How we Speak when we Speak as we Speak', *Acta Linguistica Hungarica*, 39: 225–72.

—— (1992), *Phonological Representation and Lenition Processes* (Hungarian Papers in Phonetics 24. Budapest).

SZENTGYÖRGYI, SZILÁRD (1998) 'Lowering: The Interaction of Phonology and Morphology in Hungarian', József Attila University Ph.D. dissertation, Szeged.

SZÉPE, GYÖRGY (1969), 'Az alsóbb nyelvi szintek leírása [The Description of Lower Linguistic Levels]', *Általános Nyelvészeti Tanulmányok*, 6: 359–466.

SZIGETVÁRI, PÉTER (1994), 'Coronality, Velarity and Why they are Special', in L. Varga (ed.), *The Even Yearbook 1994* (Budapest: ELTE School of English and American Studies), 185–224.

—— (1996), 'Laryngeal Contrasts and Problematic Representations', in L. Varga (ed.), *The Even Yearbook 2* (ELTE SEAS Working Papers in Linguistics), 97–110.

—— (1997), 'On Affricates', in Á. Bende-Farkas (ed.), *DOCSYMP: Graduate Students' First Linguistics Symposium* (Budapest: ELTE), 94–105.

—— (1998), 'Why [h] is Not Voiced', in E. Cyran (ed.), *Structure and Interpretation: Studies in Phonology* (Lublin: Folium), 287–301.

TÁLOS, ENDRE (1988), 'A Short Note on the Status of /x/ in Modern Hungarian', manuscript, Budapest.

TOMPA, JÓZSEF (1968), *Ungarische Grammatik* (The Hague: Mouton).

TÖRKENCZY, MIKLÓS (1989), 'Does the Onset Branch in Hungarian?', *Acta Linguistica Hungarica*, 39: 273–92.

—— (1992), 'Vowel ~ Zero Alternations in Hungarian: A Government Approach', in I. Kenesei and Cs. Pléh (eds.), *Approaches to Hungarian, 4: The Structure of Hungarian* (Szeged: JATE), 157–76.

—— (1993), 'A magyar szótag szerkezete [The Structure of the Hungarian Syllable]', Hungarian Academy of Sciences Ph.D. dissertation, Budapest.

—— (1994a), 'A szótag [The Syllable]', in Kiefer (1994), 273–392.

—— (1994b), 'Coronal Underspecification and the *CC$_i$VC$_i$ Constraint', in L. Varga (ed.), *The Even Yearbook 1994* (Budapest: ELTE School of English and American Studies), 225–32.

—— (1995), 'Underparsing and Overparsing in Hungarian: The /h/ and the Epenthetic Stems', in I. Kenesei (ed.), *Approaches to Hungarian 5: Levels and Structures* (Szeged: JATE), 321–40.

—— (1997), *Hungarian Verbs and Essentials of Grammar: A Practical Guide to the Mastery of Hungarian* (Budapest and Lincolnwood, Ill.: Corvina).

—— and SIPTÁR, PÉTER (1999), 'Hungarian Syllable Structure: Arguments for/against Complex Constituents', in H. van der Hulst and N. Ritter (eds.), *The Syllable: Views and Facts* (Berlin/New York: Mouton de Gruyter), 249–84.

TRANEL, BERNARD (1991), 'CVC Light Syllables, Geminates and Moraic Theory', *Phonology*, 8: 291–302.

VAGO, ROBERT (1974), 'Hungarian Generative Phonology', Harvard University Ph.D. dissertation.

—— (1976), 'Theoretical Implications of Hungarian Vowel Harmony', *Linguistic Inquiry*, 7: 243–63.

—— (1978a), 'Some Controversial Questions Concerning the Description of Vowel Harmony', *Linguistic Inquiry*, 9: 116–25.

—— (1978b), 'The Lengthening of Final Low Vowels in Hungarian', *Ural-Altaische Jahrbücher*, 50: 144–8.

—— (1980a), *The Sound Pattern of Hungarian* (Washington: Georgetown University Press).

—— (1980b), 'A Critique of Suprasegmental Theories of Vowel Harmony', in R. M. Vago (ed.), *Issues in Vowel Harmony* (Amsterdam: John Benjamins B. V.), 155–81.

—— (1985), 'The Treatment of Long Vowels in Word Games', *Phonology Yearbook*, 2: 329–42.

—— (1987), 'On the Status of /w/ in Hungarian Roots', *Studia Uralica*, 4: 114–30.

—— (1989), 'Empty Consonants in the Moraic Phonology of Hungarian', *Acta Linguistica Hungarica*, 39: 293–316.

—— (1991), 'Elméleti szempontok a *t* végű igék elemzésében [Theoretical Aspects of the Analysis of *t*-final Verbs]', in J. Kiss and L. Szűcs (eds.), *Tanulmányok a magyar nyelvtudomány történetének témaköréből [Studies on the History of Hungarian Linguistics]* (Budapest: Akadémiai Kiadó), 682–90.

—— (1992), 'The Root Analysis of Geminates in the Moraic Phonology of Hungarian', in I. Kenesei and Cs. Pléh (eds.), *Approaches to Hungarian 4: The Structure of Hungarian* (Szeged: JATE), 177–94.

VARGA, LÁSZLÓ (1979), 'Az ellentéti hangsúly különleges helye [An Unusual Place for Contrastive Stress]', *Magyar Nyelv*, 75: 332–4.

—— (1983), 'Hungarian Sentence Prosody: An Outline', *Folia Linguistica*, 17: 117–51.

—— (1985), 'Intonation in the Hungarian Sentence', in I. Kenesei (ed.), *Approaches to Hungarian 1: Data and Descriptions* (Szeged: JATE), 205–24.

—— (1989), 'The Stylized Fall in Hungarian', *Acta Linguistica Hungarica*, 39: 317–31.

—— (1993), *A magyar beszéddallamok fonológiai, szemantikai és szintaktikai vonatkozásai [Phonological, Semantic, and Syntactic Aspects of Hungarian Intonation]* (Nyelvtudományi Értekezések 135, Budapest: Akadémiai Kiadó).

—— (1994a), 'A hanglejtés [Intonation]', in Kiefer (1994), 468–549.

—— (1994b), 'Rhythmic Stress Alternation in Hungarian', in L. Varga (ed.), *The Even Yearbook 1994* (Budapest: ELTE School of English and American Studies), 233–54.

—— (1995), 'Stylization of the falling tone in Hungarian intonation', in J. Windsor Lewis (ed.), *Studies in General and English Phonetics. Essays in Honour of Professor J. D. O'Connor* (London: Routledge), 278–87.

—— (1996), 'Hungarian Intonation Contours', in L. Varga (ed.), *The Even Yearbook 2* (ELTE SEAS Working Papers in Linguistics), 111–44.

VENNEMANN, THEO (1988), *Preference Laws for Syllable Structure and the Explanation of Sound Change* (The Hague and Berlin: Mouton de Gruyter).

VÉRTES, O. ANDRÁS (1982), *A magyar leíró hangtan története az újgrammatikusokig [The History of Hungarian Descriptive Phonetics and Phonology to the Neogrammarian Period]* (Budapest: Akadémiai Kiadó).

VOGEL, IRENE (1988), 'Prosodic Constituents in Hungarian', in P. M. Bertinetto and M. Loporcaro (eds.), *Certamen Phonologicum. Papers from the 1987 Cortona Phonology Meeting* (Torino: Rosenberg & Sellier), 231–50.

—— (1989), 'Prosodic Constituents in Hungarian', *Acta Linguistica Hungarica*, 39: 333–51.

—— and KENESEI, ISTVÁN (1987), 'The Interface between Phonology and Other Components of Grammar: the Case of Hungarian', *Phonology Yearbook*, 4: 243–63.

—— —— (1990), 'Syntax and Semantics in Phonology', in S. Inkelas and D. Zec (eds.), *The Phonology–Syntax Connection* (Chicago: Chicago University Press), 339–63.

VOGT, HANS (1954), 'Phoneme Classes and Phoneme Classification', *Word*, 10: 28–34.

WEIJER, JEROEN VAN DE (1989), 'The Formation of Diminutive Names in Hungarian', *Acta Linguistica Hungarica*, 39: 353–71.

YIP, MOIRA (1987), 'English Vowel Epenthesis', *Natural Language and Linguistic Theory*, 5: 463–84.

—— (1988), 'The Obligatory Contour Principle and Phonological Rules: A Loss of Identity', *Linguistic Inquiry*, 19: 65–100.

—— (1989), 'Feature Geometry and Cooccurrence Restrictions', *Phonology*, 6: 349–74.

ZEC, DRAGA (1988), 'Sonority Constraints on Prosodic Structure', Stanford University Ph.D. dissertation.

ZONNEVELD, WIM (1980), 'Hungarian Vowel Harmony and the Theory of Exceptions in Generative Phonology', *Linguistic Analysis*, 8: 21–39.

ZSIGRI, GYULA (1994) 'Magyar mássalhangzószabályok [Rules for Hungarian Consonants]', JATE Ph.D. dissertation, Szeged.

—— (1997), 'Posztalveoláris összeolvadás [Postalveolar Coalescence]', in L. Büky (ed.), *Nyíri Antal kilencvenéves [A. Nyíri is 90]* (Szeged: JATE), 179–86.

—— (1998), '/h/ and /v/ in Hungarian Voicing Assimilation', in C. de Groot and I. Kenesei (eds.), *Approaches to Hungarian 6: Papers from the Amsterdam Conference* (Szeged: JATE), 89–100.

# INDEX